Constitutive Modeling of Engineering Materials

Constitutive Modeling of Engineering Materials

Theory, Computer Implementation, and Parameter Identification

Vladimir Buljak
Faculty of Mechanical Engineering, University of Belgrade
Belgrade, Serbia

Gianluca Ranzi
School of Civil Engineering, The University of Sydney
Sydney, NSW, Australia

ELSEVIER
ACADEMIC PRESS
An imprint of Elsevier

Academic Press is an imprint of Elsevier
125 London Wall, London EC2Y 5AS, United Kingdom
525 B Street, Suite 1650, San Diego, CA 92101, United States
50 Hampshire Street, 5th Floor, Cambridge, MA 02139, United States
The Boulevard, Langford Lane, Kidlington, Oxford OX5 1GB, United Kingdom

Library of Congress Cataloging-in-Publication Data
A catalog record for this book is available from the Library of Congress

British Library Cataloguing-in-Publication Data
A catalogue record for this book is available from the British Library

ISBN: 978-0-12-814696-5

For information on all Academic Press publications
visit our website at https://www.elsevier.com/books-and-journals

Publisher: Matthew Deans
Acquisitions Editor: Dennis McGonagle
Editorial Project Manager: Rafael G. Trombaco
Production Project Manager: Poulouse Joseph
Designer: Victoria Pearson

Typeset by VTeX

Contents

Preface

This book aims to provide an introduction to the basic concepts associated to constitutive models of engineering materials and to address key aspects related to their implementation in computer codes. A sound understanding of the material properties is essential for robust and effective material representations that can be used for a wide range of purposes and applications. The book places particular attention on exploiting inverse analysis techniques for the characterization of the material model parameters. Certainly, when dealing with simple constitutive models it might be easier and more efficient to rely on direct relationships that may exist between the measurements obtained from a standardized test and the sought material parameters. The use of inverse analysis techniques becomes competitive and more efficient when the aim is to minimize the amount of experimental tests carried out for calibration purposes, for example, by enriching the experimental information collected from the standard tests. Other advantages of exploiting inverse analysis techniques arise when dealing with more complex constitutive models, in which case it is generally difficult to separate the influence of different parameters from the experiment outputs and/or some parameters may not have a direct physical meaning. With these characterization techniques, it is usually possible to obtain representative parameters capable of describing the material behavior in scenarios that differ from those considered in the experiment performed for their identification. In this context, the selection of a constitutive model should aim at ensuring that the material representation is sufficiently refined to adequately capture the key features of the material response, relevant to the application considered, while keeping the model as simple as possible to simplify the identification of the material parameters.

Numerous simple examples are provided throughout the book to highlight the key challenges associated with the parameter identification for different constitutive models and how these can be addressed by either modifying or increasing the experimental results utilized in input of the inverse analysis procedure or by revising the selection of the constitutive model to limit the experimental information required for the parameter calibration while still providing an acceptable material description. Throughout the book, MATLAB® scripts are provided to assist the reader in gaining insight into the details of the computer implementations.

The main focus of the book is to deal with constitutive models that account for inelastic deformation. Such material representations are useful in a wide range of engineering problems when a material exhausts its elastic capacity. In general, the stress of a material that exhibits some degree of inelastic deformation depends on the current value of strain as well as on something else. In this book, we aim at introducing some of the possible representations that describe this something else and provide an overview of the main features exhibited by different typologies of inelastic constitutive models.

In the final part of the book, we introduce some of the key aspects associated to the modeling and characterization of particular classes of materials, such as metals, polymers, and shape memory alloys. These chapters consider simple and selected examples that aim at highlighting some of the key features related to the response of these materials. The selection of the examples does not intend to be exhaustive and reference is made to specialized textbook for a comprehensive review of the modeling of these materials. The illustrative examples presented aim at introducing sufficient information to create a connection between the areas of constitutive modeling, computer implementation, and parameter identification that are usually encountered when dealing with research and development of engineering materials and components.

Vladimir Buljak
Gianluca Ranzi

Chapter 1

Introduction

Engineering has supported human progress since ancient times and has been at the basis of its technological development. The engineering of a solution that could be a structure, a component or a device, can involve different levels of complexity and it usually starts with the conception of an idea that fulfills certain desired functionalities. This is then followed by its structural idealization and its analysis and design. Once these are finalized, prototyping and/or production/construction can take place. This is an iterative process whose stages can be repeated and arranged in different orders until the sought design engineering requirements are satisfied. Particular attention is usually placed to *engineer* a solution with suitable and optimized components, joints, and materials that, when combined together, are capable of exhibiting the sought functionalities. This is a broad definition that is applicable to a wide range of applications. Relevant to this book, examples of solutions commonly designed with this approach include buildings, bridges, dams, and tunnels in the field of civil engineering, and automotive structures, aerospace structures, boats, and internal combustion engines in the field of mechanical engineering.

A sound engineering design involves the development of a structure that is capable of maintaining its functionality, integrity, and other specific requirements throughout its design life. Such integrity is usually ensured by specifying, among the others, sufficient strength and/or stiffness to avoid failure or excessive deformations when subjected to the relevant design actions. This design and verification step makes use of structural analysis to predict the expected structural response based on information of the relevant engineering materials, the structural geometry, and the support conditions.

The main features of the structural analysis process are briefly outlined in the first part of the chapter with the assistance of two examples to better highlight the role of constitutive models and how these can influence the structural response. The fundamental concepts related to the representation of the structural behavior and to the stress analysis of deformable components (commonly referred to as *bodies* in structural mechanics) are then presented. This is followed by an introduction of the constitutive modeling for solid deformable bodies and by its implementation in numerical modeling by considering the specific case of the finite element method that is widely adopted throughout the book. The chapter closes with some remarks on structural systems that, when analyzed, undergo large deformations. The proposed chapter layout intends to provide an overview of the structural analysis process and of the role and significance played by con-

Constitutive Modeling of Engineering Materials. https://doi.org/10.1016/B978-0-12-814696-5.00006-X

stitutive modeling of engineering materials that are covered in the remaining chapters of the book.

1.1 Structural analysis: the main concept

Example 1.1

The first example considers the structure illustrated in Fig. 1.1. It consists of a simple mechanical system formed by a rigid bar that is supported vertically by three springs with known stiffnesses and is restrained horizontally at the left support. We will now determine the deformed configuration of the structure and the internal forces resisted by the three springs when subjected to the external force F.

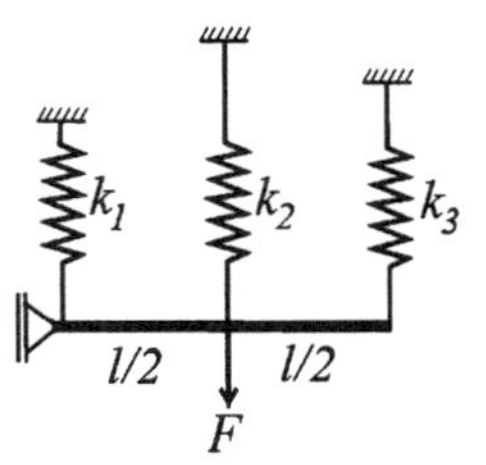

FIGURE 1.1 Simple mechanical system loaded by a single force in its undeformed configuration.

When the three springs have different stiffnesses (k_1, k_2, k_3), the bar will attain an arbitrary inclined position upon the application of the external load. This position is uniquely defined by the extensions of each of the springs (see Fig. 1.2). This example requires us to determine 6 unknowns: three elongations of the springs (Δl_1, Δl_2 and Δl_3), and three spring forces (S_1, S_2, and S_3). In order to determine their values, we will need to establish six independent algebraic equations (that do not contain additional unknowns).

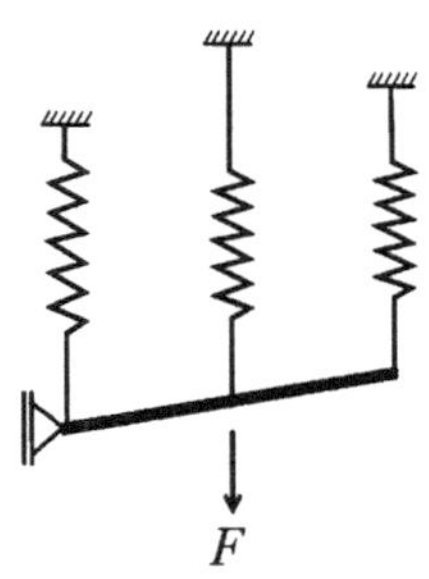

FIGURE 1.2 Possible deformed configuration of the system.

By considering a free-body diagram of the rigid bar, it is possible to identify all forces acting on it. These include the spring forces denoted as S_1, S_2, and S_3 as well as the horizontal reaction R, as shown in Fig. 1.3, and their magnitudes can be determined by enforcing vertical, horizontal, and rotational equilibrium to the free-body diagram of the bar. From the horizontal equilibrium, it is possible to evaluate that the horizontal reaction R is nil. Vertical equilibrium is obtained by summing the projections of all the forces in the vertical direction and setting their sum to zero, namely:

$$S_1 + S_2 + S_3 = F \tag{1.1}$$

The rotational equilibrium about the mid-point of the bar can then be written as:

$$S_1 \frac{l}{2} = S_3 \frac{l}{2} \tag{1.2}$$

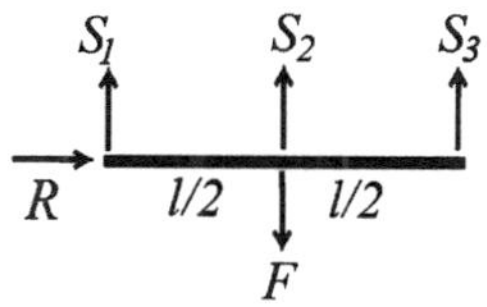

FIGURE 1.3 Free-body diagram of the bar highlighting the equilibrated system of forces.

An additional three equations can be obtained from the behavior of the springs that, under the assumption of linear response, can be written as:

$$S_1 = k_1 \cdot \Delta l_1 \tag{1.3}$$

$$S_2 = k_2 \cdot \Delta l_2 \tag{1.4}$$

$$S_3 = k_3 \cdot \Delta l_3 \tag{1.5}$$

The rigidity of the bar ensures that it remains straight before and after loading and, consequently, the vertical displacements along its length follow a linear profile. This linear variation imposes a restraint to the three elongations. Geometrically, this can be visualized by considering that the elongation of spring 2 represents the median of a trapezoid whose parallel sides are depicted by the elongations of the other two springs (see Fig. 1.4). This geometric condition can be written as:

$$\Delta l_2 = \frac{\Delta l_1 + \Delta l_3}{2} \tag{1.6}$$

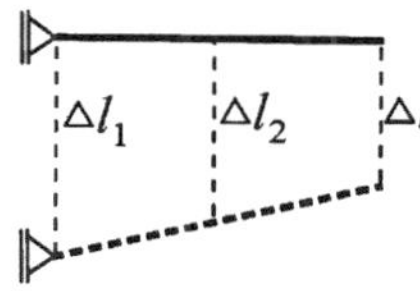

FIGURE 1.4 Kinematics of deformation for the bar elongations.

Eqs. (1.1)–(1.6) can now be solved for the unknown elongations and spring forces. In the solution presented below, Eq. (1.2) is first used to obtain an expression for S_3, while Δl_2 is obtained from Eq. (1.6). The remaining two elongations Δl_1 and Δl_3 are obtained from Eqs. (1.3) and (1.5) in terms of the unknown spring forces S_1 and S_3. The expressions determined so far can then be substituted into Eqs. (1.1) and (1.4) as a function of S_1 and S_2 as:

$$S_2 = F - 2S_1 \tag{1.7}$$

$$S_2 = k_2 \frac{\Delta l_1 + \Delta l_3}{2} = k_2 \frac{S_1}{2} \frac{(k_1 + k_3)}{k_1 \cdot k_3} \tag{1.8}$$

based on which it is possible to evaluate S_1 as follows:

$$S_1 = \frac{2F \cdot k_1 \cdot k_3}{k_2 \cdot k_3 + k_1 \cdot k_2 + 4 \cdot k_1 \cdot k_3} \tag{1.9}$$

This expression can now be substituted back in the previous equations to determine all remaining unknowns from which it can be seen that the final solution depends on the material property of the structure, here specifically expressed in terms of the spring stiffnesses.

Considering first the simplest case, where all the three springs have the same stiffness: $k_1 = k_2 = k_3 = k$, Eq. (1.9) gives the value for the force in spring 1 equal to $S_1 = \frac{F}{3}$. The forces in the remaining two springs can then be determined with Eqs. (1.1) and (1.7) and, in this case, these are the same and equal to the value of the force in spring 1, namely: $S_2 = S_3 = \frac{F}{3}$. As all springs have the same stiffness and resist the same force, they undergo identical elongations that produce a deformed configuration where the bar remains in a horizontal position (Fig. 1.5(a)).

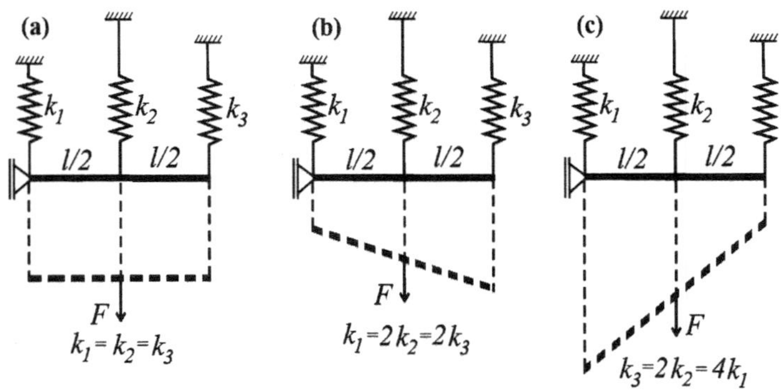

FIGURE 1.5 Deformed configurations for different values of spring stiffnesses.

Let us now consider the case for which the second and third springs have the same stiffness, while the stiffness of the first one is two times larger, namely $k_1 = 2k_2$, $k_2 = k_3$. In this case, from Eq. (1.9) the value of S_1 is $S_1 = \frac{4F}{11}$. From rotational equilibrium (Eq. (1.2)), the spring force S_3 equals S_1, while from vertical equilibrium (Eq. (1.1)), or simply from Eq. (1.7), the force in spring 2 becomes $S_2 = \frac{3F}{11}$. A qualitative representation of the consequent deformed configuration is illustrated in Fig. 1.5(b).

Let us now look at the most general case, in which all the springs have different stiffnesses. For clarity and to simplify the solution, let us assume that the coefficients are related: $k_1 = k_2/2 = k_3/4$. Following the same procedure previously adopted, the spring forces can be calculated as $S_1 = S_3 = \frac{4F}{13}$ and $S_2 = \frac{5F}{13}$. The stiffness of the first spring is 4 times smaller than the stiffness of the third one, therefore it has to elongate more to resist the same force. A possible deformed configuration for this case is shown in Fig. 1.5(c).

The solution of this example has been obtained by relying on three different groups of equations. The first set of equations enforces equilibrium, and because of this, they are commonly referred to as the *equilibrium equations*. These have been expressed in Eqs. (1.1) and (1.2) (and by the equation enforcing horizontal equilibrium that has not been previously written explicitly). The structure has only the three springs as its deformable structural components and their behavior has been assumed to be defined by a linear relationship between their elongation and force, as expressed by Eqs. (1.3)–(1.5). The equations that define the behavior of deformable materials are called *constitutive equations* and they relate internal stresses (depicted by forces for springs) to deformations (represented by elongations for springs). The assumption of the bar rigidity has been accounted for in Eq. (1.6) that represents the *kinematic equation* or *compatibility equation*, and it defines the compatibility between deformation components.

From previous observations, it is clear that by changing values of the three spring stiffnesses, considerably different responses can be obtained for this simple structure. These properties affect both the deformed shape (as outlined in Fig. 1.5) and the values of the internal forces, which may be important when quantifying the actions being transferred to surrounding structures, or when calculating the limit values of the internal actions. This simple example has highlighted the importance of adopting suitable formulations and appropriate material properties for the constitutive equations.

Constitutive modeling for the structural analysis of complex engineering structures plays a significant role because it needs to adequately capture the behavior of engineering materials that are usually described by complicated mechanisms. We will now consider another example in which the deformable structural components can exhibit a nonlinear behavior.

Example 1.2

Let us now consider a structure consisting of three bars or truss elements, as illustrated in Fig. 1.6. It is assumed that these elements can resist only axial forces and that they are connected with hinges. The aim of this example is to determine the structural response induced by the applied horizontal load by calculating the displacements of the loaded point and the internal forces resisted by each bar.

The structure is classified as statically indeterminate because the evaluation of the structural response cannot be determined only from equilibrium equations and it requires, similar to the previous example, the use of also the kinematic and constitutive equations. From a qualitative viewpoint, this is the case when a change in the material properties forming a structure lead to a change in the structural behavior. This is not the case for determinate structures, in which case the reactions and internal actions are calculated solely from equilibrium considerations and the solution can then be post-processed to obtain information on the deformations and displacements.

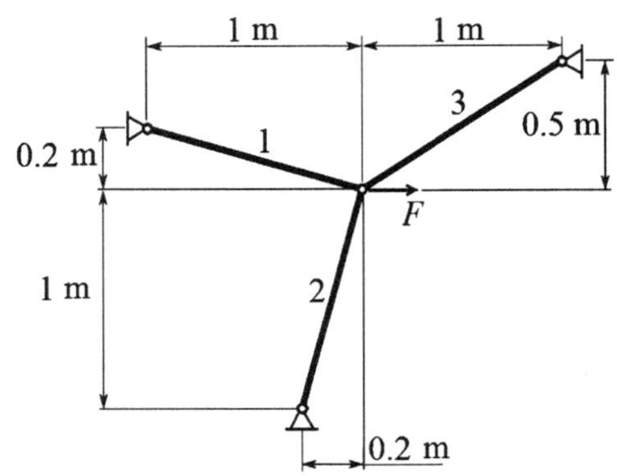

FIGURE 1.6 Statically indeterminate truss of Example 1.2.

The solution is sought in the following by applying the kinematic equations, equilibrium equations, and constitutive equations to formulate the problem in terms of displacement components as the only unknowns. This approach is known as the *displacement method* (or *stiffness method*), and is widely used to solve statically indeterminate problems, because it is suitable for computer implementation. For a detailed description of the displacement method, and the alternative *force method*, the reader should refer to books dedicated to structural analysis, e.g., [1], [2], [3], [9]. A brief overview of the displacement method is provided below for the analysis of the truss of Fig. 1.6 to highlight the role played by the constitutive equations in the solutions process.

The kinematic equations relate the displacements of the bars to their elongations. This is a slightly more complex case with respect to the previous example, where the spring elongations coincided with the vertical displacements of the structure. In this case, the only unrestrained node is the one where the load is applied, thus we can consider two unknown components of displacement for the whole structure: the horizontal displacement of the point of the application of load, denoted by u, and the vertical displacement of it, denoted by v, assumed to be positive when directed rightwards and upwards, respectively.

From geometrical considerations and within the framework of small displacements, three kinematic equations can be written by relating the elongations of each bar to the two displacement components as follows:

$$e_1 = 0.981 \cdot u - 0.196 \cdot v \tag{1.10}$$
$$e_2 = 0.196 \cdot u + 0.981 \cdot v \tag{1.11}$$
$$e_3 = -0.894 \cdot u - 0.447 \cdot v \tag{1.12}$$

where e_1 in Eq. (1.10) refers to the elongation of the first bar, e_2 in Eq. (1.11) to the second bar, and e_3 in Eq. (1.12) to the third bar.

Equilibrium conditions require that the sum of the forces acting at the point where the three bars meet equals zero. By denoting as S_i the force in the ith bar and by projecting the forces to the horizontal and vertical directions, the two equilibrium equations can be expressed as:

$$0.981 \cdot S_1 + 0.196 \cdot S_2 - 0.894 \cdot S_3 = F \tag{1.13}$$
$$-0.196 \cdot S_1 + 0.981 \cdot S_2 - 0.447 \cdot S_3 = 0 \tag{1.14}$$

The constitutive equations describe the behavior of the individual bars by relating the bar elongations to their axial force. The form of these equations depend on the mechanical behavior of the material adopted for the bars. By assuming a linear-elastic behavior for each bar (being the simplest possible response), these equations follow directly from Hooke's low, and they read:

$$e_1 = \frac{S_1 l_1}{EA_1} \tag{1.15}$$
$$e_2 = \frac{S_2 l_2}{EA_2} \tag{1.16}$$
$$e_3 = \frac{S_3 l_3}{EA_3} \tag{1.17}$$

where l_i and A_i are the length and the cross-sectional area of the ith bar, and E is the Young's modulus of the material.

Following the displacement method, the constitutive equations (1.15)–(1.17) are substituted into the equilibrium equations (1.13) and (1.14), and then further substituted into the kinematic equations (1.10)–(1.12). By rearranging these equations, the only unknown variables of the resulting system of equations are displacement components. Such combination of equations is well described if we adopt matrix notations as commonly used in the displacement method. The kinematic equations (1.10)–(1.12) written in matrix form read:

$$\begin{bmatrix} e_1 \\ e_2 \\ e_3 \end{bmatrix} = \begin{bmatrix} 0.981 & -0.196 \\ 0.196 & 0.981 \\ -0.894 & -0.447 \end{bmatrix} \begin{bmatrix} u \\ v \end{bmatrix} \rightarrow \boldsymbol{e} = \boldsymbol{B} \cdot \boldsymbol{d} \tag{1.18}$$

where $\boldsymbol{B}$ is the *kinematic matrix*. The transpose of this matrix, denoted as *static matrix*, is used in the equilibrium equations, and it relates the internal and external forces by means of $\boldsymbol{B}^T$ as described in the equilibrium equations (1.13) and (1.14). In matrix form, it becomes:

$$\begin{bmatrix} F \\ 0 \end{bmatrix} = \begin{bmatrix} 0.981 & 0.196 & -0.894 \\ -0.196 & 0.981 & -0.447 \end{bmatrix} \begin{bmatrix} S_1 \\ S_2 \\ S_3 \end{bmatrix} \rightarrow \boldsymbol{f} = \boldsymbol{B}^T \cdot \boldsymbol{s} \tag{1.19}$$

It is worth pointing out that this *duality of kinematic and equilibrium equations*, as it is usually referred to in structural mechanics, (i.e., the fact that they are written through the same matrix $\boldsymbol{B}$) is guaranteed only when the vectors $\boldsymbol{d}$ and $\boldsymbol{f}$ and the vectors $\boldsymbol{e}$ and $\boldsymbol{s}$ are selected to be work conjugate, namely when their scalar product represent energy/work. In such situations, $\boldsymbol{f}^T\boldsymbol{d}$ has the meaning of work done by the external forces, while $\boldsymbol{s}^T\boldsymbol{e}$ has the meaning of internal energy stored in the deformed body. This condition is satisfied in the present example, and generally, in the implementation of the displacement method, the variables are selected so that they form work conjugate pairs.

The constitutive equations written in matrix form are as follows:

$$\begin{bmatrix} S_1 \\ S_2 \\ S_3 \end{bmatrix} = \begin{bmatrix} \frac{EA_1}{l_1} & 0 & 0 \\ 0 & \frac{EA_2}{l_2} & 0 \\ 0 & 0 & \frac{EA_3}{l_3} \end{bmatrix} \begin{bmatrix} e_1 \\ e_2 \\ e_3 \end{bmatrix} \rightarrow \boldsymbol{s} = \boldsymbol{D} \cdot \boldsymbol{e} \tag{1.20}$$

where matrix $\boldsymbol{D}$ is called the *material stiffness matrix.*

By substituting Eq. (1.20) into Eq. (1.19) and by further eliminating $\boldsymbol{e}$ through Eq. (1.18), the system of algebraic equations derived for the displacement method can be written in terms of only displacements, expressed in matrix form as:

$$\boldsymbol{f} = \boldsymbol{B}^T \cdot \boldsymbol{D} \cdot \boldsymbol{B} \cdot \boldsymbol{d} = \boldsymbol{K} \cdot \boldsymbol{d} \tag{1.21}$$

Matrix $\boldsymbol{K} = \boldsymbol{B}^T\boldsymbol{D}\boldsymbol{B}$ is called the *stiffness matrix* and is completely defined once the geometry and the material properties of the structure are specified. The unknown displacement components $\boldsymbol{d}$ can be obtained from Eq. (1.21) by inverting the stiffness matrix. In the present example, we have included only unknown displacements in Eq. (1.21) even if it is common, when implementing the displacement method, to derive the stiffness relationship of Eq. (1.21) by considering all displacement components, both the unknown and the known ones. In the latter case, the unknown displacement components can then be determined by partitioning Eq. (1.21) and by inverting only the submatrix of the stiffness matrix that is associated to the directions (or freedoms) of the unknown displacements.

Once the unknown displacements are computed, the solution is post-processed by calculating the deformation components through Eq. (1.18) and the internal forces with Eq. (1.20).

It is clear that the properties of the solution depend, apart of the geometry and support conditions of the structure, on the constitutive behavior of the materials forming the structural members. To better illustrate this, let us see how the solution of the present example changes when considering different constitutive relations.

Let us start by assuming that the three bars are made of the same material (i.e., with the same Young's modulus E) and possess the same cross-sectional area A. In this case, the internal forces can be determined as:

$$s = F \begin{bmatrix} 0.644 \\ -0.066 \\ -0.427 \end{bmatrix} \tag{1.22}$$

The minus sign indicates that, when subjected to a positive force F, bars 2 and 3 are in compression, while bar 1 is in tension. Note that, in this case, the constitutive parameters, Young's modulus E and cross-sectional area A are eliminated from the expressions of the internal forces. Of course, the values for E and A affect the calculated bar extensions and displacement components u and v.

Eq. (1.22) provides the relationship between the applied load and the internal bar forces. It is going to be valid provided that all bars behave in a linear-elastic manner as assumed by the constitutive equation (1.20). In general, all constructive materials do have a zone in which their mechanical response can be assumed to follow linear-elastic relations. Let us denote by S_0 the limit force, equal for all three bars, up to which the relationship between the elongation and the internal force is linear. By further assuming that this value is independent of the sign of the load, we may state that the elastic behavior is valid for forces within the following bounds:

$$-S_0 < S_i < S_0, \quad \text{where} \quad i = 1, 2, 3 \tag{1.23}$$

In the present problem, bar 1 will be the first to exhaust its elastic capacity, as it carries the largest amount of load. It may be convenient to express the maximum value of the force F that the bar 1 can resist in its linear-elastic range as a function of S_0, which amounts to:

$$F_{max} = 1.553 S_0 \tag{1.24}$$

From previous considerations, it may be concluded that the structure behaves linearly up to the level of force defined by Eq. (1.24)—meaning that, for example, there will be a linear relationship between the applied force F and the horizontal displacement u of the point where the load is applied.

We will now consider what would happen if the external force F is increased beyond the value given by Eq. (1.24) as this would depend on the type of behavior that the material possesses upon exhausting the elastic capacity.

If we assume that the material cannot take any further load nor deformation upon reaching the elastic limit S_0, bar 1 would break after reaching this point. Such behavior is typical of brittle materials, such as ceramics, glass, and graphite. The constitutive behavior of the bar in this case can be graphically presented as shown in Fig. 1.7a, with loading and unloading parts (represented by arrows) both in compression and tension, following the same straight line, up to the critical force value. Upon exceeding this critical force value the material immediately breaks.

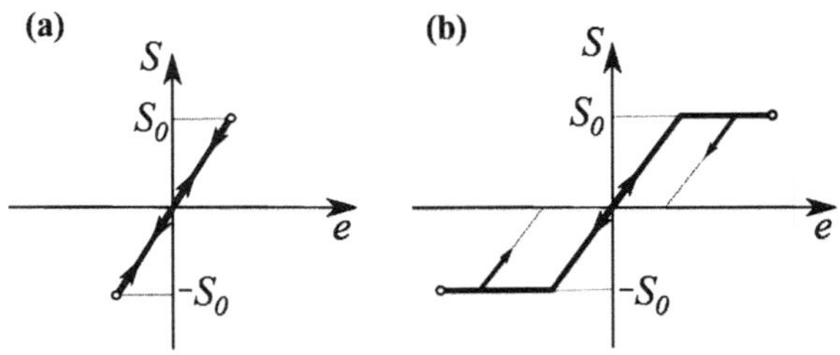

FIGURE 1.7 Constitutive behavior of a bar: (a) elastic-brittle response and (b) elastic-perfectly-plastic response.

Once bar 1 breaks, bars 2 and 3 continue to carry the external load since they have not exhausted their load carrying capacity. As the relationship of Eq. (1.22) is valid up to the point at which bar 1 reaches the critical value, it can be used to compute forces in the remaining two bars at the limiting point of failure of bar 1. By substituting $F = 1.553S_0$ into Eq. (1.22) the sought forces are obtained as:

$$S_2 = -0.102S_0 \quad \text{and} \quad S_3 = -0.663S_0 \tag{1.25}$$

Since bar 1 has now failed, the equilibrium is not satisfied anymore with these forces in bars 2 and 3, and the internal forces previously resisted by bar 1 need to be redistributed to the remaining of the structure. The values of forces in the bars can now be calculated by using only the equilibrium equations, since the structure with only two bars is statically determinate. By eliminating S_1 from the equilibrium equations, and then, by solving it for the unknowns S_2 and S_3 when $F = 1.553S_0$, the following values are obtained:

$$S_2 = -0.880S_0 \quad \text{and} \quad S_3 = -1.930S_0 \tag{1.26}$$

Clearly, bar 3 exceeds its load carrying capacity (since the force has a magnitude larger than $|-S_0|$), and it fails. As only one bar is not sufficient to resist the applied load, it is concluded that the maximum force that this structure can reach is $F_{max} = 1.553S_0$.

Let us now assume an alternative behavior to the brittle fracture upon exhausting the elastic capacity. Some materials like polymers or metals have the capability of developing, even significantly, irreversible plastic deformation prior to rapture, depicted in Fig. 1.7b by a circle at the end of lines in tension and compression. From a mechanical point of view, the development of plastic deformation does not influence (or at least not to a large extent) the load carrying capacity, which usually remains constant or slightly increases with accumulated plastic deformation. We will now assume that the bars conform to this mechanical behavior without pronounced increment in load carrying capacity after the elastic range. Such behavior is commonly referred to as elastic-perfectly-plastic and, in the present context of the axially loaded bars, it can be graphically represented as shown in Fig. 1.7b.

Since the elastic response remains the same as in the previously considered case, the results given in Eq. (1.22) are still valid until bar 1 reaches its critical force S_0. The main difference is that, in this case, upon reaching the critical force, the bar continues to carry this force while undergoing plastic (irreversible) elongation. Since the remaining two bars are still within the elastic range, they can take the additional load, so the external load F can continue to increase beyond the value $F = 1.553S_0$, but the force-displacement response of the structure will no longer follow the same straight line as it did up to this load level.

The solution given by Eq. (1.24) for the elastic part represents the value of the load at which the first load increment terminates. The corresponding values of the forces in the bars and of the external load can be marked by superscript $^{(1)}$ to highlight the fact that these values represent the end of the first load increment, namely:

$$\boldsymbol{s}^{(1)} = \begin{bmatrix} 1 \\ -0.102 \\ -0.663 \end{bmatrix} S_0 \quad \text{and} \quad F^{(1)} = 1.553S_0 \tag{1.27}$$

To determine the structural response for higher levels of load (past the first increment), we need to account that bar 1 contributes to the equilibrium with only a constant force of $S_1 = S_0$ regardless of the further growth of the external load F. It is expected that the external load increment is balanced by increasing forces in the other two bars which are still within the elastic range. It should be mentioned that this is a realistic scenario since the ultimate deformation for ductile materials (i.e., materials with pronounced plastic deformation prior to the rapture) is usually at least one order of magnitude larger than the deformation corresponding to the elastic limit. The fact that, in this example, the remaining two bars are still within the elastic range, assures through the kinematics of the structure, that the elongation of the three bars remains small, therefore preventing bar 1 of reaching large elongations and eventually breaking.

With the external load increasing beyond the value of $F = 1.553S_0$, it is more convenient to switch to the incremental form of the governing equations, as we want to develop the solution for the structural response valid for the next increment of the applied force F. The kinematics of the structure does not change (i.e., all bars are still present) and the incremental form of the kinematic equation is identical to Eq. (1.18), namely:

$$\begin{bmatrix} \Delta e_1 \\ \Delta e_2 \\ \Delta e_3 \end{bmatrix} = \begin{bmatrix} 0.981 & -0.196 \\ 0.196 & 0.981 \\ -0.894 & -0.447 \end{bmatrix} \begin{bmatrix} \Delta u \\ \Delta v \end{bmatrix} \rightarrow \dot{\boldsymbol{e}} = \boldsymbol{B} \cdot \dot{\boldsymbol{d}} \tag{1.28}$$

Note that dot sign in the previous equation is not referring to derivatives with respect to time, but merely to the increments of the variable considered. This is a common notation adopted in elasto-plastic analysis, and it is sometimes referred to as derivatives with respect to a *fictitious* time.

The fact that we express the governing equations in terms of rates (i.e., increments) of all the variables influences the constitutive equations. For the ith bar, the constitutive relation needs to be written in incremental form as follows:

$$\begin{aligned} \dot{S}_i &= \tfrac{EA_i}{l_i}\dot{e}_i \quad \text{for the elastic response} \\ \dot{S}_i &= 0 \quad \text{for the plastic response} \end{aligned} \tag{1.29}$$

that represents an elegant way to eliminate the force of the bar which started to deform plastically from the equilibrium equations written in a rate form. For the example considered, Eq. (1.29) can be written in matrix form as:

$$\begin{bmatrix} \dot{S}_1 \\ \dot{S}_2 \\ \dot{S}_3 \end{bmatrix} = \begin{bmatrix} 0 & 0 & 0 \\ 0 & \frac{EA_2}{l_2} & 0 \\ 0 & 0 & \frac{EA_3}{l_3} \end{bmatrix} \begin{bmatrix} \dot{e}_1 \\ \dot{e}_2 \\ \dot{e}_3 \end{bmatrix} \rightarrow \dot{\boldsymbol{s}} = \boldsymbol{D}_T^{(1)} \cdot \dot{\boldsymbol{e}} \tag{1.30}$$

in which $\boldsymbol{D}_T^{(1)}$ is the material tangent stiffness matrix valid for the second load increment and it accounts for the material properties written in a rate form.

By considering the rate form expressions of the constitutive equations given by Eq. (1.29) the equilibrium equations can be derived in a rate form as:

$$\dot{F} = 0.196\dot{S}_2 - 0.894 \cdot \dot{S}_3 \tag{1.31}$$

$$0 = 0.981 \cdot \dot{S}_2 - 0.447 \cdot \dot{S}_3 \tag{1.32}$$

based on which, once bar 1 reaches its elastic limit, incrementally speaking, the structure behaves as if bar 1 is removed. The increments $\dot{S}_2$ and $\dot{S}_3$ can be determined from the equilibrium equations that can be written for this example in matrix form as follows:

$$\begin{bmatrix} \dot{F} \\ 0 \end{bmatrix} = \begin{bmatrix} 0.981 & 0.196 & -0.894 \\ -0.196 & 0.981 & -0.447 \end{bmatrix} \begin{bmatrix} \dot{S}_1 \\ \dot{S}_2 \\ \dot{S}_3 \end{bmatrix} \rightarrow \dot{\boldsymbol{f}} = \boldsymbol{B}^T \cdot \dot{\boldsymbol{s}} \tag{1.33}$$

It is worth pointing out that Eq. (1.33) contains the three variables associated to the incremental internal forces of the bars ($\dot{S}_1$, $\dot{S}_2$, and $\dot{S}_3$), while Eqs. (1.31) and (1.32) include only the incremental forces of the bars still in the elastic range ($\dot{S}_2$ and $\dot{S}_3$). In reality, Eq. (1.33) is equivalent to Eqs. (1.31) and (1.32) because the latter equations can be obtained by substituting $\dot{S}_1 = 0$ (in accordance with the definition of the constitutive equations (1.30)) into the former one.

The solution of this example can still be obtained by following the displacement method which, expressed in the incremental form, can be calculated based on the following system of equations:

$$\dot{\boldsymbol{f}} = \boldsymbol{K}_T^{(1)} \cdot \dot{\boldsymbol{d}} \tag{1.34}$$

where $\boldsymbol{K}_T^{(1)} = \boldsymbol{B}^T \boldsymbol{D}_T^{(1)} \boldsymbol{B}$.

Considering that the kinematic equation do not change by switching to the rate form, matrix $\boldsymbol{B}$ can still be computed through (1.18). Matrix $\boldsymbol{K}_T^{(1)}$ is usually referred to as the *tangent stiffness matrix*.

Following the same procedure adopted in the previous load step, it is possible to calculate the increments in the internal bar forces as a function of the external load increment as:

$$\Delta \boldsymbol{s}^{(2)} = \begin{bmatrix} 0 \\ -0.567 \\ -1.242 \end{bmatrix} \Delta F^{(2)} \tag{1.35}$$

By recalling that the two bars are still within the elastic limit, it can be seen that bar 3 is resisting a larger force and, consequently, it is expected to mark the end of the second increment when it is going to reach the elastic limit. It is therefore possible to determine the increment in the external force $\Delta F^{(2)}$ that can be applied before the elastic limit is reached for bar 3.

The updated values of the internal forces in the bars, considering their values at the end of the first increment from Eq. (1.27) and the computed "jump" given by Eq. (1.35), are:

$$\boldsymbol{s}^{(1-2)} = \begin{bmatrix} 1 \\ -0.102 \\ -0.663 \end{bmatrix} S_0 + \begin{bmatrix} 0 \\ -0.567 \\ -1.242 \end{bmatrix} \Delta F^{(2)} \tag{1.36}$$

where superscript $^{(1-2)}$ indicates that this relation is valid throughout the whole second increment, while bars 2 and 3 are still within the elastic limit. By equating the force in the third bar to the one at the elastic limit the increment for the external load to reach the end of the second increment becomes:

$$-0.663 S_0 - 1.242 \Delta F^{(2)} = -S_0 \quad \rightarrow \quad \Delta F^{(2)} = 0.271 S_0 \tag{1.37}$$

while the external load at the end of the second increment is obtained after updating:

$$F^{(2)} = 1.553S_0 + 0.271S_0 = 1.824S_0 \tag{1.38}$$

After the external load reaches this value, bar 3 starts to deform plastically. As there is only one bar that remains elastic, the structure cannot take any further increment in load. From this point, the structure starts to behave as a kinematic mechanism, since an arbitrary displacement of the loading point in the direction of the applied load is possible, with the elastic bar keeping a constant length, while the other two bars are further deforming but with no increments in the internal forces. The displacement would continue until the bars fracture after undergoing extensive deformation, at the point depicted at the end of the tension line with a circle in Fig. 1.7b. Based on these observations, the force computed with Eq. (1.38) represents the maximum load that the structure can take.

By adopting a Young's modulus of E = 200,000 MPa, a cross-sectional area of A = 600 mm^2 and a force at the elastic limit of S_0 = 300 kN, the structural response can be described by the force-displacement curve as shown in Fig. 1.8. The displacement here is normalized by dividing it with its maximal value. There is a clear change in slope of the curve which corresponds to the point when bar 1 starts exhibiting a plastic behavior. For the first loading case analyzed in this example (when the bars possessed an elastic-brittle response), the curve would terminate here as, at this point, bar 1 reaches its elastic limit and the structure is not able to carry any additional load. The maximum load that the structure can take for this case is $F = 465.9$ kN.

For the elastic-perfectly-plastic response, due to the fact that the plastically deformed bar is still able to carry a constant force and that the remaining two elastic bars can resist additional load increments, there is an overall increment of almost 20% in the collapse load of the structure.

This example shows how the constitutive behavior influences the global structural response. When only the elastic part of the response is required, such calculations can be rather simple. Indeed, in the present case there is no difference in the results obtained up to the elastic limit for the two material models considered. For the calculation of the collapse load of a structure, it is important to know what happens beyond the elastic limit as the post-elastic behavior defined by the constitutive model has an important influence on the calculated maximum load that a structure can resist.

Inelastic analyses of structures are also important for the determination of local stress distributions, as the redistribution of internal loads as plastic deformations taken place in one part of the structure can affect the internal actions resisted in other locations of the structure. This aspect has also been observed in the proposed example.

Let us consider an alternative case, where bar 3 starts to deform plastically once it reaches the limit force of $|S_3| = 150$ kN, while the remaining two bars still have the elastic limits defined as $|S_0| = 300$ kN. Clearly, the elastic part of

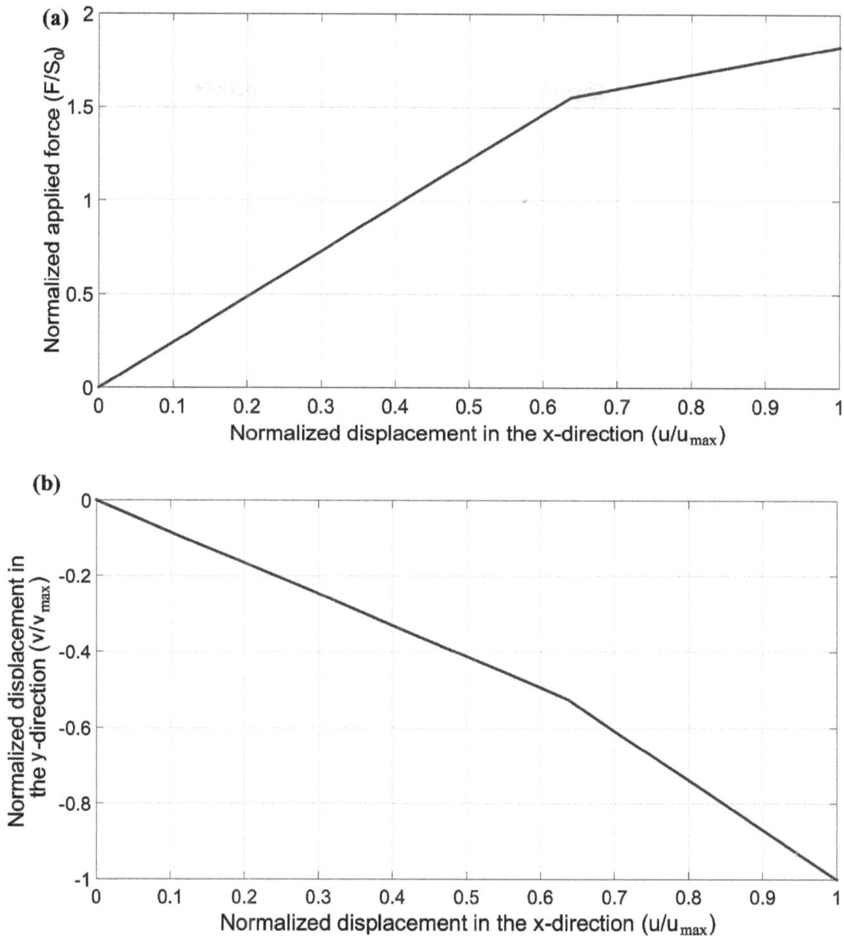

FIGURE 1.8 Structural response for Example 1.2: (a) force-displacement curve and (b) displacement curve of loaded point.

the response will not change, and the values of forces in the bars are still going to be those given by (1.22). In this case, it will be bar 3 that will first exhaust its elastic capacity and start to deform plastically. From this point, incrementally speaking, we can remove it from the structure and consider the remaining system of two bars, in order to calculate the second increment of the applied force. In the absence of bar 3, both bars 1 and 2 would be stretched for a given load direction. Indeed, for the second increment of load, once bar 3 starts to deform plastically, bar 2 will start to unload, considering that the new increment in the load produces positive values for both internal forces, while at the end of the first increment bar 2 was in compression.

Keeping the remaining constants the same as in the previous case, we can calculate the values of the internal forces in the bars corresponding to the collapse load to be equal to:

$$\boldsymbol{s}^{(2)} = \begin{bmatrix} 300\ \text{kN} \\ -8.4\ \text{kN} \\ -150\ \text{kN} \end{bmatrix} \tag{1.39}$$

The value of the internal forces at the end of the first increment for this case can be obtained by simply rescaling the solution given by (1.22) so that the value in bar 3 reaches its elastic limit:

$$\boldsymbol{s}^{(1)} = \begin{bmatrix} 226.2\ \text{kN} \\ -23.2\ \text{kN} \\ -150\ \text{kN} \end{bmatrix} \tag{1.40}$$

By comparing the two values, we can note that the force in the bar 2 is actually reduced by almost three times, when the external load was increased from increment one to increment two. The additional load is balanced by the increment in the force of bar 1 due to the redistribution of the loads that took place once bar 3 started to deform plastically. Such solution can be obtained by a proper elasto-plastic analysis, but at least at the first glance, may look unexpected and is different from what could be estimated by rescaling the elastic solution given by (1.40).

From the above considerations, it is clear that the proper constitutive modeling is of a crucial importance if the structural response is to be evaluated accurately, particularly within the nonlinear range. We saw that, even in this rather simple example quite different results are obtained when different post-elastic behaviors are assumed. Within the considered case the internal forces and deformation are quantified by a single value, namely the force in the bar S_i and the elongation e_i, respectively. For more complex, three-dimensional structures, these measures are represented by *stress* and *strain*. The concept remains the same, so the constitutive equations have still to relate the internal forces to the deformation. In order to establish a framework for further discussion of different constitutive behaviors, a short outline of the stress and strain measurements in a deformable three-dimensional body is presented in the following.

1.2 Stress and strain measurements in a deformable body

When a deformable body is subjected to external actions, it reacts by *deformation*: it changes both shape and volume in the most general case. The body deforms while the internal actions develop to equilibrate the external actions. The way in which these deformations relate to the internal actions is a consequence of the mechanical behavior of the material. To describe mathematically this behavior, it is important to establish the measurements for both the deformation and the internal actions that do not depend on the geometry and dimensions of the body. These measurements are provided through the *analysis of strains and stresses* and are briefly introduced in the following.

1.2.1 Kinematics of a deformable body: strain analysis

When the relative positions between points of a body are altered, we say that the body is deformed or strained. The change in the relative positions of points is referred to as deformation. The deformation, in the general case, may vary from point to point within the strained body and, in order to quantify it over the whole body, it is required to know the displacements of all the material points within the body.

The continuum hypothesis implies the definition of a displacement as a field quantity, therefore having its value defined at each point. Using the appropriate geometrical interpretations, particular deformation measurements can be constructed from the displacement field, leading to the formulation of the strain tensor. It is convenient to establish a relationship between displacements and deformation, as the displacements can be measured physically.

Although there are different approaches to relate displacements to the deformation, or strain measurements, one practical approach stems from geometrical considerations. Let us look at an infinitesimally small domain with sides $dx \times dy$, within a strained body as shown in Fig. 1.9. As a result of the deformation, the initially rectangular domain turns into a general quadrilateral, by having the points $ABCD$ moved into the new positions $A'B'C'D'$.

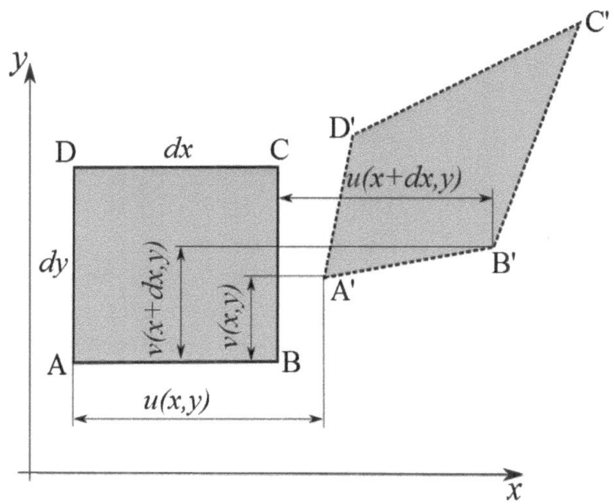

FIGURE 1.9 Two-dimensional domain with the initial and deformed configurations within an infinitesimal small domain.

When moving from the initial to the final configuration, not all of the displacements contributed to the deformation. We now want to extract just the part of the displacements that quantifies the relative change in the position between the points. We can therefore define a relative change in the length measured along the x-direction, defined as a change in length per unit length. This variable represents the *normal strain* or *extensional strain* for the direction x, which is positive if the fibers increase the length, or negative for shortening fibers.

By considering that the orientation is such that direction $A - B$ coincides with the x-axis in the initial configuration (i.e., before the deformation), the normal strain for the x-direction can be formulated as:

$$\varepsilon_x = \frac{\overline{A'B'} - \overline{AB}}{\overline{AB}} \tag{1.41}$$

where $\overline{\times\times}$ indicates the distance between the two points specified under the line.

We now want to express the relation depicted in Eq. (1.41) in terms of displacements. For this purpose, we can consider the displacement components u and v measured along the $x-$ and $y-$axes, respectively, and assume them to be continuous functions over the domain of interest. By taking the corner point A as a reference point with coordinates (x, y), the point B can be defined by coordinates $(x + dx, y)$. The displacements along both directions of point A are described by:

$$\begin{aligned} &u(x, y) \\ &v(x, y) \end{aligned} \tag{1.42}$$

and, for point B, they can be written as a Taylor expansion, while keeping only the first derivatives, as:

$$\begin{aligned} u(x + dx, y) &\approx u(x, y) + \frac{\partial u(x, y)}{\partial x} dx \\ v(x + dx, y) &\approx v(x, y) + \frac{\partial v(x, y)}{\partial x} dx \end{aligned} \tag{1.43}$$

Strictly speaking, we had to used the sign $\approx$ since we dropped terms of higher order than the first derivatives. For ease of notation, we will denote $\partial u(x, y)/\partial x$ and $\partial v(x, y)/\partial x$ as $\partial u/\partial x$ and $\partial v/\partial x$, respectively, in the following.

Considering that dx is infinitesimally small, points A and B are neighboring points and, therefore, the linearization of displacement functions $u(x, y)$ and $v(x, y)$ over the infinitesimal domain $dx \times dy$ gives generally accurate results.

From the geometrical considerations presented in Fig. 1.9, we can write that the length $A'B'$ is equal to:

$$\overline{A'B'} = \sqrt{\left(dx + \frac{\partial u}{\partial x} dx\right)^2 + \left(\frac{\partial v}{\partial x} dx\right)^2} \tag{1.44}$$

Consistently with small deformation theory, higher order terms can be neglected, and the change in length can be expressed as the difference between the projections of AB and $A'B'$ along the x-axis. While the initial direction AB is coaligned with the x-axis, thus having this projection equal to dx, by replacing the length $A'B'$ with its projection to the x-axis an approximation is made, namely:

$$\overline{A'B'} \approx \left(1 + \frac{\partial u}{\partial x}\right) dx \tag{1.45}$$

By comparing Eqs. (1.44) and (1.45), it can be observed that the latter equation is obtained when, in the former one, it is assumed that $\left(\frac{\partial v}{\partial x} dx\right)^2 \approx 0$. Since $\left(\frac{\partial v}{\partial x} dx\right)^2$ is of the same order as $\left(\frac{\partial u}{\partial x} dx\right)^2$, by assuming that the former term can be neglected hypothesizes that the displacement component u undergoes a more pronounced change along the x-direction than v. This condition is true if there are small angular deformation and small rigid-body rotations, since $\frac{\partial v}{\partial x}$ is produced by the rotation of AB. Small angular deformations are generally ascertained within small deformation theory, while large rotations can occur even for small or null deformations. For such problems the strain measure defined here performs poorly and generates fictitious strains, therefore requiring, in these instances, the use of alternative strain measures.

By substituting Eq. (1.45) into Eq. (1.41) and by recalling that $\overline{AB} = dx$, the normal strain in the x-direction can be expressed as:

$$\varepsilon_x = \frac{\partial u}{\partial x} \tag{1.46}$$

As previously applied for point B in Eq. (1.43), the displacement components of point D can be written through a truncated Taylor series expansion as:

$$\begin{aligned} u(x, y + dy) &\approx u(x, y) + \frac{\partial u(x, y)}{\partial y} dy \\ v(x, y + dy) &\approx v(x, y) + \frac{\partial v(x, y)}{\partial y} dy \end{aligned} \tag{1.47}$$

In the following, we will denote, for ease of notation, $\partial u(x, y)/\partial y$ and $\partial v(x, y)/\partial y$ as $\partial u/\partial y$ and $\partial v/\partial y$, respectively. Based on similar geometrical considerations (i.e., this time applied to the direction $A - D$), the normal strain in the y-direction can be written as:

$$\varepsilon_y = \frac{\partial v}{\partial y} \tag{1.48}$$

With these two components of strain, the relative change of length is defined in the x and y-directions in the neighborhood of the reference point A. Through these components, we can quantify how much the domain stretches (or compresses) along the two directions. In order to uniquely describe the change in shape, we need to evaluate a third strain component: the *shear strain.*

The shear strain is defined as the change in the angle between two originally orthogonal directions. It has a positive (negative) sign if the right angle decreases (increases). The shear strain for the x- and y-directions in the neighborhood of point A, that is taken as the reference point, is thus equal to (see Fig. 1.10):

$$\gamma_{xy} = \frac{\pi}{2} - \angle B'A'D' = \alpha + \beta \tag{1.49}$$

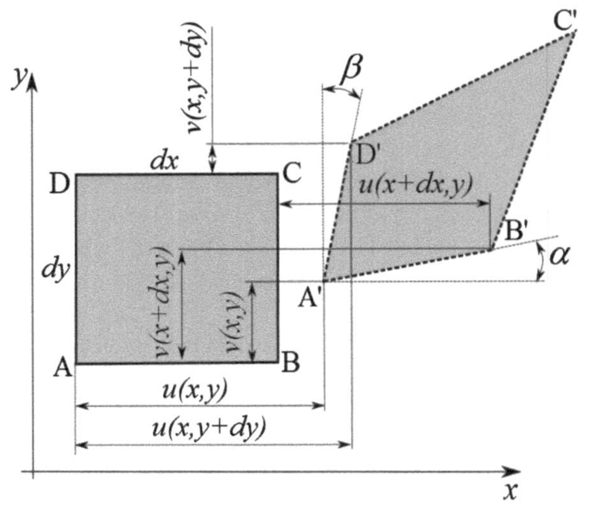

FIGURE 1.10 Two-dimensional domain for the definition of the shear strain.

For small deformations, the angles α and β are also small, and we can assume that $\tan\alpha \approx \alpha$ and $\tan\beta \approx \beta$. The shear strain component can then be expressed in terms of displacements as:

$$\gamma_{xy} = \frac{v(x+dx,y) - v(x,y)}{dx + u(x+dx,y) - u(x,y)} + \frac{u(x,y+dy) - u(x,y)}{dy + v(x,y+dy) - v(x,y)} \tag{1.50}$$

By substituting Eqs. (1.43) and (1.47) into Eq. (1.50) the following expression is obtained:

$$\gamma_{xy} = \frac{\frac{\partial v}{\partial x}dx}{dx + \frac{\partial u}{\partial x}dx} + \frac{\frac{\partial u}{\partial y}dy}{dy + \frac{\partial v}{\partial y}dy} \tag{1.51}$$

By neglecting higher order terms (i.e., by adopting $\frac{dx}{dx+\frac{\partial u}{\partial x}dx} \approx 1$ and $\frac{dy}{dy+\frac{\partial v}{\partial y}dy} \approx 1$), we obtain the following simplified expression for the shear strain component valid for small displacements:

$$\gamma_{xy} = \frac{\partial u}{\partial y} + \frac{\partial v}{\partial x} \tag{1.52}$$

Since, by definition, the shear strain component is a change in angle between two originally orthogonal directions (given as indexes of the strain component), the order of indexes plays no role and we can write that $\gamma_{xy} = \gamma_{yx}$.

By applying similar geometrical considerations for planes $y - z$ and $x - z$, it is possible to extend this concept to a three-dimensional body and obtain the following results:

$$\begin{aligned} \varepsilon_x = \frac{\partial u}{\partial x}, \varepsilon_y = \frac{\partial v}{\partial y}, \varepsilon_z = \frac{\partial w}{\partial z} \\ \gamma_{xy} = \frac{\partial u}{\partial y} + \frac{\partial v}{\partial x}, \gamma_{xz} = \frac{\partial u}{\partial z} + \frac{\partial w}{\partial x}, \gamma_{yz} = \frac{\partial v}{\partial z} + \frac{\partial w}{\partial y} \end{aligned} \tag{1.53}$$

Eqs. (1.53) formulate strain-displacement relations, important for finite element implementations, where the problem is usually formulated in terms of unknown displacements. The three normal strains quantify the dilation of the deformed body, while the three shear strains depict the distortion of the body. Together, they form the *engineering strain tensor* that defines the deformation of the body by making a reference to the initial configuration, while assuming small deformations and displacements.

From previous considerations, we have seen that, in a single point of a deformed body, the value of the normal strain (i.e., dilation) changes when computed along different directions. If we know the dilation for any arbitrary direction from a particular point as well as the shear deformation for any orthogonal pair of directions passing through the same point, we can say that we know the *state of deformation* in that point. It can be easily demonstrated that this is sufficient to determine the strain tensor at that point, based on which we can then compute either dilation or shear strain for any arbitrary direction or pair of directions.

Let us consider a case where we know the six components of the strain tensor at point O of a deformed body, and let us assume that we want to calculate the dilation along a direction passing through point O, defined by vector $\vec{OP}$ (see Fig. 1.11a). The displacement of point P can be computed on the basis of the known strain components with respect to the $x - y - z$ coordinate system. Considering that the deformations are small, we may apply superposition to consider the displacements produced along each direction of the relevant strain components. For example, the three strain components that affect the displacement of point P along the x-direction are ε_{xx}, γ_{xy} and γ_{xz}. Based on this, the infinitesimal displacement of point P along the x-direction du_P can be determined as follows (see Fig. 1.11b):

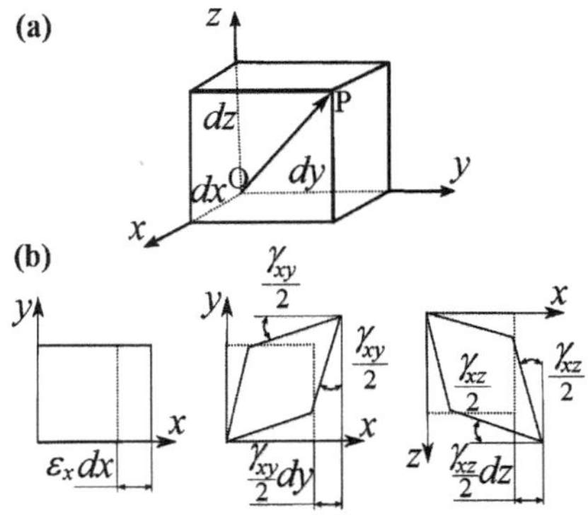

FIGURE 1.11 Dilation for an arbitrary direction expressed in terms of engineering strain components.

$$du_{\mathrm{P}} = \varepsilon_{xx}dx + \frac{1}{2}\gamma_{xy}dy + \frac{1}{2}\gamma_{xz}dz \tag{1.54}$$

It should be noted that the complete angle of shear deformation γ_{xy} is divided into two halves, one placed on each side of the angle considered. By definition, the shear deformation is the angle that quantifies how much an originally right angle changes. These rotations are the same for a pure shear deformation, which is the case that we are considering here, as we are superimposing separate effects.

From similar geometrical considerations applied along the other two directions, we can obtain the displacement components of point P along the y- and z-axes:

$$dv_{\mathrm{P}} = \frac{1}{2}\gamma_{yx}dx + \varepsilon_{yy}dy + \frac{1}{2}\gamma_{yz}dz \tag{1.55}$$

$$dw_{\mathrm{P}} = \frac{1}{2}\gamma_{zx}dx + \frac{1}{2}\gamma_{zy}dy + \varepsilon_{zz}dz \tag{1.56}$$

Vector $\vec{OP}$ can be expressed in terms of its components as $\vec{OP} = \boldsymbol{dr} = [dx, dy, dz]^T = dr[n_x, n_y, n_z]^T$, with $\boldsymbol{n} = [n_x, n_y, n_z]^T$ being a unit vector along the $\vec{OP}$ direction. The displacements components of point P can therefore be expressed as:

$$\begin{aligned} du_{\mathrm{P}} &= \varepsilon_{xx}drn_x + \frac{1}{2}\gamma_{xy}drn_y + \frac{1}{2}\gamma_{xz}drn_z \\ dv_{\mathrm{P}} &= \frac{1}{2}\gamma_{yx}drn_x + \varepsilon_{yy}drn_y + \frac{1}{2}\gamma_{yz}drn_z \\ dw_{\mathrm{P}} &= \frac{1}{2}\gamma_{zx}drn_x + \frac{1}{2}\gamma_{zy}drn_y + \varepsilon_{zz}drn_z \end{aligned} \tag{1.57}$$

Since we consider the case in which point O is fixed, the displacement of point P, defined by the displacement vector $[du_P, dv_P, dw_P]^T$, is due to pure deformation (i.e., there is no translation in it). If we divide this vector by the initial length between the points O and P (i.e., by dr), we can get the *deformation vector*, which represents the quantification of the deformation at point P. The components of the deformation vector are therefore equal to:

$$\begin{aligned}
d_x^{(n)} &= \varepsilon_{xx} n_x + \frac{1}{2}\gamma_{xy} n_y + \frac{1}{2}\gamma_{xz} n_z \\
d_y^{(n)} &= \frac{1}{2}\gamma_{yx} n_x + \varepsilon_{yy} n_y + \frac{1}{2}\gamma_{yz} n_z \\
d_z^{(n)} &= \frac{1}{2}\gamma_{zx} n_x + \frac{1}{2}\gamma_{zy} n_y + \varepsilon_{zz} n_z
\end{aligned} \tag{1.58}$$

which written in matrix form reads:

$$\boldsymbol{d}^n = \begin{bmatrix} d_x^{(n)} \\ d_y^{(n)} \\ d_z^{(n)} \end{bmatrix} = \begin{bmatrix} \varepsilon_{xx} & \frac{1}{2}\gamma_{xy} & \frac{1}{2}\gamma_{xz} \\ \frac{1}{2}\gamma_{yx} & \varepsilon_{yy} & \frac{1}{2}\gamma_{yz} \\ \frac{1}{2}\gamma_{zx} & \frac{1}{2}\gamma_{zy} & \varepsilon_{zz} \end{bmatrix} \begin{bmatrix} n_x \\ n_y \\ n_z \end{bmatrix} = \boldsymbol{E} \cdot \boldsymbol{n} \tag{1.59}$$

The deformation vector $\boldsymbol{d}^n$ is a scaled vector of displacement of point P in a view of being multiplied by scalar $\frac{1}{dr}$. Therefore it has an arbitrary direction and represents a measure of deformation as it contains the relative change in length. If we project it on the direction $\boldsymbol{n}$, we will get the dilation for this direction:

$$\varepsilon_n = \boldsymbol{n}^T \boldsymbol{E} \boldsymbol{n} \tag{1.60}$$

By projecting it on an arbitrary perpendicular direction to vector $\boldsymbol{n}$, defined by vector $\boldsymbol{s}$, a shear deformation is obtained, which reads:

$$\gamma_{ns} = \boldsymbol{s}^T \boldsymbol{E} \boldsymbol{n} \tag{1.61}$$

What we just demonstrated is that in a particular point of a deformed body, it is enough to know the quantity:

$$\boldsymbol{E} = \begin{bmatrix} \varepsilon_{xx} & \frac{1}{2}\gamma_{xy} & \frac{1}{2}\gamma_{xz} \\ \frac{1}{2}\gamma_{yx} & \varepsilon_{yy} & \frac{1}{2}\gamma_{yz} \\ \frac{1}{2}\gamma_{zx} & \frac{1}{2}\gamma_{zy} & \varepsilon_{zz} \end{bmatrix} = \begin{bmatrix} \varepsilon_{xx} & \varepsilon_{xy} & \varepsilon_{xz} \\ \varepsilon_{yx} & \varepsilon_{yy} & \varepsilon_{yz} \\ \varepsilon_{zx} & \varepsilon_{zy} & \varepsilon_{zz} \end{bmatrix} \tag{1.62}$$

to be able to compute dilation for any arbitrary direction $\boldsymbol{n}$, and shear strain for the two perpendicular directions $\boldsymbol{n} - \boldsymbol{s}$ ($\boldsymbol{s}$ being an arbitrary direction perpendicular to $\boldsymbol{n}$ direction). Second-order tensor $\boldsymbol{E}$ represents the *engineering strain tensor*. It is worth noting that out-of-diagonal terms in the engineering strain tensor are one-half of the shear deformation $\varepsilon_{ij} = \frac{1}{2}\gamma_{ij}$. Since, by definition, there is only one shear deformation describing the change in angle between two directions the strain tensor is a symmetric tensor and it is usually represented, for ease of notation, as a column vector:

$$\boldsymbol{\varepsilon} = \begin{bmatrix} \varepsilon_{xx} \\ \varepsilon_{yy} \\ \varepsilon_{zz} \\ \varepsilon_{xy} \\ \varepsilon_{xz} \\ \varepsilon_{yz} \end{bmatrix} \quad (1.63)$$

This representation is called the Voigt notation, in which there is a reduction in notation for all symmetric tensors. In this particular case, the second-order deformation tensor is represented as a vector. It is important to note that tensor properties are lost with this reduction, so the vector given by Eq. (1.63) cannot be treated as a first-order tensor.

1.2.2 Equilibrium considerations and stress analysis

Let us consider a deformable body subjected to an arbitrary system of equilibrated external actions (i.e., with the resultant equal to zero) as depicted in Fig. 1.12.

The external actions acting on a deformable body are transferred through it in the form of internal forces. The continuum hypothesis implies that the internal actions are distributed continuously within the solid. In order to study the nature of this distribution, let us consider a deformable body subjected to an arbitrary system of equilibrated external actions (i.e., with the resultant equal to zero) depicted in Fig. 1.12.

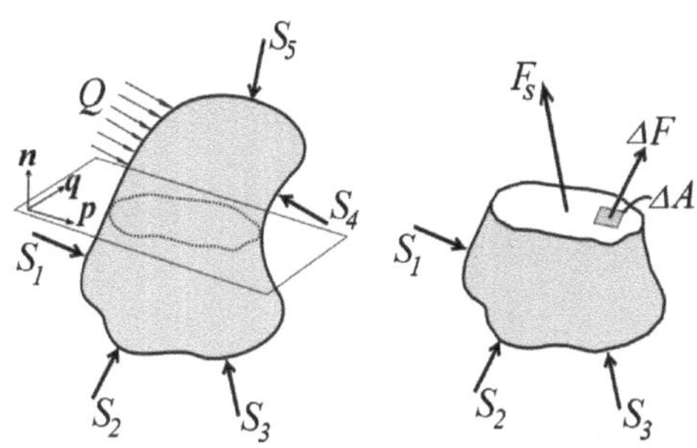

FIGURE 1.12 Deformable body subjected to an equilibrated system of actions.

We need to cut the body to determine the internal actions. This can be carried out by means of an imaginary plane, marked by $\boldsymbol{p} - \boldsymbol{q}$ with the normal $\boldsymbol{n}$ on the left diagram of Fig. 1.12. Let us then consider only one part of the body as shown on the right diagram of Fig. 1.12. At the cut surface a resultant surface force $\boldsymbol{F}_s$ has been included to account for the internal actions that consider the influence of the removed part. By considering a small area ΔA of the section the resultant surface force acting on it can be denoted as $\Delta \boldsymbol{F}$, based on which we can define the *traction vector* or *stress vector* as follows:

$$\boldsymbol{t} = \lim_{\Delta A \to 0} \frac{\Delta \boldsymbol{F}}{\Delta A} \tag{1.64}$$

In an arbitrary loading context, the traction varies from point to point over a section and its resultant can be computed from:

$$\boldsymbol{F}_s = \int_S \boldsymbol{t}(p, q) ds \tag{1.65}$$

The adopted definition of stress or traction vector is convenient for the quantification of the internal actions as it is expressed in terms of a force per unit area and, therefore, it does not depend on the dimensions of the body. The direction and intensity of this vector is defined by the external actions applied to the removed part. It is usually more convenient to define the traction vector in terms of its projections along the normal of the intersecting plane $\vec{n}$ and of the two axes defining the intersecting plane, namely:

$$\boldsymbol{t} = \sigma_n \vec{n} + \tau_{np} \vec{p} + \tau_{nq} \vec{q} \tag{1.66}$$

where the projections are referred to as the *normal stress component* σ_n and the *shearing stress components* τ_{pn} and τ_{qn}, for the fact that they are acting along the normal and within the intersecting plane, respectively.

By definition, the traction vector depends on the location of the point and on the intersecting plane considered. In fact, for a particular point, the traction vector can vary as a function of the orientation of the intersecting surface normal. If we are capable of formulating the traction (or stress) vector at a given point for any arbitrary plane, we can then claim to know the *stress state* at that particular point. The minimum amount of data required to define the stress state in a point can be deduced from equilibrium considerations.

Let us assume that we cut from the loaded body an infinitesimally small pyramidal element as depicted in Fig. 1.13. The pyramid is enclosed by three orthogonal planes $x-y$, $x-z$, and $z-y$, and an arbitrary plane defined by the normal $\vec{n}$, which, in a matrix notation, is defined by $\boldsymbol{n} = [n_x, n_y, n_z]^T$ with n_x, n_y and n_z being its projections on the x-, y-, and z-axes, respectively.

Since the original body is in equilibrium, the pyramidal element has to be subjected to an equilibrated system of actions. The actions acting on the element are determined by integrating the traction vectors over the surface areas. In particular, the tractions acting on the three orthogonal surfaces can be expressed in terms of their components as:

$$\begin{aligned} \boldsymbol{t}_x &= [\sigma_x, \tau_{xy}, \tau_{xz}]^T \\ \boldsymbol{t}_y &= [\tau_{yx}, \sigma_y, \tau_{yz}]^T \\ \boldsymbol{t}_x &= [\tau_{zx}, \tau_{zy}, \sigma_z]^T \end{aligned} \tag{1.67}$$

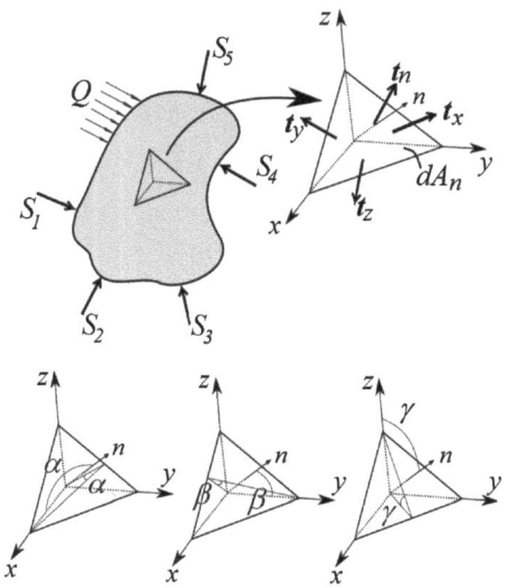

FIGURE 1.13 Stress vector acting on an arbitrary plane defined by the normal.

After projecting all the forces to the x-, y-, and z-directions, equilibrium can be enforced along the three directions as follows:

$$\begin{aligned} t_n^x dA_n &= \sigma_x dA_x + \tau_{yx} dA_y + \tau_{zx} dA_z \\ t_n^y dA_n &= \tau_{xy} dA_x + \sigma_y dA_y + \tau_{zy} dA_z \\ t_n^z dA_n &= \tau_{xz} dA_x + \tau_{yz} dA_y + \sigma_z dA_z \end{aligned} \tag{1.68}$$

where dA_n is the area of the triangle normal to $\boldsymbol{n}$, while dA_x, dA_y, and dA_z denote the areas of the remaining three triangles. Considering that $n_x = \cos\alpha$, $n_y = \cos\beta$, and $n_z = \cos\gamma$, where α, β, and γ are the angles between vector $\boldsymbol{n}$ and the x, y, and z axes, respectively, the relationship between the orthogonal surfaces and the one defined by $\boldsymbol{n}$ can be written as (see Fig. 1.13):

$$dA_x = dA_n n_x, \quad dA_y = dA_n n_y, \quad dA_z = dA_n n_z. \tag{1.69}$$

By substituting Eq. (1.69) into Eq. (1.68), the following relations are obtained:

$$\begin{aligned} t_n^x &= \sigma_x n_x + \tau_{yx} n_y + \tau_{zx} n_z \\ t_n^y &= \tau_{xy} n_x + \sigma_y n_y + \tau_{zy} n_z \\ t_n^z &= \tau_{xz} n_x + \tau_{yz} n_y + \sigma_z n_z \end{aligned} \tag{1.70}$$

that can be written in matrix form as:

$$\begin{bmatrix} t_n^x \\ t_n^y \\ t_n^z \end{bmatrix} = \begin{bmatrix} \sigma_x & \tau_{yx} & \tau_{zx} \\ \tau_{xy} & \sigma_y & \tau_{zy} \\ \tau_{xz} & \tau_{yz} & \sigma_z \end{bmatrix} \begin{bmatrix} n_x \\ n_y \\ n_z \end{bmatrix} \rightarrow \boldsymbol{t}_n = \boldsymbol{S} \cdot \boldsymbol{n} \tag{1.71}$$

The last equation is usually referred to as the *Cauchy equilibrium equation* and it highlights how the stress vector $\boldsymbol{t}_n$ can be calculated for any arbitrary plane defined by a normal $\boldsymbol{n}$ if the tractions (i.e., stress vectors) for three orthogonal planes are known. Based on this consideration, the state of the stress at a single point is defined once we know the 9 components in matrix $\boldsymbol{S}$, referred to as the *stress tensor.*

It can be further demonstrated that there are only 6 components of the stress tensor that are independent because, due to the symmetry, we have $\tau_{ij} = \tau_{ji}$. It is common therefore also for the stress tensor to adopt Voigt notation and express the stress tensor as a vector with six stress components as follows:

$$\sigma = \begin{bmatrix} \sigma_x \\ \sigma_y \\ \sigma_z \\ \tau_{xy} \\ \tau_{xz} \\ \tau_{yz} \end{bmatrix} \tag{1.72}$$

For completeness, symmetry of the stress tensor can be demonstrated based on equilibrium considerations by writing the moment equilibrium with respect the x-axis for an infinitesimally small cube $dx \times dy \times dz$, with the origin of the reference system taken in the center of the small cube. After neglecting the higher order terms, the following equation is obtained:

$$(\tau_{yz} dx dz) dy - (\tau_{zy} dx dy) dz = 0 \rightarrow \tau_{yz} = \tau_{zy} \tag{1.73}$$

The symmetry of the other two index pairs can be derived by enforcing moment equilibrium with respect to the y- and z-axes.

The loss of tensor properties with Voigt reduced notation is easily demonstrated within stress (or strain) analysis. When representing second order stress tensor in full order notation (i.e., as a matrix) principal stresses are computed as the eigenvalues of that matrix. Clearly, the eigenvalues of a column vector cannot be computed and, for this operation, it is required to re-express stress tensor in a full order representation.

Certain important tensor calculations can be preserved by slight modifications within the reduced Voigt notation. Here, we will just briefly make a reference to the internal energy, stored within the deformed body in the context of Voigt notation. This quantity is going to be explained and used more extensively within Chapter 2. By considering stress and strain as second-order tensors, the internally stored energy is computed as:

$$U = \boldsymbol{S} : \boldsymbol{E} \tag{1.74}$$

where ":" is the contraction operator that implies $\boldsymbol{S} : \boldsymbol{E} = S_{ij} \cdot E_{ij}$, in which summation is implied over the range of the repeated indexes i and j.

To obtain the same result in the reduced notation, the out-of-diagonal terms have to be multiplied. It is therefore common that a strain vector in Voigt notation is represented as:

$$\boldsymbol{\varepsilon} = \begin{bmatrix} \varepsilon_{xx} \\ \varepsilon_{yy} \\ \varepsilon_{zz} \\ 2\varepsilon_{xy} \\ 2\varepsilon_{xz} \\ 2\varepsilon_{yz} \end{bmatrix} \tag{1.75}$$

so that the last three terms represent engineering shear strain (i.e., γ_{ij}). In such way the scalar product $\boldsymbol{\varepsilon}^T \cdot \boldsymbol{\sigma}$ gives the same result, therefore the internally stored energy, as the tensor double-dot product given by (1.74). Indeed, it is generally common to represent the strain tensor through Eq. (1.75) instead of Eq. (1.63).

1.3 Constitutive modeling for solid deformable bodies

Constitutive theory studies the mechanical behavior of materials subjected to external actions. The nature of such response represents the physical property of the material and is generally deduced from experiments. For modeling purposes, a mathematical model is needed to transfer into analytical terms the major phenomenological characteristics of the material observed in the experiments. Such models are called the *constitutive models*.

For deformable solids, the governing equations of the constitutive models relate stress components to strain components. Most commonly, stresses are expressed as a function of strains while, in more complex deformation scenarios, they may depend also on the strain rate (i.e., the velocity of deformation), the strain history, the temperature, other material properties, or environmental conditions.

Deformation mechanisms that develop under the application of loads are determined by the material structure at the atomic, molecular or crystal scales. Such scales are much smaller than usual dimensions of the structural components, where commonly the material is treated as a homogeneous continuum. The constitutive models which tend to capture and describe mechanisms at the former scale belong to *micro-mechanical approaches*. The *phenomenological approaches* address the latter scale, and define governing equations by studying the response of a representative volume of material.

Phenomenological constitutive models are usually better suited for structural applications as micro-mechanical approaches are generally more demanding, being intrinsically multiscale. In some situations, inclusions of microscopic information may become crucial to describe the mechanical response of a material, and purely phenomenological model may be insufficient to adequately describe the material response. In such cases, recourse should be made to an appropriate multiscale model.

None of the constitutive models is general enough to describe the material response under all possible circumstances, which makes the selection of an appropriate model an important task in the modeling process. The correct choice clearly depends on the material type, but also on the loading context that is going to be considered. To illustrate these aspects, let us consider a simple tensile test, commonly performed to obtain stress–strain curves on the basis of measured tensile force applied during the test and corresponding elongation of the specimen (see Fig. 1.14a).

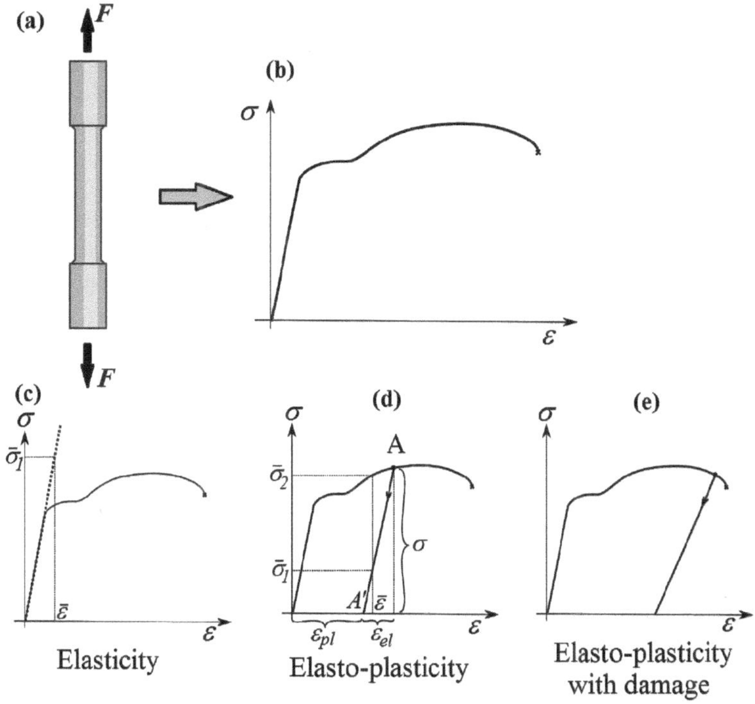

FIGURE 1.14 Stress–strain curves obtained from tensile tests and with different constitutive models.

The stress–strain curve illustrated in Fig. 1.14b is obtained from a tensile test and it is typical for ductile materials, such as metals and polymers. In such cases, the stress depends linearly on the strain for initially small values of applied load (and induced stress) and the deformations are fully elastic, meaning that they would totally disappear upon removal of the applied load. In order to model the material response for this range of loads, it is sufficient to consider a *linear-elastic* constitutive model, where the stress depends only on the current deformation. The curve shows the uniaxial representation of this model, while for a general multidimensional state of stress, each stress component is taken to be linearly related to the strain components. Such relationship is usually expressed in a matrix form as:

$$\begin{bmatrix} \sigma_x \\ \sigma_y \\ \sigma_z \\ \tau_{xy} \\ \tau_{xz} \\ \tau_{yz} \end{bmatrix} = \begin{bmatrix} D_{11} & D_{12} & . & . & . & D_{16} \\ D_{21} & D_{22} & . & . & . & . \\ . & . & . & . & . & . \\ . & . & . & . & . & . \\ . & . & . & . & . & . \\ D_{61} & . & . & . & . & D_{66} \end{bmatrix} \begin{bmatrix} \varepsilon_x \\ \varepsilon_y \\ \varepsilon_z \\ \gamma_{xy} \\ \gamma_{xz} \\ \gamma_{yz} \end{bmatrix} \rightarrow \boldsymbol{\sigma} = \boldsymbol{D} \cdot \boldsymbol{\varepsilon} \tag{1.76}$$

where matrix $\boldsymbol{D}$ is referred to as *material elasticity matrix.* The entries of this matrix are related to the elastic properties of the material. The 36 members of matrix $\boldsymbol{D}$ are not all independent and, for the simplest elasticity constitutive model based on linear, isotropic elasticity, they are reduced to 2 elastic constants. Detailed aspects of these constitutive representations are not further considered here as they are discussed in later parts of the book.

Another important remark here is related to the representation of matrix $\boldsymbol{D}$. Considering that both stress and strain are tensors of second order, what linearly relates the latter to the former must be a tensor of fourth order. Indeed, the material elasticity tensor $\boldsymbol{D}$ is a fourth-order tensor, with formally 81 entries. However, due to the symmetries—both the inner and the outer symmetries such that $D_{ijkl} = D_{klij} = D_{jikl} = D_{ijlk}$, within Voigt notation it is reduced to a symmetric matrix.

The linear relationship between stress and strain is usually exhausted after certain levels of loads are achieved. For an uniaxial stress state, the material starts to develop an irreversible plastic deformation once the stress reaches a threshold value—called the *yield limit.* Beyond this point, predictions based on a linear-elastic constitutive model would lead to an overestimation of the stress (see Fig. 1.14c), and plasticity theory has to be considered to develop a suitable *elasto-plastic* constitutive model. This further increases the complexity of the model as the stress now cannot be related to the complete value of the strain (called the *total strain*) but only to its recoverable part—the *elastic strain* component (see Fig. 1.14d).

Another restatement of this complexity is that the stress starts to depend on the *history* of previous deformation, apart on its current value. As it may be observed from Fig. 1.14d, the same value of strain $\bar{\varepsilon}$ can correspond to different values of stresses, denoted in the figure as $\bar{\sigma}_1$ and $\bar{\sigma}_2$, that depend on the previous loading history of the material. For example, stress $\bar{\sigma}_2$ can be achieved upon first loading cycle, while the material would exhibit a stress of $\bar{\sigma}_1$ if it had been first loaded up to point A, completely unloaded, and then further reloaded up to $\bar{\sigma}_1$. An effective way to avoid cumbersome formulations of constitutive models as a function of the history of certain variables is to introduce *internal state variables* that relate to the dissipative part of the energy (i.e., the energy which is not recovered upon the removal of loads). In this manner, the state of stress depends only on current values (and not the history) of certain state variables. Elasto-plastic constitutive models usually consider the accumulated plastic strain as an internal state variable, so the stress state can be uniquely determined by knowing its current value and the elastic constants.

The unloading branch of the curve ($A - A'$) depicted in Fig. 1.14d follows the line which is parallel to the initial linear-elastic line up to the first yield limit σ_y. In the elasto-plastic constitutive models the elastic constants (identical to those also used in the linear elastic models), are adopted to compute these unloading stresses.

When the material is subjected to increasing levels of loads, other phenomena start to develop, such as the occurrence of internal damage, that macroscopically is manifested by the drop of the elastic constants and by the depart from the linear elastic behavior of the stress–strain curve in the loading stage (see Fig. 1.14e). In the unloading stage, the extent of occurred damage influences the slope of the unloading branch. To describe this behavior a more refined constitutive model has to be employed which accounts for *damage with plasticity* and incorporates a larger number of state variables.

At this point, let us emphasize the important difference between damage and plasticity phenomenological models. These models may look quite similar when considering the loading part, where, in fact, they may be formulated to give exactly the same response. This may be evidenced by considering a simple uniaxial nonlinear response depicted in Fig. 1.15. If the nonlinear part is to be modeled by a plasticity constitutive model, the stress should be related to the elastic part of the strain, namely:

$$\sigma = E \cdot (\varepsilon^{tot} - \varepsilon^{pl}) = E\varepsilon^{tot} - E\varepsilon^{pl} \tag{1.77}$$

where ε^{tot} represents total strain and ε^{pl} its plastic (i.e., irreversible) part.

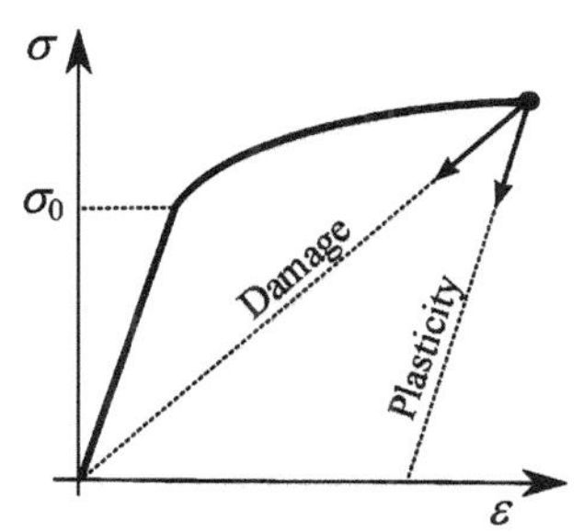

FIGURE 1.15 Uniaxial stress–strain curve modeled with a plasticity or a damage model.

The damage model assumes that the inelastic part of the deformation is mainly due to the development of microcracks, macroscopically manifested as the reduction in elastic constants. These models are usually formulated employing, as the internal state variable, the *damage variable D*, with value growing from 0 (for an undamaged material) to 1 (for a completely damaged material), thus having the stress–stain relation expressed as:

$$\sigma = E(1 - D)\varepsilon^{tot} = E\varepsilon^{tot} - ED\varepsilon^{tot} \tag{1.78}$$

The substantial difference between the two phenomena is that plasticity introduces irreversible deformation, while, with the damage model, the deformation is totally reversible, but the stiffness is reduced. This behavior cannot be observed during loading as the two models give the same response in the loading stage if they are "tuned" so that $E\varepsilon^{pl} = ED\varepsilon^{tot}$, which is possible by postulating appropriate evolution laws for the plastic strain in one case and for the damage variable in the another case.

During unloading ε^{tot} clearly decreases, therefore the satisfaction of the condition $E\varepsilon^{pl} = ED\varepsilon^{tot}$ implies that either ε^{pl} drops or D increases. None of these two solutions is feasible because the former does not satisfy the fundamental requirement for the plastic deformation to remain upon removal of the applied load, while the latter is not acceptable from a physical viewpoint as it would require the damage of the material to grow during the removal of the external loads. These considerations highlight the fact that the differences between the two phenomena become evident during the unloading of the material, as they follow different unloading curves (see Fig. 1.15).

The previous discussion has demonstrated that, even for the same material, the loading requirements and scenarios influence the choice of the suitable constitutive models to be used in the modeling. In particular, we have seen that constitutive models with different complexities can be successfully used, provided that they are selected, considering the purpose of the analysis, to well predict the material response for the expected typology and range of loading.

In general, depending on the loading scenario, the complexity of the material response can increase well beyond the one discussed in the previous cases. For example, if the application of the load can occur at different temperatures, significantly different from room temperature, it is expected that the elastic constants would vary for most of the materials with temperature. In such situations, it becomes essential to express the governing model parameters as a function of temperature. The material response might also undergo significant changes at different temperatures. At higher temperatures, the long-term behavior of some materials may no longer be accurately computed by linear elasticity when subjected to even small deformation, and time-dependent effects, such as creep and stress relaxation, might need to be accounted for because crucial for an accurate representation of the material response.

The above discussion corroborates the importance of a careful and adequate selection of a constitutive model that well suits the nature of the material (e.g., ductile, brittle, and others) and the expected loading context of the particular application. Once a constitutive model is selected, the reliability of the subsequent analyses rests on the accuracy of the governing material parameters entering into the constitutive equations. For this reason, there are two problems that are central for an accurate structural analysis: (i) the selection of an adequate constitutive model and (ii) the associated quantification of the governing material constants that quantify the mechanical properties of the material. Both of these problems are going to be treated in subsequent chapters by considering different structural materials and by relying on the finite element method for the numerical implementations, briefly presented in the following section.

1.4 Finite element modeling

In the previous sections of this chapter, we have highlighted that the use of structural analysis enables the quantification of the deformations and of the internal forces exhibited by the structure when subjected to external actions. In the more general case of deformable solids, this can be seen as a boundary value problem in which different field equations, that relate stress, strain, and displacement fields, are combined with boundary conditions to formulate the complete problem.

In structural analysis, the boundary conditions commonly describe how the structure is supported and/or loaded, therefore resulting in prescribed displacements and/or loads (i.e., tractions) applied to the boundaries of the structure. The solution of the problem is expressed in terms of stress, strain, and displacement fields that satisfy the specified boundary conditions and the remaining field equations. It is worth pointing out that such a solution can be achieved following different strategies.

The preferred solution strategy considered in this book is the one based on the finite element method (FEM) that relies on a displacement-based formulation of the problem. With this method of analysis, the constitutive equations relating stresses to strains and the strain-displacement relations are used to eliminate other field unknowns from the equilibrium equations to describe the problem only in terms of unknown displacements. Once the problem is solved for the unknown displacements, the field equations are then applied to post-process (or "revolve-back") the solution to calculate the remaining fields of strains and stresses.

Typical displacement-based FE formulations follow this procedure, even when they deal with nonlinear constitutive models. There are applications of mixed term formulations (see, e.g., [4], [5]) in which different fields, e.g., displacement and stress fields, can be separately approximated.

The single field displacement-based FEM in its classical formulation represents a form of *weighted residuals method*, and we now seek its solution in the approximate form as:

$$\boldsymbol{u}(\boldsymbol{x}) \approx \hat{\boldsymbol{u}}(\boldsymbol{x}) = \sum_{i=1}^{n} \boldsymbol{a}_i \cdot \phi_i(\boldsymbol{x}) \tag{1.79}$$

where $\boldsymbol{x} = [x, y, z]^T$ is a spatial coordinate vector, $\boldsymbol{u} = [u, v, w]^T$ is a vector of displacements, measured along the three coordinate axes, with $\hat{\boldsymbol{u}}$ being its approximation. The approximate solution is written as a linear combination of *a priori* selected functions $\phi_i(\boldsymbol{x})$, denoted as the *weighting functions*, with coefficients of this combination $\boldsymbol{a}_i$ being the main unknowns. These coefficients typically represent displacements of certain particular points within the deformable body, called *nodes* or *nodal points*. Consequently, the finite element approximated solution should be looked at as the interpolation between nodal displacements that consist of the unknown parameters of the problem.

The interpolation is performed locally over a small region of the regular shape—called *finite element*. With this approach the continuum is first divided by imaginary lines and surfaces into a finite number of smaller elements. This process is usually referred to as *meshing*. These elements are connected at a discrete number of nodal points, and the displacement field is interpolated within each element based on the values of the displacement components of its nodes. The functions used for this interpolation are referred to as *shape functions*. The approximate solution of the displacement within an element can take the following form:

$$\boldsymbol{u}(\boldsymbol{x}) \approx \hat{\boldsymbol{u}}(\boldsymbol{x}) = \sum_{i=1}^{m} \boldsymbol{d}_i \cdot N_i(\boldsymbol{x}) \tag{1.80}$$

where $\boldsymbol{d}_i$ is the vector of displacement for the ith node, m is the number of nodes that define the element, and N_i are the adopted shape functions and these can be obtained following different procedures. For example, if these functions are chosen to be weighting functions, then this specific representation of the weighted residuals defining the problem is denoted as the *Galerkin method* and it constitutes the displacement-based finite element method.

To simplify the interpolation of displacements over the element, it is desirable to use the simplest possible shapes for the elements. For two-dimensional problems the simplest possible shape is a triangular element with three nodes. For convergence reasons the most convenient form of the shape functions is a polynomial function. The typical interpolation of displacements through this type of function would take the following form:

$$u(x, y) = a_1 + a_2 x + a_3 y + \ldots \tag{1.81}$$

which is a horizontal displacement component at an arbitrary point within the element defined by coordinates x and y.

Clearly, terms of higher order than linear cannot be taken into account for a three-node element, considering that the unknown interpolation coefficients a_i should be determined from known nodal displacements (see Fig. 1.16a), namely:

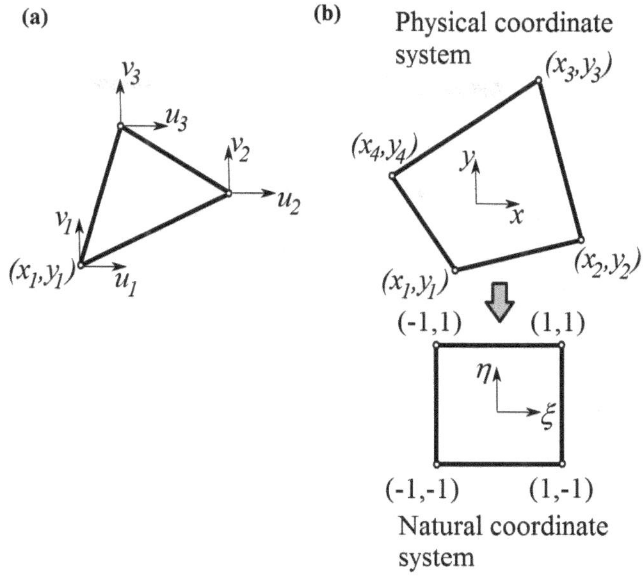

FIGURE 1.16 Continuum 2D finite elements with three and four nodes.

$$\begin{aligned} a_1 + a_2 x_1 + a_3 y_1 &= u_1 \\ a_1 + a_2 x_2 + a_3 y_2 &= u_2 \\ a_1 + a_2 x_3 + a_3 y_3 &= u_3 \end{aligned} \tag{1.82}$$

Given the strain-displacement relations defined by Eq. (1.53), with three-node triangular elements only a constant strain field can be modeled over the element, as the coordinates will vanish after the differentiation of the displacements formulated through Eq. (1.81). This limitation can be eliminated by employing higher order elements, obtained by either inserting internal nodes within the edges or elements, or by including more nodes, e.g., leading to a four-node quadrilateral element.

A general quadrilateral element with unparalleled edges enhances the complexity related to the integration of stiffness and mass matrices as the Gauss numerical integration cannot be directly employed. A remedy commonly used within FEM analysis is to map the element into a *natural coordinate system* $\xi - \eta$ (see Fig. 1.16b). In this system the element has a square shape, and domain defined by $[-1, 1]$ for both coordinates, and is well suited for Gauss numerical integration.

Isoparametric finite elements are usually used to solve both the mapping and the interpolation of the displacements over the element. Within this formulation the same shape functions are used to interpolate coordinates from physical to natural coordinate system, and for the interpolation of displacements. According to this scheme a point (ξ, η) of the natural system is mapped into the point (x, y)

of the physical system through:

$$x(\xi,\eta)=\sum_{i=1}^{4}N_i x_i$$
$$y(\xi,\eta)=\sum_{i=1}^{4}N_i y_i \tag{1.83}$$

where x_i and y_i are the coordinates of the four element nodes, and N_i are the shape functions associated to each node. Eq. (1.83) written in matrix form reads:

$$\begin{bmatrix} x \\ y \end{bmatrix}=\begin{bmatrix} N_1 & 0 & N_2 & 0 & N_3 & 0 & N_4 & 0 \\ 0 & N_1 & 0 & N_2 & 0 & N_3 & 0 & N_4 \end{bmatrix}\begin{bmatrix} x_1 \\ y_1 \\ x_2 \\ y_2 \\ x_3 \\ y_3 \\ x_4 \\ y_4 \end{bmatrix} \to \boldsymbol{X}=\boldsymbol{N}\cdot\boldsymbol{x}_e \tag{1.84}$$

Shape functions N_i are selected to be polynomials and they should have the property that the interpolation at corner nodes has to be exact (i.e., corner points from natural system will be mapped into corner nodes of physical system). The corner coordinates for the reference element in natural system are $(-1,-1)$, $(1,-1)$, $(1,1)$, and $(-1,1)$ (see Fig. 1.16b). It can be easily verified that the shape functions given in the form:

$$N_i(\xi,\eta)=\frac{1}{4}(1+\xi_i\xi)(1+\eta_i\eta) \tag{1.85}$$

satisfy the above mentioned property. For more detailed derivation of shape functions for quadrilateral isoparametric finite elements, reference should be made to dedicated finite element textbooks, e.g., [6] and [7].

As already mentioned the displacements are interpolated through the same shape functions, therefore the displacement components in an arbitrary point within the element are given by:

$$\boldsymbol{u}(\xi,\eta)=\boldsymbol{N}(\xi,\eta)\cdot\boldsymbol{d}_e \tag{1.86}$$

where $\boldsymbol{N}$ is matrix of shape functions given by (1.84) and $\boldsymbol{d}_e$ is vector of nodal displacements of the four node element.

Stiffness matrix for each finite element relates the nodal displacement components to the nodal forces. It is usually derived by relying on some energy theorem like the principle of minimum potential energy. For details of this derivation the reader should refer to [4], [6], [7], and [8]. Considering that the initial geometry is first divided into a finite number of smaller elements, by assembling these elements' stiffness matrices, a global stiffness matrix is formed. The structural problem under consideration is further expressed as a set of algebraic equations that can be written in matrix form as:

$$\boldsymbol{K}\boldsymbol{d} = \boldsymbol{F} \tag{1.87}$$

where $\boldsymbol{K}$ is a global stiffness matrix, $\boldsymbol{d}$ is a vector of nodal displacements and $\boldsymbol{F}$ is a vector of nodal forces. The starting continuum with an infinite number of degrees of freedom is thus transformed (i.e., *discretized*) into an equivalent system with finite number of degrees of freedom, namely the displacements of nodal points, collected in vector $\boldsymbol{d}$.

To complete the problem formulation, boundary conditions have to be imposed, which in a discretized system are represented by prescribed values of nodal displacements and/or nodal forces given to certain nodes (i.e., entries to the vectors $\boldsymbol{d}$ and $\boldsymbol{F}$). After the system (1.87) is solved, all components of displacements, representing the main field of approximation are known. These values are them fed into the displacement-strain relations to compute the components of strain. This is resolved at an element basis and thus the implementation varies depending on the type of finite element used.

Adopting a matrix notation, typical for FEM implementation, and limiting ourselves to two-dimensional problems, the strain-displacement relations can be expressed as:

$$\begin{bmatrix} \varepsilon_{xx} \\ \varepsilon_{yy} \\ \gamma_{xy} \end{bmatrix} = \begin{bmatrix} \frac{\partial}{\partial x} & 0 \\ 0 & \frac{\partial}{\partial y} \\ \frac{\partial}{\partial y} & \frac{\partial}{\partial x} \end{bmatrix} \cdot \begin{bmatrix} u \\ v \end{bmatrix} \rightarrow \boldsymbol{\varepsilon} = \boldsymbol{L} \cdot \boldsymbol{u} \tag{1.88}$$

where $\boldsymbol{u}$ is the vector of displacement at an arbitrary point. In general, strain components can vary from point to point, and this variation, as discussed previously, can be modeled within a quadrilateral element. Considering that the displacements within these elements are interpolated with Eq. (1.86) the corresponding strain components can be computed within the element as:

$$\boldsymbol{\varepsilon} = \boldsymbol{L}\boldsymbol{N} \cdot \boldsymbol{d}_e = \boldsymbol{B} \cdot \boldsymbol{d}_e \tag{1.89}$$

where matrix $\boldsymbol{B}$ is called the *strain matrix* and, for a quadrilateral element, it has the following form:

$$\boldsymbol{B} = \begin{bmatrix} \frac{\partial N_1}{\partial x} & 0 & \frac{\partial N_2}{\partial x} & 0 & \frac{\partial N_3}{\partial x} & 0 & \frac{\partial N_4}{\partial x} & 0 \\ 0 & \frac{\partial N_1}{\partial y} & 0 & \frac{\partial N_2}{\partial y} & 0 & \frac{\partial N_3}{\partial y} & 0 & \frac{\partial N_4}{\partial y} \\ \frac{\partial N_1}{\partial y} & \frac{\partial N_1}{\partial x} & \frac{\partial N_2}{\partial y} & \frac{\partial N_2}{\partial x} & \frac{\partial N_3}{\partial y} & \frac{\partial N_3}{\partial x} & \frac{\partial N_4}{\partial y} & \frac{\partial N_4}{\partial x} \end{bmatrix} \tag{1.90}$$

with N_i being the shape functions.

It is worth noting that shape functions (1.85) are defined as a function of the natural coordinates, while for strain calculation it is required to find derivatives with respect to physical coordinates. The relationship is established through the chain rule for derivation, namely:

$$\begin{aligned}\frac{\partial N_i}{\partial \xi} &= \frac{\partial N_i}{\partial x}\frac{\partial x}{\partial \xi} + \frac{\partial N_i}{\partial y}\frac{\partial y}{\partial \xi}\\ \frac{\partial N_i}{\partial \eta} &= \frac{\partial N_i}{\partial x}\frac{\partial x}{\partial \eta} + \frac{\partial N_i}{\partial y}\frac{\partial y}{\partial \eta}\end{aligned} \tag{1.91}$$

that in matrix form becomes:

$$\begin{bmatrix}\frac{\partial N_i}{\partial \xi}\\ \frac{\partial N_i}{\partial \eta}\end{bmatrix} = \boldsymbol{J}\begin{bmatrix}\frac{\partial N_i}{\partial x}\\ \frac{\partial N_i}{\partial y}\end{bmatrix} \tag{1.92}$$

where $\boldsymbol{J}$ is the Jacobian matrix of the coordinate transformation defined by:

$$\boldsymbol{J} = \begin{bmatrix}\frac{\partial x}{\partial \xi} & \frac{\partial y}{\partial \xi}\\ \frac{\partial x}{\partial \eta} & \frac{\partial y}{\partial \eta}\end{bmatrix} \tag{1.93}$$

Since physical coordinates are computed from nodal values through Eq. (1.84), the Jacobian can be written as:

$$\boldsymbol{J} = \begin{bmatrix}\frac{\partial N_1}{\partial \xi} & \frac{\partial N_2}{\partial \xi} & \frac{\partial N_3}{\partial \xi} & \frac{\partial N_4}{\partial \xi}\\ \frac{\partial N_1}{\partial \eta} & \frac{\partial N_2}{\partial \eta} & \frac{\partial N_3}{\partial \eta} & \frac{\partial N_4}{\partial \eta}\end{bmatrix}\begin{bmatrix}x_1 & y_1\\ x_2 & y_2\\ x_3 & y_3\\ x_4 & y_4\end{bmatrix} \tag{1.94}$$

The outlined scheme to compute derivatives, together with Eq. (1.89), provides a mathematical framework to compute strain components in any arbitrary point of the finite element.

When forming the element stiffness matrix, the following integral has to be determined over the element:

$$\boldsymbol{K} = \int_V \boldsymbol{B}^T \boldsymbol{D}\boldsymbol{B}\,dV \tag{1.95}$$

with $\boldsymbol{D}$ being a material elasticity matrix. Within FEM procedures, this integral is solved numerically by the Gauss quadrature rule, which requires calculation of the integrand function value only in a limited number of points within the element domain, called *Gauss points*. The required number depends on the type of interpolation functions used and desired accuracy. For bilinear quadrilateral element with displacement interpolation given by Eq. (1.86), four Gauss points could be used for integration, which results in an accurate integral if quadrilateral element has internal angles equal to 90^0. It is therefore common practice that also within the post-processing of the results, strain components are computed only of these Gauss points.

Once the strains are calculated, constitutive relations are invoked to compute stresses. For the particular case of plane stress and by assuming a linear-elastic response of the material the stress vector is equal to:

$$\begin{bmatrix} \sigma_{xx} \\ \sigma_{yy} \\ \sigma_{xy} \end{bmatrix} = \boldsymbol{D} \cdot \begin{bmatrix} \varepsilon_{xx} \\ \varepsilon_{yy} \\ \gamma_{xy} \end{bmatrix} \tag{1.96}$$

For an isotropic plane-stress problem, matrix $\boldsymbol{D}$ takes the following form:

$$\boldsymbol{D} = \frac{E}{1-\nu^2} \begin{bmatrix} 1 & \nu & 0 \\ \nu & 1 & 0 \\ 0 & 0 & \frac{1-\nu}{2} \end{bmatrix} \tag{1.97}$$

which is uniquely defined by two elastic constants: Young's modulus E and Poisson's coefficient ν.

Calculating stresses from strains depends directly on the adopted type of material constitutive behavior. For the elastic case, strain components, recovered from the displacements through Eq. (1.88) are going to be treated as the elastic, and used as such in Eq. (1.96) to evaluate stresses. For inelastic constitutive models, more complex relations need to be employed. As an example, for the elasto-plastic constitutive models, usually additive decomposition of strains into the elastic and plastic parts have to be performed, with additional governing equations regarding the yield criterion and plastic flow rules required to compute the plastic part of the strains. Once the plastic strains are subtracted from the total strains, Eq. (1.96) is used to calculate the stresses from the elastic part of the strains.

An additional remark here has to be made regarding the system of algebraic equations (1.87) resulting from the discretization of the system. In the case when all the entries to the stiffness matrix $\boldsymbol{K}$ are constants, computed before the solution is found, the resulting system represents a set of linear algebraic equations. On the contrary, if the stiffness matrix entries depend on the solution, the system of algebraic equations is nonlinear and the corresponding analysis belongs to the group of nonlinear FEM problems. Such problems have to be solved with some iterative strategies.

When dealing with nonlinear problems an additional step is required within the operative sequence of procedures invoked by the FEM solution strategy. Upon the calculation of stresses, equivalent nodal forces have to be computed on the basis of equilibrium conditions imposed to each element. These equivalent nodal forces are then compared with the externally applied forces and used

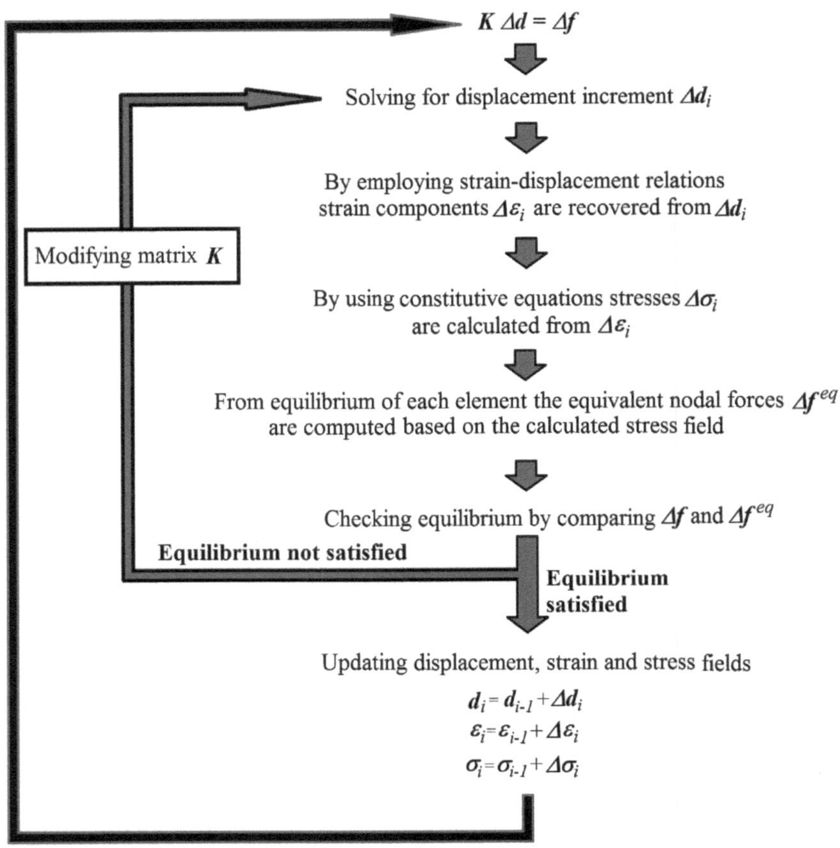

FIGURE 1.17 Overview of the sequence of operations within typical nonlinear solution strategy.

to check global equilibrium, which in common nonlinear problems is obtained after a few iterations. Without entering into a detailed description of each sequence at this stage, a schematic representation of a complete FEM analysis is visualized in Fig. 1.17. To be more consistent with general procedures applied to the nonlinear problems, the sequence presented in the figure considers an incremental formulation, almost exclusively applied within nonlinear FEM analysis.

One of the main reasons for dealing with a nonlinear system is to account for material nonlinearity, and thus for constitutive modeling. In the context of structural analysis, following a solution strategy typically adopted within a displacement formulation, the constitutive model should provide the mathematical formulation that relates stress components to strain components. Within the finite element implementation, these relations are inserted in two different stages of the operative procedure, i.e., when formulating the stiffness matrix and when computing the stresses from the strains. Different strategies of these implementations within finite element modeling techniques, with references to diverse types of constitutive models, are going to be discussed in more detail in the following chapters.

1.5 Some comments regarding large deformation problems

The material mechanical behavior can be such as to exhibit large deformation. For example, most ductile metals have the elastic deformation fairly small, but once they enter into plasticity they can develop significant plastic deformations, which can be larger by couple of orders of magnitude than the elastic ones. A similar and even more pronounced effect is encountered with certain polymers. On the other hand, rubbers develop usually fairly large deformation while still remaining within the elastic range. Therefore an additional complexity, apart from the nonlinear stress–strain relationship is encountered when the material undergoes large deformation.

Within the formulation of the structural analysis problem, one of the three groups of equations used consisted of the equilibrium equations. When formulating these equations, a reasonable simplification is achieved by assuming that the deformation are small when compared to the body size, thus the equilibrium can be written with respect to the initial configuration of the body—the one that does not depend on the solution. If the considered material mechanical behavior allows large deformation, this assumption may no longer be justified, and change in the geometry due to large displacements and large deformations may have a significant effect on the load-displacement characteristics of the structure being analyzed. In such cases, the equilibrium equations need to be written with respect to the deformed configuration, which consequently would require to distinguish between different measures of strains and stresses. Such formulation represents what is called a *geometrically nonlinear* problem, as the deformed configuration, which here is the part of the problem formulation, clearly depends on the solution.

For the completeness of the material presented in this book the main difficulties, and ways to treat them, related to large deformation and large displacements are separately outlined in Chapter 6, while they formally belong to the material presented in this chapter, namely to the finite element modeling techniques. Considering that certain constitutive models discussed in further chapters of this book involve large deformation, in this section we simply intend to highlight that significant differences exist between what is outlined in this chapter and the problems that involve large deformation.

While the displacement field is uniquely defined within the deformed body, there are different strain measurements that can be used for the evaluation of the deformation. The difference between them may in certain situations become significant, and the appropriate selection of a suitable strain measurement can be crucial when dealing with large deformation problems. Based on the fact that the strain energy, which is used to formulate the solution within displacement-based FEM, depends on the product of stress and strain, the selection of the strain measurement conditions the selection of the appropriate stress measurement. Thus the "work conjugate" pairs of stress and strain that produce the same strain energy must be selected. In Sections 1.1 and 1.2 the terms "engineering stress" and "engineering strain" were formulated to represent work conjugates, since their product results in correct strain energy. However, they are defined under the assumption of small deformation and small displacements by making a reference to the initial configuration. Such measurements need to be changed in certain constitutive models when dealing with large deformation.

Geometrically, nonlinear problems represent a rather broad subject and their detailed treatment is not included in this book. For further readings, reference should be made to [8], [10], [11], and [12]. A dedicated chapter (i.e., Chapter 6) is included in this book to outline the main concept related to the implementation of constitutive models associated to large deformation.

References

[1] E.F. Bruhn, Analysis and Design of Flight Vehicle Structures, Jackob Publishing, Indianapolis, 1973.

[2] M. Jirasek, Z. Bazant, Inelastic Analysis of Structures, Wiley, West Sussex, 2002.

[3] I. Karanovsky, O. Lebed, Advanced Methods of Structural Analysis, Springer - Verlag, Heidelberg, 2010.

[4] O. Zienkiewicz, R. Taylor, Finite Element Method, vol. 1, fifth edition, Butterworth-Heinemann, Woburn, 2000.

[5] O. Zienkiewicz, R. Taylor, Finite Element Method, vol. 2, fifth edition, Butterworth-Heinemann, Woburn, 2000.

[6] S. Quek, G. Liu, Finite Element Method: A Practical Course, Butterworth-Heinemann, Oxford, 2003.

[7] T. Hughes, Finite Element Method. Linear Static and Dynamic Finite Element Analysis, Dover Publications INC, New York, 2000.

[8] K.J. Bathe, Finite Element Procedures, Prentice Hall, New Jersey, 1996.

[9] G. Ranzi, R.I. Gilbert, Structural Analysis: Principles, Methods and Modelling, CRC Press, Boca Raton, 2015.

[10] J.N. Reddy, An Introduction to Nonlinear Finite Element Analysis, Oxford University Press, Oxford, 2004.

[11] J.N. Reddy, Principles of Continuum Mechanics: Conservation and Balance Laws with Applications, Oxford University Press, Oxford, 2012.

[12] P. Wriggers, Nonlinear Finite Element Methods, Springer - Verlag, Heidelberg, 2010.

Chapter 2

Calibration of constitutive models

The quantification of the internal stresses and deformations induced in a structure is performed by applying the principles of structural analysis that have been briefly introduced in the previous chapter. In a design scenario, this quantification is necessary to ensure that the predicted response satisfies the *margin of safety* required for the structure under consideration. For example, this can be carried out by comparing the calculated stresses with some reference values. Traditionally, admissible stress criteria have been adopted for safety analysis, where stresses are compared against their values at the limit of elastic deformation. Over the last few decades, developments in structural mechanics have led to increasing number of situations in engineering practice in which the desired safety margin is computed with respect to collapse loads. Relevant design standardization usually reflects such approach in their specifications. The underlying numerical calculations are supported by computational procedures, such as the finite element method, implemented through robust computer codes. This shift (i.e., expressing the safety margin in terms of collapse loads) has increased the need for quantifying the inelastic properties of the materials that are, in general, more difficult to characterize.

The mechanical properties of a material are generally assessed based on the measurements of experiments. It is common to encounter scenarios in which a sought property cannot be directly measured, and for this reason, it is evaluated on the basis of some other experimentally measurable quantity. Standard tests commonly used for the material characterization involve simple specimen shapes and loading conditions to provide simple procedures that translate the measured quantities to the sought material properties. Simple tests have also the advantage of limiting the influence of other experimental factors that might influence the variability of the measured quantities.

One of the most widely used tests is the uniaxial tensile test [1]. In this test a standardized specimen is subjected to a gradually increasing tensile load up to its final rapture. This test is performed in order to establish the stress–strain diagram that describes the main mechanical features of the tested material in both linear and nonlinear regimes. Testing equipment used to perform the test is commonly equipped with precise instrumentation to measure both the applied load, and the elongation of the specimen. The engineering stress during a test can be obtained by dividing the measured load by the area of the sample's

Constitutive Modeling of Engineering Materials. https://doi.org/10.1016/B978-0-12-814696-5.00007-1

cross-section (measured in its unloaded condition). The engineering strain is determined by dividing the elongation exhibited by the sample during the test by its initial length. With this simple transformation, the experimentally measured quantity, i.e., force *vs.* elongation curve, is expressed in a more meaningful material representation, i.e., the *stress–strain curve.*

For simple experiments the transition from experimentally measured quantity to material sought parameters is easily achieved. For more complex deformation mechanisms, establishing this transition can become fairly difficult, if at all possible, through simple mathematical calculations. The main reason for this difficulty is that it is not easy to separate the influence of a single property from the overall measured response. This aspect is now illustrated by considering the case of an instrumented indentation test.

The indentation experiment has been traditionally used for the evaluation of the material hardness, where an indenter of a standardized shape made of some material harder than the one of the sample, is pushed against the surface of the tested specimen with the aim of leaving a permanent imprint on it (see Fig. 2.1). Modern indentation instruments are equipped with tools to provide digitalized indentation curves, namely indenter tip penetration *vs.* force applied on it during the test. Fig. 2.2 highlights the difference in the measured indentation curves by considering, for illustrative purposes, two curves produced with the same indenter for two different materials.

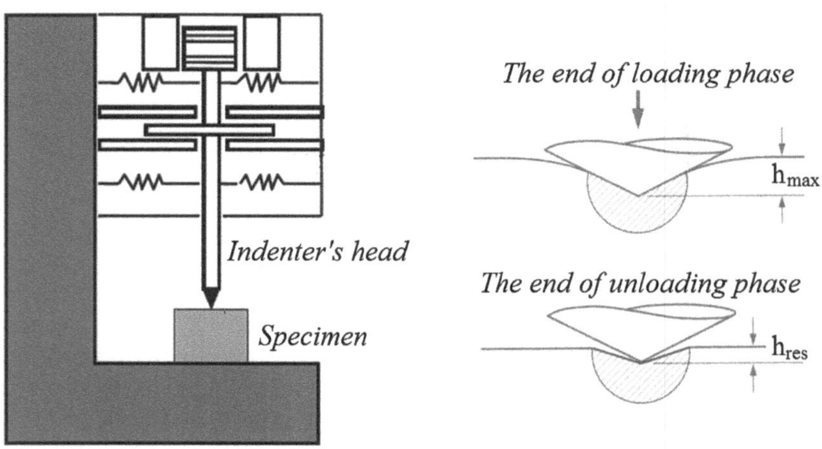

FIGURE 2.1 Schematic representation of the indentation test.

As it can be intuitively expected, the material properties influence the shape of the indentation curve and, because of this, the measured curve can be considered as an experimentally measured quantity to represent and assess material properties. In this case, it is difficult to relate indentation curve directly to specific material properties, as the curve profile is influenced simultaneously by different material properties, such as the Young's modulus, the yield limit, and

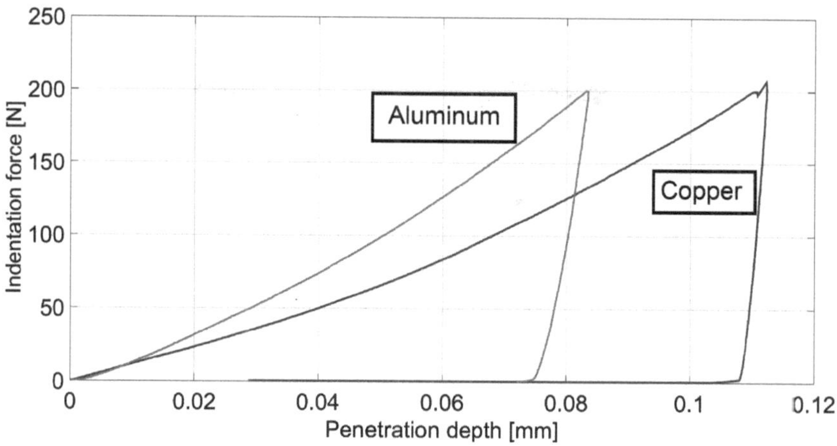

FIGURE 2.2 Curves resulting from Rockwell indentation performed on two different materials.

the hardening properties. In the literature, it has also been observed that the same indentation curve can be obtained by different materials (see, e.g., [2]), for example, when considering two materials one could possess a comparably lower value of the yield limit and a pronounced hardening while the other could exhibit a higher yield limit followed by a perfect plasticity (i.e., increase in plastic deformation at constant stress).

To avoid this inconvenience, the approach pursued by standardized tests is to establish simplified experiments, where a measured quantity can be related with sufficient accuracy to a specific sought material mechanical property. For complex material models, this may not be the most effective strategy, as it may require a large number of experiments. An alternative advantageous strategy for material calibration is offered by the *inverse analysis* where, contrary to standardized tests, the experiment is selected to be complicated enough to "activate" all sought material parameters. In order to emphasize the difference in the two approaches, we will first give a short overview of some commonly used procedures to assess different material properties.

2.1 Mechanical characterization of materials—some standard procedures

Probably the most basic test that can be performed on a material is the tensile test. It subjects the material to what it can be assumed to be a simple, uniaxial state of stress, and it provides a complete tensile profile of the material both in the elastic and in inelastic range. A typical curve resulting from the test, corresponding to a ductile material (e.g., metals or polymers), is illustrated in Fig. 2.3. Most materials have an initial linear relationship between stress and strain that can be used to determine the Young's modulus. As the state of stress is defined by only one component different from zero, it stems directly from Hooke's law that the Young's modulus represents the slope of the diagram, measured at the starting of the linear part.

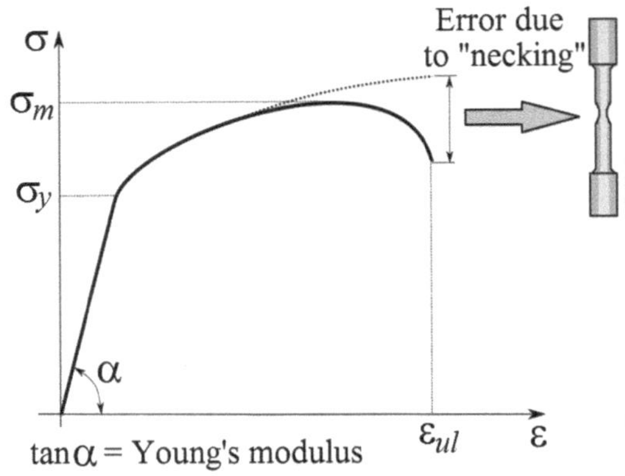

FIGURE 2.3 Stress–strain curve resulting from tensile test.

The Young's modulus is certainly not the only meaningful quantity that can be assessed from a tensile test, as the test provides insight also into the material inelastic behavior. The initial linear response of the stress–strain curve ends at a point referred to as the *proportionality limit*. All of the deformation that the material has experienced up to this point would vanish upon the removal of the load. For this reason, the response of this initial part of the curve is denoted as linear-elastic. By increasing the load beyond this point, some permanent deformation would eventually start to appear, therefore marking the *elastic limit* of the material. Even though there are materials with different proportionality and elastic limits, for most of them these points are very close and, therefore, these can be approximated by the same point.

For most of engineering materials, the elastic deformation are fairly small, thus requiring very precise equipment for the identification of the elastic limit (or proportionality limit). In order to make the measurement less dependent on the accuracy of the instrument the elastic limit has been defined to depict the value of stress that corresponds to some value of permanent deformation. This value was originally set to be equal to 0.001% at The Fourth Congress of the International Association for Testing Materials, held in Brussels in 1906. It was later increased to 0.01% and it corresponds to today's most commonly used value [3].

Another point of the stress–strain curve that is important for ductile materials is the one that marks the onset of large plastic deformation. This point is called the *yield limit*, with corresponding value of stress usually denoted by σ_y as shown in Fig. 2.3. The behavior of different materials in this zone can vary. Some materials, such as aluminum and copper, have this point well identified, as illustrated in Fig. 2.3. Mild steel, frequently used in civil engineering applications, can exhibit an initial drop in stress followed by an increase in plastic deformation. There are many other materials for which it is difficult to identify exactly the stress value at which plastic deformation starts to occur. For such materials, the yield limit is usually taken as the stress needed to induce a permanent plastic deformation of 0.2%.

Yield stress is used as the reference value when sizing engineering structures according to the requirement that no permanent deformation can occur within service loads. If, however, it is required to calculate failure of the structure, then the material behavior beyond the yield limit should be considered. Such behavior can also be captured from a stress–strain curve, with the most important quantities being the ultimate strain that the material can withstand prior to failure and the maximum stress (also called the *tensile strength*) denoted in Fig. 2.3 as ε_{ul} and σ_m, respectively.

If the tested sample has a pronounced ductility, the plastic deformation exhibited during the test can be significant and it might contribute to changes in the cross-sectional area that might not be neglected. The stress–strain curve is usually obtained from a tensile test and the relevant stress is determined by dividing the measured load by the initial cross-sectional area. For high levels of deformation, this calculation can lead to an underestimation of stresses, since the reduction in cross-sectional area undergone by the material is not accounted for in the stress calculation. The reduction becomes more pronounced prior to failure with plastic deformation localized in a small region of the material and resulting in the so-called *necking* of the specimen. If, throughout the test, the load is divided by the true area exhibited by the sample, the *true stress* would continue to increase up to the failure of the specimen (see Fig. 2.3). The tensile strength is usually computed on the basis of the initial cross-sectional area, thus leading to an erroneous result but acceptable because conservative (i.e., on the safe side) because assuming the material to fail upon reaching a stress smaller than the real failure stress.

The tensile test is simple and inexpensive, and it does provide information about both the elastic and inelastic material response. It is also a destructive test that, in some applications, should be avoided. If the aim is to assess the material mechanical properties of a structure that is in service, it is generally required, and certainly preferable, to avoid extracting specimens from it for the laboratory testing. In such situations, the elastic properties can be effectively quantified on the basis of measuring the sound velocity within the structure. The equipment used for this purpose measures the propagation speed of longitudinal and shear waves because these depend on the material elastic properties (i.e., Young's modulus and shear modulus) and density. Readouts are obtained through the application of straightforward analytical equations that can evaluate the values of the elastic properties. For such measurements, the equipment is usually portable (see Fig. 2.4) and applicable to a wide range of materials that exhibit a linear relationship between stress and strain, thus excluding, for example, rubber-like materials [4]. Even though ultrasonic methods are fast and powerful, they have the limitation of assessing only the elastic properties.

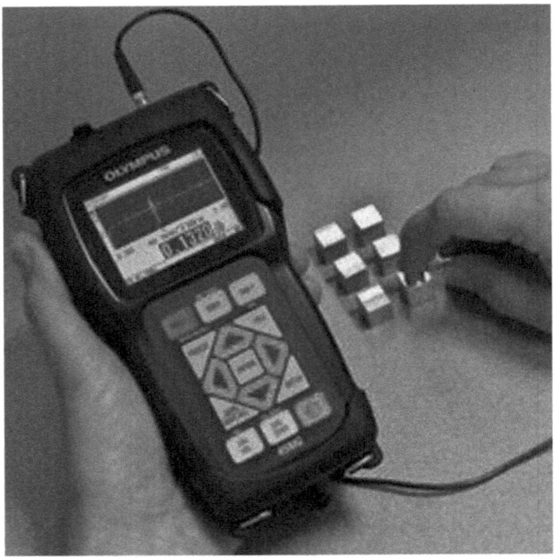

FIGURE 2.4 Portable equipment for the quantification of the elastic properties of materials based on ultrasound measurements.

The behavior of materials such as high carbon steels, ceramics, or rock, is characterized by insignificant plastic deformation prior to rapture. For this reason, these materials are denoted as brittle or quasi-brittle, since fracturing of the material takes place without appreciable plastic flow. For these materials, it is important to assess the value of the failure stress, while its mechanical behavior, in some cases, can be modeled as linear-elastic up to failure. Since the fracture initiates at large defects or internal cracks, compressive loading is beneficial as it will tend to close the existing cracks. Thus such materials usually have significantly larger compressive then tensile strength. When testing them, it is more pertinent to perform a compressive rather than a tensile test.

Samples specified for compressive tests are usually shorter than those adopted for tensile tests to prevent any possible instability and buckling phenomena during testing. For example, length to diameter ratio of cylindrical specimens typically falls in the range of $1 \div 2$. It is common to assume that the stress distribution is uniform over the sample and, therefore, the compressive strength is calculated by dividing the maximum load carried by the specimen during the test by its cross-sectional area. In reality, the stress distribution varies significantly through the volume of the sample because it is influenced by the restraints provided, through friction, by the supporting plates of the testing equipment to the end surfaces of the sample. This lateral deformation restraint tends to induce a beneficial state of stress during the test. Some standardized procedures (see, e.g., [5]) define correction coefficients which decrease the measured compressive strength to account for this effect. Another way to remove the influence of the end restraints is to adopt particular end supports, as proposed in [3], that are produced in a conical shape with an angle that matches the friction angle to eliminate its influence and to produce a uniform stress distribution over the sample (see Fig. 2.5).

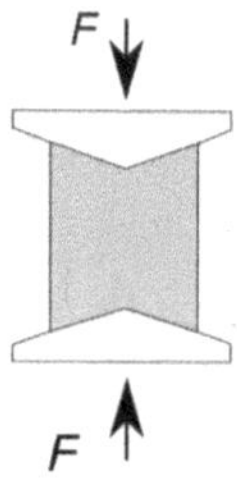

FIGURE 2.5 Compressive test of a cylindrical sample with conical ends.

For certain brittle materials, such as rocks or ceramics, it may be also important to evaluate the tensile strength. An indirect tensile test is commonly used for this purpose based on a standardized procedure [6], known as the Brazilian test. It is a test in which a disc is put in diametrical compression by applying concentrated forces in two points on its radial surface (see Fig. 2.6a). The load pattern is such that a tensile zone is created along the diameter, perpendicular to the direction of the forces. With this setup, the principal tensile stress component of largest intensity is located in the cylinder center and it corresponds to the place where the crack initiates followed by the immediate failure of the specimen, thus allowing the measurement of the material tensile strength (see Fig. 2.6b). Based on these test results the tensile strength of the material is computed from the measured failure load F_f as follows:

$$\sigma_t = \frac{2F_f}{\pi Dh} \tag{2.1}$$

which stems directly from the theoretical solution of the stress distribution within the sample, and D and h depict the sample diameter and height, respectively. For a detailed derivation of the stress distribution over the disc when subjected to diametrical compression, reference should be made to textbooks of theory of elasticity (see, e.g., [7]).

In all previous examples, the sought material property was directly related to the quantity measured in the test. The experiment was kept simple, so that the state of the stress remained as close as possible to the theoretical solution. Such approach can be used to assess a limited number of material properties. If more complex constitutive models are employed, it may become difficult, or even impossible, to quantify certain material properties by directly relating them to experimentally measured data. In such situations an advantageous strategy is offered by the use of inverse analysis theory. Within an inverse analysis approach, there is no need to establish a direct transition between a measurement and a material property, and the experiment can be rather complicated. It also offers

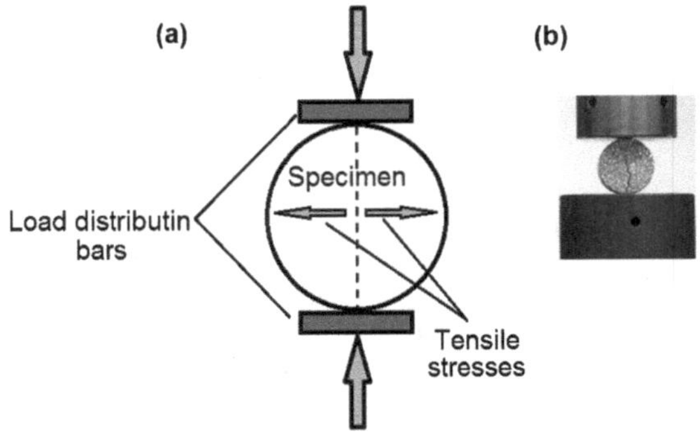

FIGURE 2.6 Indirect tensile (Brazilian) test: a) schematic representation and b) sample with the crack after testing.

the advantage of testing and characterizing materials with more realistic stress states.

2.2 Inverse analysis for parameter quantification problems

Let us introduce the concept of inverse analysis by considering a general problem of modeling natural phenomena that take place within a system. The conceptual and mathematical model of a system is developed by using the laws of physics, and it results in algebraic or differential equations that relate the quantities of interest. These models are commonly used to simulate the physical events by providing the solution to the governing set of equations. For example, the solution of a mathematical model described by partial differential equations requires the knowledge of the initial conditions, boundary conditions and all constants entering into the governing equations. Such solution can frequently be obtained through a straightforward solution strategy (analytically or numerically) and this scenario is typically referred to as the "forward" problem.

The "inverse" problem, on the other hand, is the problem in which, the solution is known, and the aim of the computation is to identify the conditions provided in input to solve the full forward problem (i.e., initial conditions, boundary conditions and parameter values). In this context, the inverse problem can be characterized as the problem in which we aim at identifying "something" that is missing. Depending on the data that is missing, different types of inverse problems can be formulated. Inverse problems that are of interest for the material characterization problems belong to the group of *parameter identification* problems, since they are formulated such that sought material properties are parameters entering into the governing equations of the adopted constitutive models.

Contrary to the forward problems, it is not possible to use a single straightforward solution strategy when solving an inverse problem because the latter requires the combination of experiments with modeling techniques and mathematical programming. This can be exemplified for a parameter identification inverse problem that is of primary interest for the material characterization problems discussed further in this book.

The starting point of a parameter identification inverse problem is the known material response. This response is generally known from the experiment. The scheme for solving the inverse problem, the so-called *inverse operator*, inherently incorporates the experimental technique. The considered experiment has to be modeled, traditionally through some of the commonly accepted numerical methods, such as the finite element method. The numerical model of the test should provide the computed counterpart of the experimentally measured quantities. In the context of material characterization, a suitable constitutive model is selected and specified in the test simulations, thus making the numerically computed quantities dependent on the sought parameters. Further on, a suitable discrepancy function is formulated that quantifies the difference between the measured quantities and their computed counterparts. In a subsequent phase, this function is minimized with respect to the sought parameters. The parameters for which the simulated response ideally matches the measured one, or in a view of various modeling and measuring errors, is the closest to it, is regarded as the solution to the inverse problem.

The schematic representation of the inverse operator for the parameter identification problems is presented in Fig. 2.7 and it consists of the experimental techniques (depicted by the vector of measurable quantities $\boldsymbol{u}^{exp}$), the numerical techniques (included with the vector of computed variables $\boldsymbol{u}^{com}(\boldsymbol{p})$), and the discrepancy function $\omega(\boldsymbol{p})$ that is expressed as a function of the sought parameters collected in vector $\boldsymbol{p}$ and minimized through the mathematical programming.

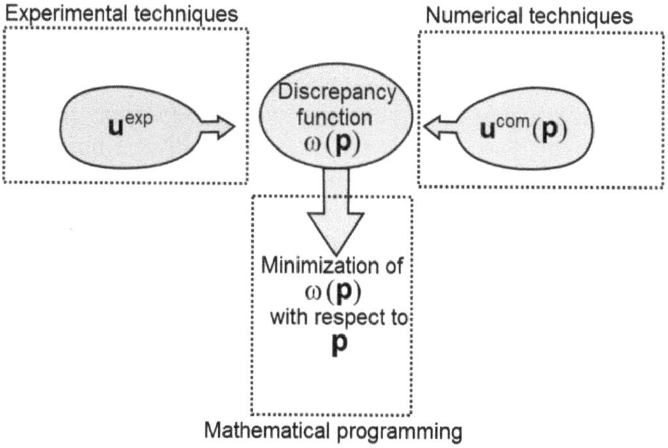

FIGURE 2.7 Schematic representation of the inverse problem expressed as a parameter identification problem.

It may be anticipated that finding the solution for an inverse problem is generally more complex than solving the direct problem. Aside from the inherent multidisciplinary nature of the inverse operator, it is possible for an inverse problem not to be in a condition to guarantee the uniqueness of the solution that it finds (this class of problems is usually referred to as ill-posed) or the identification of the sought minimum values, generally produced by non-convex discrepancy functions. From a practical viewpoint, the finding of the solution in an inverse problem often faces various computational challenges.

Despite the above mentioned difficulties, the inverse analysis methodology is gaining popularity in the material characterization because offering significant advantages. The numerical model of the test can be used to estimate the error of the assessed parameters, to quantify the statistical nature of parameters, starting from the uncertainty in the measured experimental data. Simulations of the test can also be used for the selection of suitable experimentally measured quantities, by choosing them on the basis of their sensitivity with respect to the sought quantities. Such sensitivity can be ascertained by separately perturbing sought parameters and quantifying the changes in measured quantities. With this approach, it is possible to optimize the experiment for particular calibration purposes. When the number of constitutive parameters is elevated, there is a larger possibility to have the compensation between the effects of different parameters on the measured quantities. This difficulty, if present, cannot be evidenced through sensitivity tests, as it is occurring when all the parameters influence the measurable quantities. The inverse analysis, because established on mathematical concepts, represents an effective tool to avoid these situations. Some of the strategies regarding the main concepts of well-posing the inverse problems (i.e., defining an inverse problem in which the uniqueness of the solution can be guaranteed), and various difficulties encountered when solving them, are discussed in the following chapter.

2.3 Summary

Constitutive models represent a mathematical framework to describe the material mechanical behavior in an arbitrary context. Depending on the required modeling complexity, these models can depend on a diverse number of parameters that usually represent some mechanical characteristics of the material. For reliable numerical simulations, it is crucial to adequately quantify these parameters (i.e., constants) entering into the governing equations of the adopted constitutive model. Such model calibration represents the mechanical characterization of the material.

Mechanical characterization is performed through experiments. It is usually impossible to measure all sought material parameters directly and, therefore, these are then computed through some other experimentally measured quantity. For a successful material characterization, it is therefore essential to clearly define the transition from what can be measured within the experiment to what is sought by the identification process.

For certain parameters related to simple constitutive models, standardized tests may be used with direct transition from a measurable quantity to the sought parameter, usually through a straightforward analytical relation. For more complex constitutive models, it is generally difficult to separate the influence of diverse parameters from the experiment outputs, thus a more systematic way, offered by the inverse analysis should be adopted.

It is important to emphasize that the material calibration through inverse analysis is different from fitting the experimentally measured results. Proper material calibration should result in *representative parameters* capable of describing the material behavior in the contexts that differ from the experiment performed for the characterization. For this purpose, a *representative experiment* has to be devised and carried out for a successful inverse analysis. The selection of an experiment is particularly important when calibrating constitutive models with an elevated number of parameters. The case of adopting an overly simple experiment might lead to the situation in which the particular solution is applicable to only the specific loading conditions of the adopted experiment. A proper material calibration can be regarded as a characterization in which the material response is adequately captured and can be applied to loading conditions that differ from those of the characterization experiment. Instead of fitting a single experimentally measured result a proper material calibration can be illustratively expressed as the one that is fitting all possible experiments. In this context the inverse analysis offers a mathematical framework to reach this goal as it is presented in the following chapters for different constitutive models.

References

[1] Standard test methods for tension testing of metallic materials, 2013, ASTM International, ASTM E8/E8M-13a.
[2] X. Chen, N. Ogasawara, M. Zhao, N. Chiba, On the uniqueness of measuring elastoplastic properties from indentation: the indistinguishable mystical materials, Journal of the Mechanics and Physics of Solids 55 (8) (2007) 1618–1660.
[3] S. Timoshenko, Strength of Materials Part II: Advanced Theory and Problems, Van Nostrand, Princeton, New Jersey, 1958.
[4] P. Moore, Nondestructive Testing Handbook, vol. 7, American Society for Nondestructive Testing, 2007.
[5] ASTM standard test method for compressive strength of cylindrical concrete specimens, ASTM International.
[6] ASTM standard test method for indirect tensile testing, ASTM International.
[7] Martin Saad, Elasticity: Theory, Applications, and Numerics, Elsevier Academic Press, 2003.

Chapter 3

Inverse analysis

3.1 Introduction

In Chapter 1, we provided a brief overview of the basis of structural analysis to highlight the role played by the constitutive material models in influencing the structural behavior. For example, in Fig. 1.5 we considered the response of a simple structural system and noted the significant differences in the deformed configurations obtained by specifying different elastic stiffnesses for the springs supporting the rigid bar. We then investigated the response of the truss depicted in Fig. 1.6 by considering the constitutive representations of Fig. 1.7 that accounted for the possible fracture of the material and for its yielding. In both of these cases, we aimed at determining the structural response by starting from the complete knowledge of the structure expressed in terms of its geometry, its material properties, its support conditions, and the applied loads. This information was specified as the input of the analysis and was used to determine the deformations and/or stresses induced in the structure. An overview of this process is outlined in Fig. 3.1 and this type of analysis is commonly denoted as a *forward* analysis to highlight the fact that we start from the knowledge of the "system," e.g., the structure and its support and loading conditions, to determine its structural response, e.g., displacements, deformations, and stresses.

In such an analysis, we usually have a unique relationship between the input parameters and the structural response, as depicted graphically in Fig. 3.2. Such an analysis can be carried out by means of different methods of analysis, the most popular being the stiffness method and the finite element method. For a successful modeling, the constitutive models need to provide a satisfactory representation of the material behavior. An introduction to the calibration of constitutive models was previously provided in Chapter 2. For example, we have seen that it is common to characterize the material response of metals from the stress–strain curves obtained from tensile tests (see Fig. 2.3).

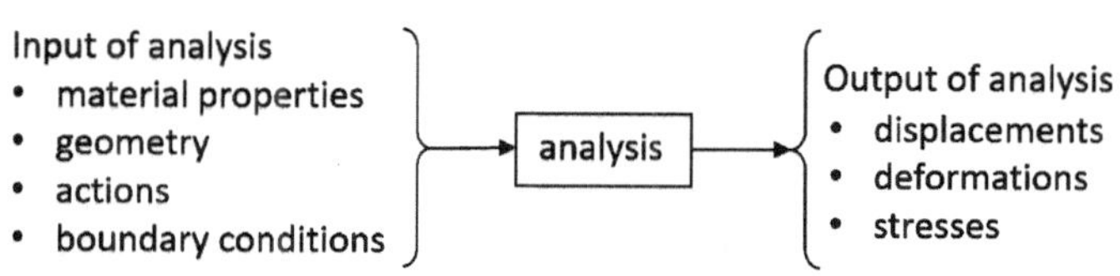

FIGURE 3.1 Example of input and output information specified in a *forward* analysis.

Constitutive Modeling of Engineering Materials. https://doi.org/10.1016/B978-0-12-814696-5.00008-3

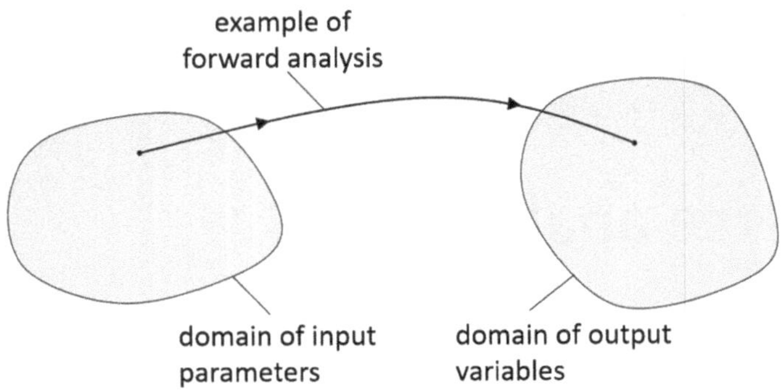

FIGURE 3.2 Example of the unique relationship that can exist between the input parameters and the structural response (expressed in terms of output variables).

The amount of refinement to be introduced for the material description depends, among the other factors, on the type of analysis to be performed. For example, if the purpose of the analysis is to simulate the response of a steel truss subjected to very low levels of load, all structural components could be assumed to remain in their linear-elastic range and it may be sufficient to model the structural system with simple truss (line) elements and to adopt a suitable elastic modulus for the steel. In the case the analysis requires the evaluation of the response for higher levels of loads, it might be necessary to ensure that other properties, such as the parameters describing yielding or hardening, are included in the analysis.

When moving to more complex material models, the evaluation of all material parameters may not be easy and it could require extensive testing. In some cases, there may be no simple direct characterization of specific material parameters from experimental testing and other identification strategies should be considered. Other identification approaches could also be explored when trying to minimize the number of tests to be carried out, to move away from destructive testing (commonly used for the direct characterization of material properties) for the exploitation of the information collected from non-destructive tests and to enhance the efficiency and robustness of the overall identification procedure.

In this context, the *inverse analysis* technique provides a useful support for the characterization of material parameters. For example, in Chapter 2 we considered the use of an indentation test (Fig. 2.1) for the identification of the material properties. This approach has been applied for the material characterization of a wide range of materials, including metals whose response could be expressed in terms of the Young's modulus, the yield stress and the strain-hardening coefficient.

When performing the indentation test, we obtain load-indentation curves such as those shown in Fig. 2.2. As it is not possible to directly extrapolate the three sought material parameters from these curves, it could be argued that a tensile test may be a better choice to obtain these parameters experimentally. Certainly this would be the case if other material characterization strategies would not be available. In reality and relevant to this book, it has been shown that it is possible to establish suitable inverse analysis procedures for the identification of these parameters that make use of the experimental information obtained from the load-indentation curve and from the mapping of the deformation left by the indenter on the material sample. The main advantage of such an approach is to rely on a non-destructive testing protocol and to move away from the conventional (destructive) tensile tests.

For the purpose of this book, an inverse analysis procedure aims at identifying material properties possessed by a structure or component based on information obtained from the structural response. Such an approach is illustrated in Fig. 3.3. Depending on the problem considered, it may be possible to identify the sought material parameters based on the available input information and monitored (i.e., measured) structural response. Such a scenario is usually referred to as a *well-posed* problem to depict the fact that the identification process is capable of establishing a unique relationship between the available information and the sought parameters (see well-posed inverse problem in Fig. 3.3).

There is another class of inverse problems, commonly denoted as *ill-posed* problems, for which it is not possible to uniquely characterize the sought parameters. These scenarios may occur when different combinations of sought parameters lead to the same structural response or when the inverse analysis algorithm is not able to distinguish the parameters (see ill-posed inverse problem in Fig. 3.3).

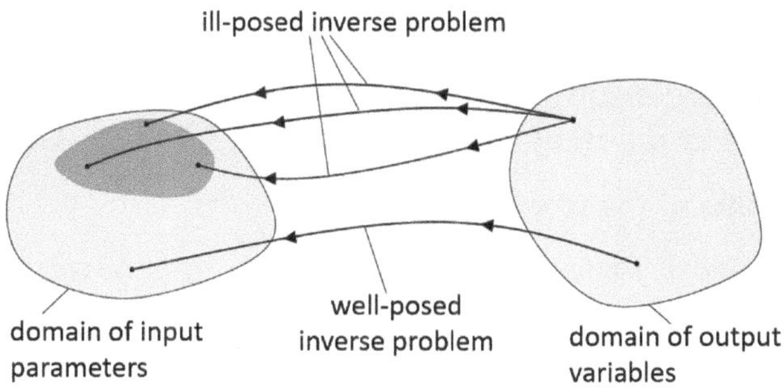

FIGURE 3.3 Scenarios of well-posed and ill-posed inverse analysis problems.

An inverse analysis aims at identifying the sought material parameters by running different forward analyses while varying the unknown material parameters (i.e., the material parameters that we want to identify) and refining their predictions to better fit the experimental measurements. This procedure is expressed mathematically in the form of a minimization problem and, therefore, the ability of the inverse analysis to capture the appropriate material properties depends on the way the minimization problem is setup. Usually, it involves the definition of an *objective function*, also known as a *discrepancy function* or *cost function*, that needs to be minimized and the ability of an inverse analysis to successfully recognize the sought material parameters highly depends on the shape of this function. For example, a convex objective function would ensure the existence of only one global minimum and we would expect, in this case, the sought parameters to be well identifiable. In a more difficult scenario, the objective function may have a complex shape with several local and even global minimum points, therefore leading to a situation in which the sought parameters cannot be uniquely identifiable and/or the identified values could depend on the way the inverse analysis is implemented and performed. In such situations, it might be possible to improve the performance of the inverse analysis process by selecting a suitable minimization algorithm and/or by revising the experimental data to be used in input for the inverse analysis, for example, by adding new experimental measurements.

There are several numerical strategies that have been used to date for solving the minimization problem of an inverse analysis procedure. While it is beyond the scope of this book to provide a comprehensive review and description of these techniques (see dedicated textbooks, e.g., [1–4]), a brief introduction to these numerical strategies is provided at the end of the chapter.

Our interest in the use of an inverse analysis procedure relies on its ability to characterize material parameters that are used in the governing equations of constitutive models. With such objective, the setup of the inverse analysis procedure is highly dependent on the type of parameters that we are trying to identify. In the remainder of the chapter, we consider two illustrative examples to present some of the key steps involved in the design and implementation of an inverse analysis process.

3.2 Illustrative example 1—simple structure formed by springs and a rigid bar

3.2.1 Outline of the structure

The first illustrative example is based on the spring system supporting a rigid bar loaded at mid-span previously introduced in Fig. 1.1 and reproduced here for ease of reference in Fig. 3.4.

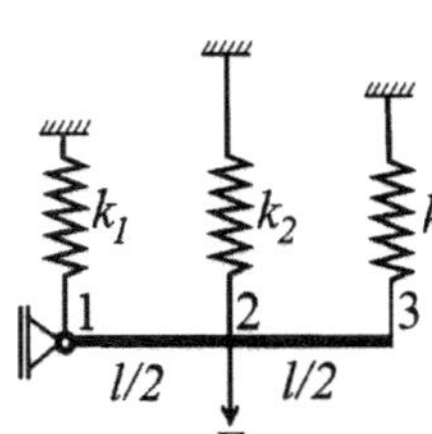

FIGURE 3.4 Illustrative example 1 consisting of the spring system supporting a rigid bar loaded at mid-span (previously introduced in Fig. 1.1).

In Chapter 1, we derived a closed-form solution for the structural response (Eq. (1.9) defines force S_1 resisted by spring 1) and we now rearrange those equations to obtain the expressions for the vertical displacements exhibited at the connection points between the bar and the springs, here denoted as v_1, v_2, and v_3 (that are equal to the corresponding spring elongations Δl_1, Δl_2 and Δl_3, respectively) as follows:

$$v_1 = \Delta l_1 = \frac{2 \cdot F \cdot k_3}{k_1 \cdot k_2 + 4 \cdot k_1 \cdot k_3 + \cdot k_2 \cdot k_3} \tag{3.1}$$

$$v_2 = \Delta l_2 = \frac{F \cdot (k_1 + k_3)}{k_1 \cdot k_2 + 4 \cdot k_1 \cdot k_3 + \cdot k_2 \cdot k_3} \tag{3.2}$$

$$v_3 = \Delta l_3 = \frac{2 \cdot F \cdot k_1}{k_1 \cdot k_2 + 4 \cdot k_1 \cdot k_3 + \cdot k_2 \cdot k_3} \tag{3.3}$$

Let us now describe the response of the structure of Fig. 3.4 by considering the following properties (these are selected for illustrative purposes):

- the applied force F is taken as 1000 N;
- the spring stiffnesses are assumed to be $k_3 = 2k_2 = 4k_1 = k$; and
- $k = 400$ N/mm.

Based on these properties, the deflections calculated at the nodes 1, 2, and 3 (i.e., the connection points between the springs and the rigid bar) are:

$$v_1 = 3.07 \text{ mm} \tag{3.4}$$

$$v_2 = 1.92 \text{ mm} \tag{3.5}$$

$$v_3 = 0.77 \text{ mm} \tag{3.6}$$

We will use these results as reference in the following discussions related to the inverse analysis approach.

3.2.2 Identification of one spring stiffness

Let us now assume that, for the structure of Fig. 3.4, we know the spring stiffnesses k_1 and k_2, and we want to identify the stiffness of the third spring k_3.

We will now go through the process of setting up an inverse analysis procedure while keeping the complexity of the problem and numerical calculations to a minimum. In reality, these steps are performed in an iterative fashion until an effective inverse analysis procedure is obtained.

The first step that we need to carry out when setting up an inverse analysis procedure is to define the type of experiment that we want to perform to make sure that we can collect sufficient experimental data for the parameter identification. Let us assume that, as a first attempt in the setting up of the inverse analysis approach, we consider a simple experiment in which we apply a force at the mid-span of the rigid bar and measure the corresponding displacement taking place at the loaded node (i.e., at the node at which the force is applied). At this point in time, we do not really know (even if, of course, we could guess) whether or not the selected measurement could enable a robust identification of the sought parameter (i.e., k_3) and we will need to go through the next steps to determine this and, if necessary, revise the approach. In general, a good experiment is the one that is sufficiently simple to be easily performed while still being able to exhibit and capture the influence of the sought material properties.

The second step in the setting up of the inverse analysis process is to determine the numerical model to be used for simulating the experiment defined in the first step. The numerical model should be carefully selected because it needs to possess a sufficient level of accuracy and details while being computationally efficient, as computationally demanding numerical models could significantly increase the overall duration of the inverse analysis. The latter remark is particularly important when seeking the identification of several material parameters. For the purpose of our illustrative example, we will keep the model as simple as possible and make use of the analytical solutions previously introduced to perform the forward analyses required by the inverse analysis procedure and, in particular, we make use of Eq. (3.2) for the evaluation of v_2.

The third step focuses on the sensitivity analysis to evaluate whether the selected experimental measurements (those that will be recorded from the experiment) are sufficiently sensitive to variations in the sought material parameters. This is usually carried out by investigating the influence of small variations in the sought material parameters on the measured data. As one would expect, if a variation of a parameter does not affect the measured data it is unlikely that the inclusion of this information in the input data of the inverse analysis will enable the characterization of the parameters under consideration. For example, if we want to determine the yield strength of a material we need to make sure to apply sufficient loading to the material sample to exhibit yielding in the response. If the loading is too small and the material remains in the linear-elastic range, the sensitivity study would highlight the inability of the proposed experiment (in this case due to the low level of load) to detect the yield strength. In such a case, the experiment should be changed, for example, by applying a higher level of load.

For the purpose of the illustrative example, we perform a sensitivity study to determine the influence of varying the sought parameter k_3 on the mid-span deflection v_2. This is usually carried out with the help of sensitivity indices that evaluate the partial derivatives of the measured variables with respect to the sought material parameters. For the problem under consideration, we consider the calculation of the following sensitivity index S_{v_2,k_3}:

$$S_{v_2,k_3} = \frac{\partial v_2}{\partial k_3} \cdot \frac{F}{v_2^2} \tag{3.7}$$

where the additional term F/v_2^2 has been introduced to non-dimensionalize S_{v_2,k_3}. Other variables and ratios could have been used for this purpose.

The main advantage of dealing with a non-dimensionalized sensitivity index is to limit its changes due to, for example, the level of applied load F or other variables that affect the magnitude of the calculated values without varying the influence of k_3 and this would not enable a direct differentiation and comparison between values calculated for sensitivity interpretations. By exploiting the analytical model available to describe the response of the structure, we evaluate S_{v_2,k_3} in closed-form and numerically. In general, the partial differentiation specified in Eq. (3.7) is performed numerically (i.e., by applying finite difference technique) because the forward analysis is commonly carried out with a numerical model.

We now start to determine S_{v_2,k_3} analytically by partially differentiating the expression for v_2 (Eq. (3.2)) with respect to k_3 and substituting it in Eq. (3.7) as follows:

$$S_{v_2,k_3} = \frac{F_{(ref)}}{k_0} \cdot \left[1 - \frac{(4 \cdot k_1 + k_2) \cdot (k_1 \cdot k_2 + k_2 \cdot k_{3(ref)})}{k_2 \cdot k_0}\right] \cdot \frac{F_{(ref)}}{v_{2(ref)}^2} \tag{3.8}$$

where the level of the force considered in the evaluation of the sensitivity index is denoted as $F_{(ref)}$ and it produces a deflection $v_{2(ref)}$ based on a reference stiffness $k_{3(ref)}$ for spring 3, and $k_0 = k_1 k_2 + 4k_1 k_{3(ref)} + k_2 k_{3(ref)}$.

The numerical evaluation of S_{v_2,k_3} is carried out by considering the variation of the response obtained with $k_{3(ref)}$ (with corresponding deflection $v_{2(ref)}$) and the behavior exhibited by a "varied" stiffness parameter $k_{3(var)}$ that leads to a displacement $v_{2(var)}$. The variation of the material property for the evaluation of the sensitivity index is kept to the smallest possible value and it is usually in the order of 0.01%, even if this value is problem-specific and varies for different identification scenarios. Based on the notation introduced, the sensitivity index S_{v_2,k_3} of Eq. (3.7) can be determined numerically as:

$$S_{v_2,k_3} = \frac{v_{2(var)} - v_{2(ref)}}{k_{3(var)} - k_{3(ref)}} \cdot \frac{F_{(ref)}}{v_{2(ref)}^2} \tag{3.9}$$

that yields values equivalent to those determined with Eq. (3.8) as long as adequate increments are specified when considering the "varied" properties.

The variation of S_{v_2,k_3} for different values of k_3 is plotted in Fig. 3.5 and highlights how the response for v_2 is more sensitive for lower values of k_3 in the domain of the response solution considered here (while considering all other input parameters constant). This is a consequence of the fact that changes in the k_3 values have a more pronounced variation on the calculated deflection v_2 for lower values of k_3 as shown in Fig. 3.6.

At this stage, we cannot draw any conclusions on the identifiability of the material parameter k_3 as this depends on a number of factors. The main purpose of the sensitivity analysis is to ensure that the sought parameter produces variations in the selected measurement and, in the case several types of measurements are available, it can assist in the selection of measurements that maximize the sensitivity of the material parameters in the domain of interest. The non-dimensionalized sensitivity indices become particularly useful in these cases, when we have to deal and compare the impact of different types of measurements.

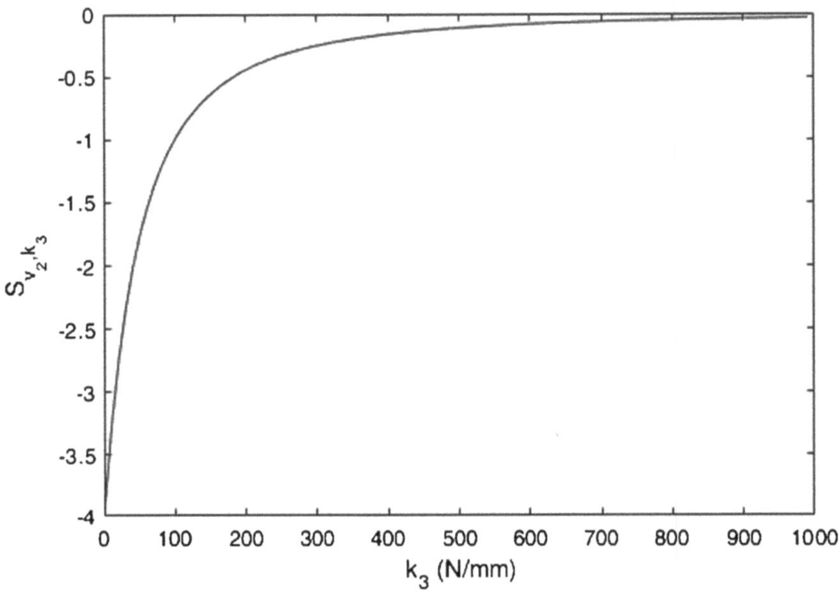

FIGURE 3.5 Sensitivity analysis performed for the illustrative example 1.

The fourth step of the inverse analysis procedure setup deals with the selection of a suitable objective function and of an adequate optimization algorithm. The objective function $f(x)$ drives the inverse analysis procedure because it consists of the function that is minimized in the domain of interest to identify the sought parameters. Such a function is usually expressed in terms of some selected residuals that can be written in the following form:

$$f(x) = \frac{1}{2}\sum_{i=1}^{n_R} r_i^2(x) \tag{3.10}$$

in which r_i denotes the ith residual and represents the difference between the computed estimate of a measurement and the actual recorded measurement, while n_R denotes the number of residuals considered (i.e. number of measurable quantities) in the evaluation of the objective function. The writing of the objective function is an important step in the definition of an inverse analysis because it needs to contain sufficient information for a robust identification of the parameters sought. It is possible, even if not explicitly shown in Eq. (3.10), to apply different weighted coefficients w_i to the residuals r_i in the case the uncertainty possessed by the different measurements is not consistent, for example, due to variations of the measurements performed under different conditions or

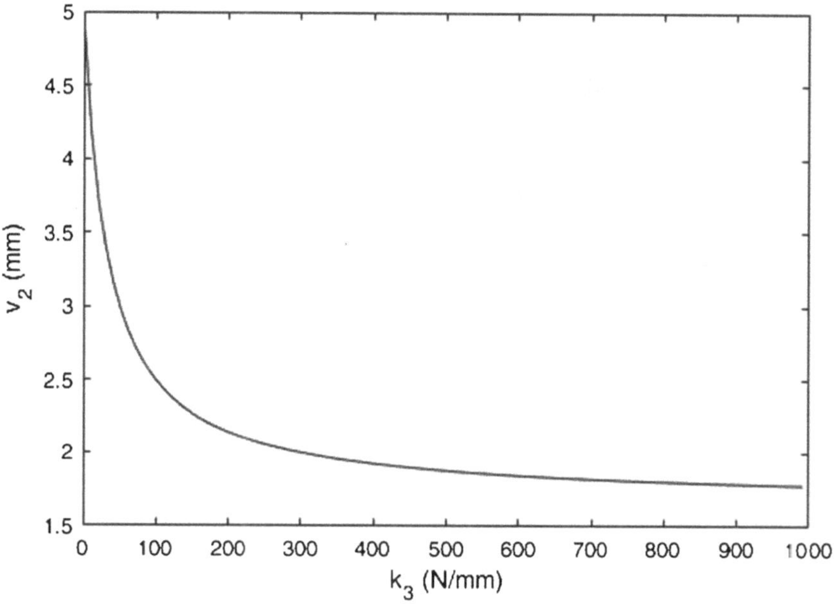

FIGURE 3.6 Variation of the mid-span deflection v_2 with respect to the stiffness k_3 for the illustrative example 1.

locations. A measurement with a higher weight has a higher influence in the identification process in moving toward the identified values because regarded to be more reliable.

For our illustrative example, the objective function of Eq. (3.10) is calculated by considering only the vertical displacement at node 2 at a known level of applied load F and the corresponding (only) residual, here referred to as r_1, is obtained as follows:

$$r_1 = v_2^{com} - v_2^{exp} \tag{3.11}$$

where v_2^{exp} depicts the measured displacement and v_2^{com} denotes the computed value.

The selection of an adequate optimization algorithm is vital for the successful implementation of an inverse analysis, especially considering the fact that the minimization problems commonly considered in material parameter identifications are constrained as some of these parameters represent physical quantities that can only vary within arbitrary values. The selected algorithm has also to deal with the possible complexity of the objective function that, in most cases, is not convex and, sometimes, even not continuous.

To minimize the complexity of the solution, we can express the objective function analytically by relying on Eq. (3.2) for the calculation of v_2 and, by rearranging the expression, to compute k_3 based on the experimental value of the mid-span deflection as follows:

$$k_3^{comp} = \frac{F \cdot k_1 - k_1 \cdot k_2 \cdot v_2^{exp}}{4 \cdot k_1 \cdot v_2^{exp} - F + k_2 \cdot v_2^{exp}} \tag{3.12}$$

that is valid for all values except those that cause the denominator to be equal to zero. The latter limitation is acceptable for the purposes of this illustrative example.

The fifth step in the setting up of an inverse analysis procedure relies on the validation of the approach that should be carried out against real experimental data. Since it is very unlikely to have access to sufficient data for a comprehensive experimental validation, it is convenient and recommended to initially determine the effectiveness of the inverse analysis procedure based on computer generated results, usually denoted as *pseudo-experimental* results. One of the main advantages in testing inverse analysis techniques with pseudo-experimental data is that the numerical results are generated from known and defined material parameters and, therefore, the accuracy and robustness of the procedure can be carefully evaluated. We could even verify the influence of introducing different levels of noise to account for the possible uncertainty and variability that is usually embedded in real-world measurements.

Let us start by testing the inverse analysis procedure by considering the value for v_2 (=1.92 mm) provided in Eq. (3.5) as our pseudo-experimental measurement (obtained with the spring stiffness parameter value of $k_3 = 400$ N/mm and for the externally applied force $F = 1000$ N).

Fig. 3.7 provides an overview of the shape of the objective function calculated by varying the values for k_3 within the range of interest (that is arbitrary taken to be within 100 N/mm $\leq k_3 \leq$ 1000 N/mm). In this case, the problem is regarded as *well posed* because the shape of the objective function is convex within the range of parameters considered and, therefore, guarantees the existence of only one minimum value which corresponds to $k_3 = 400$ N/mm (initially used to generate the pseudo-experimental measurement). Other considerations might be required when moving outside this domain of interest.

Before we continue with variations on our illustrative example, it is useful to summarize the steps that we have just introduced for the setting up of an inverse analysis procedure in Table 3.1. Such steps intend to provide an overview of the entire process and do not intend to be prescriptive as different inverse analysis scenarios may require additional and/or different steps.

3.2.3 Identification of two spring stiffnesses

In this section, we revisit the five steps introduced in Table 3.1 to setup the inverse analysis procedure for the identification of the two spring stiffnesses k_2 and k_3 of the illustrative example of Fig. 3.4. This time we assume to know only spring stiffness k_1.

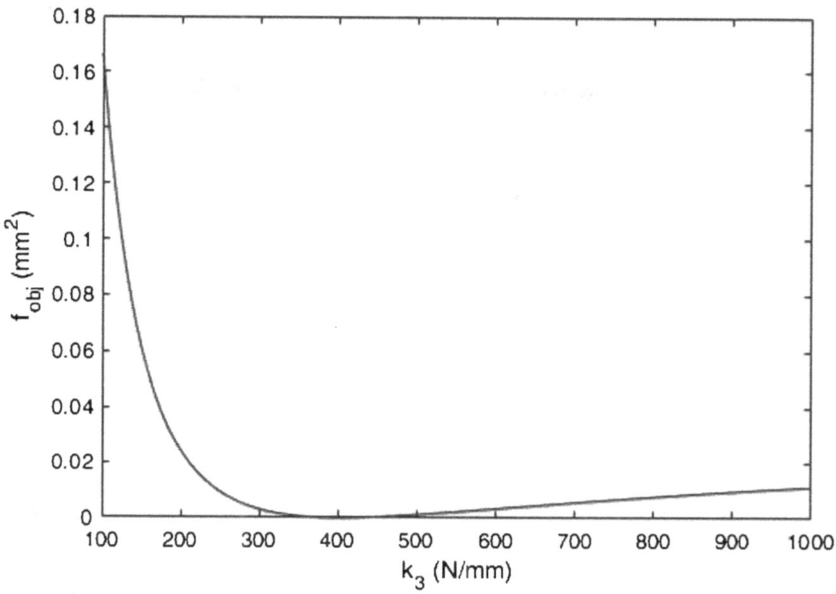

FIGURE 3.7 Variation of the objective function with respect to the stiffness k_3 for the illustrative example 1 considering the identification of one spring stiffness.

TABLE 3.1 Typical steps implemented for the setup of an inverse analysis procedure.

Step	Description of the task
1	Selection of experiment
2	Determination of the numerical model to be used for the forward analysis
3	Sensitivity analysis
4	Selection of suitable objective function and adequate optimization algorithm
5	Validation of inverse analysis performance (e.g., based on pseudo-experimental data)

Step 1: selection of experiment

We start by reconsidering the same experiment previously adopted for the identification of the spring stiffness k_3. We therefore measure the mid-span deflection v_2 at a known level of applied load F.

Step 2: determination of the numerical model to be used for the forward analysis

For the forward analysis, we make use of the analytical solution for the mid-span deflection v_2, previously introduced in Eq. (3.2), that is written as a function, among the others, of the sought parameters k_2 and k_3.

Step 3: sensitivity analysis

The sensitivity study aims at establishing the influence of both spring stiffnesses k_2 and k_3 on the mid-span deflection. This is carried out numerically based on the following expression:

$$S_{v_2,k_j} = \frac{v_{2(var)} - v_{2(ref)}}{k_{j(var)} - k_{j(ref)}} \cdot \frac{F_{(ref)}}{v_{2(ref)}^2} \tag{3.13}$$

in which $k_{j(var)}$ and $k_{j(ref)}$ depict the varied and reference values used in the calculations associated to the jth spring (with $j = 2, 3$). This expression is identical to Eq. (3.9) that considered $j = 3$.

Figs. 3.8 and 3.9 provide an overview of the variation of the sensitivity index introduced in Eq. (3.13) for k_2 and for the deflection v_2, respectively, in the domain of the reference solution considered here (while considering all other input parameters constant). The sensitivity study for k_3 was performed previously and it is not repeated here (see Fig. 3.5). As expected, the deflection v_2 is sensitive to variations in both material parameters, i.e., k_2 and k_3. At this point, it is not possible to draw conclusions on the identifiability of the parameters and this is explored in the following steps.

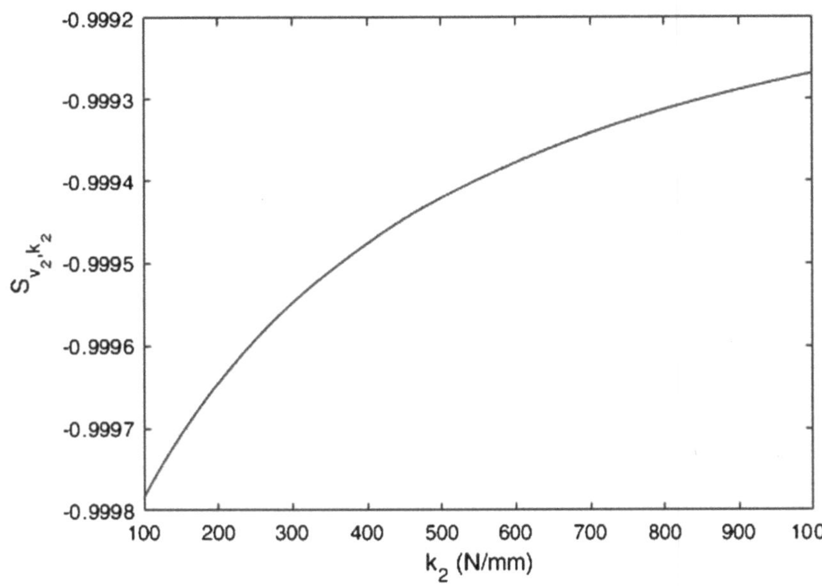

FIGURE 3.8 Sensitivity analysis performed for the spring stiffness k_2 for the illustrative example 1 considering the identification of two spring stiffnesses.

Step 4: selection of a suitable objective function and an adequate optimization algorithm

For the purpose of this example, we use the objective function previously utilized and specified in Eq. (3.10) with the residual of Eq. (3.11).

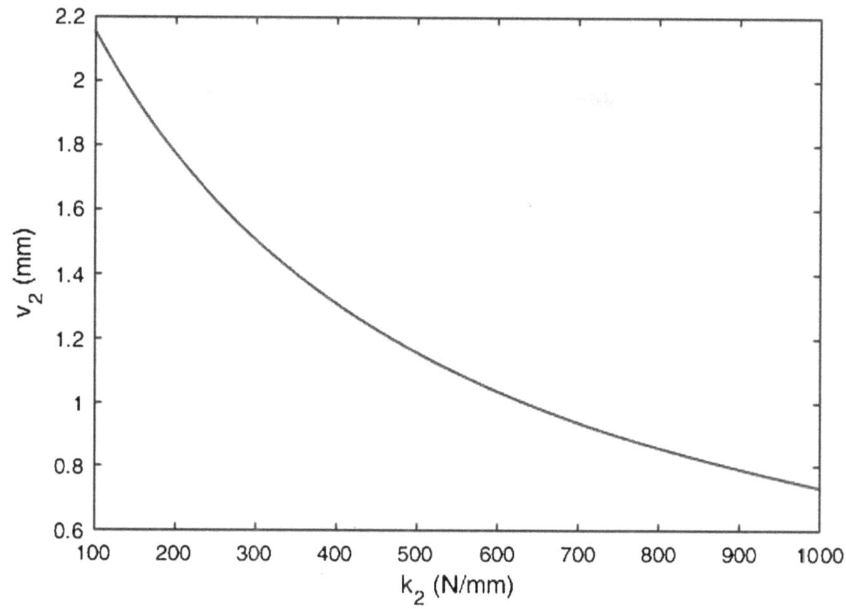

FIGURE 3.9 Variation of the mid-span deflection v_2 with respect to the stiffness k_2 for the illustrative example 1 considering the identification of two spring stiffnesses.

The identification process relies on analytical expressions of the model, if available, or alternatively, on a numerical solution. The idea is to exploit closed-form solutions if available as carried out in the previous example (in which k_3 was estimated based on the experimental measurement of v_2 and on the given values for k_1 and k_2 with Eq. (3.12)).

Step 5: validation of the inverse analysis procedure based on pseudo-experimental data

This step focuses at the evaluation of the inverse analysis procedure based on pseudo-experimental results generated with the expressions previously derived in Eqs. (3.5) and (3.6).

By inspecting Eq. (3.12) (used in the previous example), it is possible to note that different solutions for k_3 can be obtained for a given value of experimental mid-span v_2 while varying k_2. As a consequence, it is not possible to determine a unique set of k_2 and k_3 for each measured v_2. This is because different sets of these parameters can produce the same mid-span deflection. For example, the sets of k_2 and k_3 that produce v_2 =1.92 mm are plotted in Fig. 3.10 for the domain of interest considered here. Such a situation implies that there is no unique set of k_2 and k_3 values that leads to the considered pseudo-experimental measurement. This is also depicted by the variation of the objective function shown in Fig. 3.11 that highlights how the minimum region of the function extends over different sets of k_2 and k_3 (that, as expected, follows the trend of the sets of parameters depicted in Fig. 3.10).

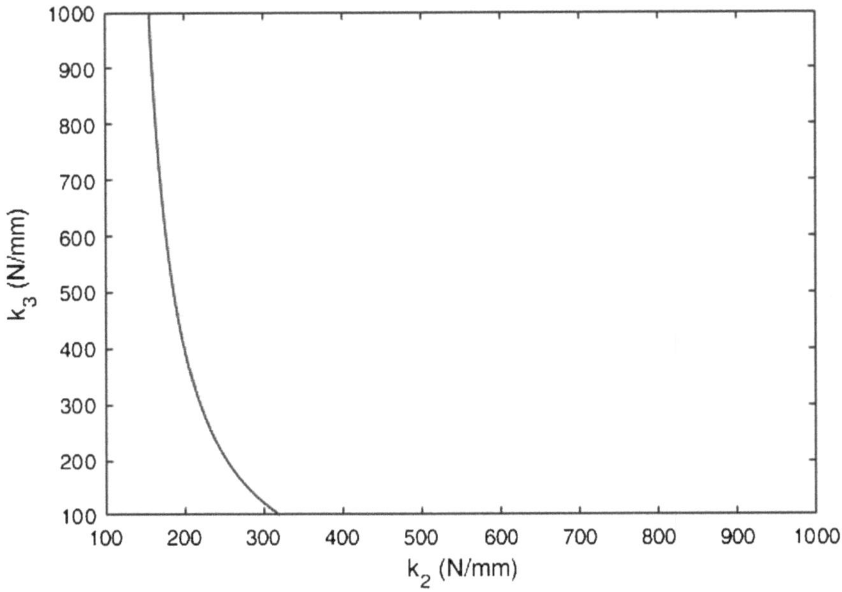

FIGURE 3.10 Set of values for k_2 and k_3 that leads to a mid-span deflection v_2 of 1.92 mm for the illustrative example 1 considering the identification of two spring stiffnesses.

To avoid this ill-posed condition, it is necessary to modify the objective function and to include additional experimental information. For this purpose, a second residual is included in the objective function that accounts for the variation of the deflection under node 3, i.e., v_3, defined as:

$$r_2 = v_3^{com} - v_3^{exp} \tag{3.14}$$

By including the residuals r_1 and r_2 defined in Eqs. (3.11) and (3.14), respectively, the objective function develops a pronounced minimum value as shown in Fig. 3.12 in the domain of the stiffnesses k_2 and k_3. In this case, its minimum value occurs for the sought values for k_2 and k_3 that were 200 N/mm and 400 N/mm, respectively. These were the values initially used for the determination of the pseudo-experimental deflection used in input in the inverse analysis procedure and, therefore, the identification process can be regarded to have been completed successfully.

When starting the inverse analysis procedure initial *guess* values for the sought parameters need to be specified. For the purpose of this example, the inverse analysis procedure has been working for all initial values specified in the trial runs. Initial values considered here were selected within $\pm$ 10% of the pseudo-experimental measurements. Different ranges could be considered depending on the type of identification problem considered.

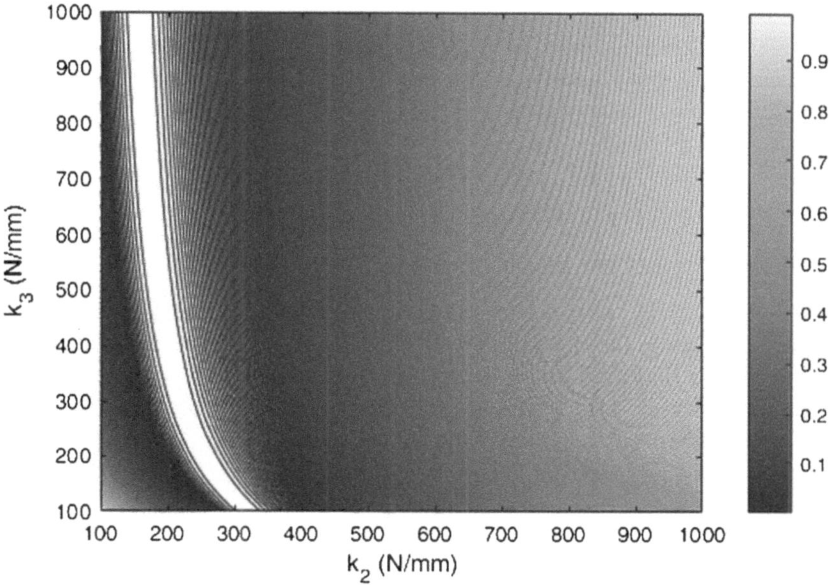

FIGURE 3.11 Variation of the objective function when considering only the mid-span deflection v_2 as a function of k_2 and k_3 for the illustrative example 1 aiming at the identification of the two spring stiffnesses.

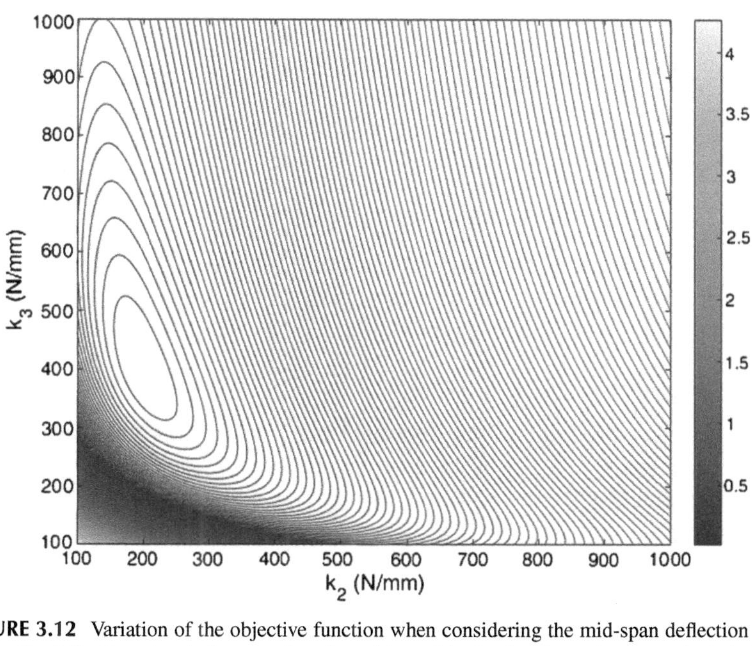

FIGURE 3.12 Variation of the objective function when considering the mid-span deflection v_2 as well as the deflection at node 3 v_3 as a function of k_2 and k_3 for the illustrative example 1 aiming at the identification of the two spring stiffnesses.

This example has provided, while minimizing the complexity of the solution, an overview of the iterative process that is usually required when setting up an inverse analysis procedure. In the next section, we consider a slightly more complex problem to provide more details on the key aspects of the numerical implementation.

3.3 Illustrative example 2—simple truss system

3.3.1 Outline of the structure

The second illustrative example considered in this chapter consists of the truss system previously introduced in Fig. 1.6 and reproduced here for ease of reference in Fig. 3.13. In the next subsections, we introduce the setting up of the inverse analysis approach followed by the identification of two case studies that rely on a linear-elastic response of the truss members.

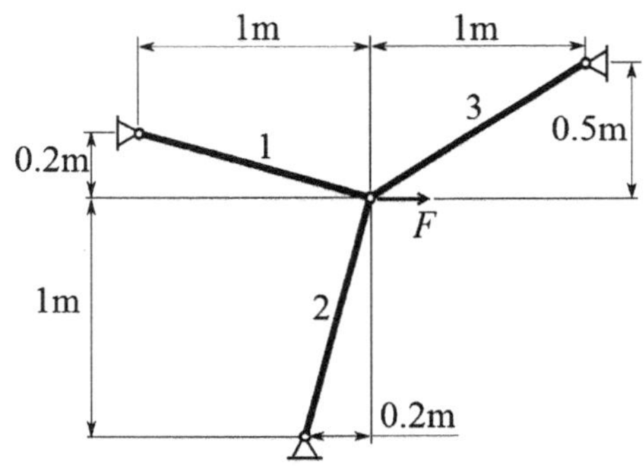

FIGURE 3.13 Illustrative example 2 consisting of the truss system introduced in Fig. 1.6.

3.3.2 Overview of the identification procedure

The setting up of the inverse analysis procedure follows the steps introduced in Table 3.1. Depending on the case study considered, one or more of the material parameters of the truss members are assumed to be unknown.

Step 1: selection of experiment

We consider an experiment that resembles the loading condition and support conditions depicted in Fig. 3.13. The measurements that are recorded consist of the displacements of the unrestrained node and the corresponding level of force. These values are collected, for ease of notation, in the following vector:

$$\mathbf{p}^{exp} = \begin{bmatrix} u_1^{exp} & v_1^{exp} & F_1^{exp} & \dots & u_n^{exp} & v_n^{exp} & F_n^{exp} \end{bmatrix}^T \tag{3.15}$$

where the terms u_i^{exp} and v_i^{exp} depict the horizontal and vertical displacements, respectively, that occur at the load level F_i^{exp} for the ith measurement recorded for input in the inverse analysis approach. In this representation, n depicts the number of measurements. Other representations of the experimental measurements could have also been considered.

Step 2: determination of the numerical model to be used for the forward analysis

The forward analysis specified at this step is implemented numerically with the stiffness method to highlight the numerical solution of the inverse problem. The analysis can be performed based on the MATLAB® script provided below that is called, as required, by the inverse analysis procedure. For clarity, the MATLAB® script has been written to incorporate the information related to the geometry and support conditions of the truss system considered.

```
function [u_comp,v_comp,F_comp] = MATLAB_script_3_1(inputParameter,soughtParameter
      ,F_exp)

% All units to be in mm and N.

i1=0;
for(i2=1:length(inputParameter))
    if(inputParameter(i2)==0)
        i1=i1+1;
        E(i2)=soughtParameter(i1);
    elseif(inputParameter(i2)~=0)
        E(i2)=inputParameter(i2);
    end
end

% Input geometry and loading
A=600;
bar1_x1=1000*0.0; bar1_y1=1000*1.2; bar1_x2=1000*1; bar1_y2=1000*1;
bar2_x1=1000*0.8; bar2_y1=1000*0.0; bar2_x2=1000*1; bar2_y2=1000*1;
bar3_x1=1000*2.0; bar3_y1=1000*1.5; bar3_x2=1000*1; bar3_y2=1000*1;
F=F_exp; % applied force

% Stiffness matrix and loading vector related to unrestrained freedoms
Lx1=bar1_x2-bar1_x1;  Lx2=bar2_x2-bar2_x1;  Lx3=bar3_x2-bar3_x1;
Ly1=bar1_y2-bar1_y1;  Ly2=bar2_y2-bar2_y1;  Ly3=bar3_y2-bar3_y1;
L1=sqrt(Lx1^2+Ly1^2); L2=sqrt(Lx2^2+Ly2^2); L3=sqrt(Lx3^2+Ly3^2);
l1=Lx1/L1; m1=Ly1/L1; l2=Lx2/L2; m2=Ly2/L2; l3=Lx3/L3; m3=Ly3/L3;
lF=1; mF=0; eq_matrix=[-l1 -l2 -l3 lF; -m1 -m2 -m3 mF];
B=[l1 l2 l3 ; m1 m2 m3]';
D = [E(1)*A/L1 0 0 ; 0 E(2)*A/L2 0 ; 0 0 E(3)*A/L3 ];
K = B' * D * B;
f = [ F_exp(1) ; 0 ];
d = inv(K)*f; e = B * d; s = D * e;
sol_F=[ [0;0] f ]; sol_D=[ [0;0] d ];
F_comp=sol_F(1,:); u_comp=sol_D(1,:); v_comp=sol_D(2,:);
```

Step 3: sensitivity analysis

The sensitivity study aims at establishing the influence of the material parameters on the structural response expressed in terms of horizontal and vertical displacements of the unrestrained node. For this purpose, the following sensitivity index is introduced to depict the influence of the jth material parameter k_j on the displacement d:

$$S_{d,k_j} = \frac{d_{(var)} - d_{(ref)}}{k_{j(var)} - k_{j(ref)}} \cdot \frac{F_{(ref)}}{d^2_{k(ref)}} \tag{3.16}$$

where d can be replaced by the horizontal displacement u, by the vertical one v or by a combination of the two as required by the analysis.

The sensitivity analysis is performed by determining S_{d,k_j} for the domain of interest and by varying the sought material parameters. In this case, as expected, the truss node displacements are influenced by the geometry of the truss and by its elastic properties as they enter in the stiffness formulation of the forward analysis.

It is common to perform the sensitivity study by varying one material parameter at the time. Depending on the problem considered, it is possible for the same material parameter to produce different sensitivity responses in different locations of the solution domain and, therefore, the sensitivity analysis should be performed by varying the reference parameters of the study when helpful in informing the setting up of the inverse analysis approach.

Step 4: selection of a suitable objective function and an adequate optimization algorithm

The objective function considered for the truss example is based on Eq. (3.10) and the associated residuals are calculated based on the expression format already introduced in Eqs. (3.11) and (3.14).

For the purpose of the numerical calculations, the optimization algorithm is implemented in the MATLAB® script provided below. In the script, the selection of the residuals to be included in the calculation of the objective function is specified in the input file to enable the consideration of different cases.

```
function [f_obj] = ...
    MATLAB_script_3_2(data_comp,data_exp,flag_f_obj)

r_1=1/2*(data_comp(1)-data_exp(1))^2;
r_2=1/2*(data_comp(2)-data_exp(2))^2;

if(flag_f_obj=="10") f_obj=r_1;
elseif(flag_f_obj=="12") f_obj=r_1+r_2;
end
```

The optimization algorithm is implemented with the gradient based line search method to take advantage of its simplicity (so that it can be easily coded in MATLAB®) and its robustness (i.e., its typical ability to identify sought material parameters in characterization problems). The MATLAB® script provided below presents the main code driving the inverse analysis process that includes the optimization algorithm [1].

```

clc
clear

% Setting up options for optimisations
% guess=[210000], inputParameter=[0,200000,200000] % for case study 1
guess=[210000,210000], inputParameter=[0,200000,0] % for case study 2

pert=0.001; % perturbation for first derivative
lengthStep=0.5; % Initial length step
resMIN=5e-6; % Residual value for termination check

% Definition of objective function
% flag_f_obj="10" % for case study 1
flag_f_obj="12" % for case study 2

F_pseudo_exp=300000
u_pseudo_exp=1.5765
v_pseudo_exp=-0.4864
data_exp=[u_pseudo_exp v_pseudo_exp,F_pseudo_exp]
F_exp=[F_pseudo_exp ]

% Optimisation cycle
res=10; iter=0;
vect_disp=[2];
length_guess=length(guess);
while res>resMIN
    itiner(iter+1,1:length_guess)=guess';
    [u_comp,v_comp,F_comp]=MATLAB_script_3_1(inputParameter,guess,F_exp);
    data_comp=[u_comp(vect_disp) v_comp(vect_disp) F_comp(vect_disp)];
    e=MATLAB_script_3_2(data_comp,data_exp,flag_f_obj);
    itiner(iter+1,length_guess+1)=e;
    iter=iter+1;
    for iCount=1:size(guess,2)
        guessp=guess;
        guessp(iCount)=guessp(iCount)*(1+pert);
        [u_comp,v_comp,F_comp]=MATLAB_script_3_1(inputParameter,guessp,F_exp);
        data_comp=[u_comp(vect_disp) v_comp(vect_disp) F_comp(vect_disp)];
        e1=MATLAB_script_3_2(data_comp,data_exp,flag_f_obj);
        grad(iCount,1)=(e1-e)/pert;
    end
    stpdsc=-grad'/norm(grad);
    % Trying different step lengths
    guess1=guess+lengthStep*stpdsc;
    guess2=guess+(lengthStep/4)*stpdsc;
    guess3=guess+(lengthStep*2)*stpdsc;
    [u_comp1,v_comp1,F_comp1]=MATLAB_script_3_1(inputParameter,guess1,F_exp);
```

```
    data_comp1=[u_comp1(vect_disp) v_comp1(vect_disp) F_comp1(vect_disp)];
    e1tr=MATLAB_script_3_2(data_comp1,data_exp,flag_f_obj);
    [u_comp2,v_comp2,F_comp2]=MATLAB_script_3_1(inputParameter,guess2,F_exp);
    data_comp2=[u_comp2(vect_disp) v_comp2(vect_disp) F_comp2(vect_disp)];
    e2tr=MATLAB_script_3_2(data_comp2,data_exp,flag_f_obj);
    [u_comp3,v_comp3,F_comp3]=MATLAB_script_3_1(inputParameter,guess3,F_exp);
    data_comp3=[u_comp3(vect_disp) v_comp3(vect_disp) F_comp3(vect_disp)];
    e3tr=MATLAB_script_3_2(data_comp3,data_exp,flag_f_obj);
    best=min([e1tr,e2tr,e3tr]);
    if e1tr==best
        guess=guess1;
        step=1;
    end
    if e2tr==best
        guess=guess2;
        step=2;
        lengthStep=lengthStep/4;
    end
    if e3tr==best
        guess=guess3;
        step=3;
        lengthStep=lengthStep*2;
    end
    step;
    res=best; % Computing residual
    itiner(iter+1,1:length_guess)=guess';
    itiner(iter+1,length_guess+1)=best;
end

guess
best

if(length_guess==1)
    plot(itiner(:,1),itiner(:,2),'MarkerSize',10,'Marker','o','LineWidth',3,'
     LineStyle','--','Color',[0 0 0])
    xlabel('E_1 (N/mm^2)')
    ylabel('f_{obj} (mm^2)')
elseif(length_guess==2)
    plot3(itiner(:,1),itiner(:,2),itiner(:,3),'MarkerSize',10,'Marker','o','
     LineWidth',3,'LineStyle','--','Color',[0 0 0])
    xlabel('E_1 (N/mm^2)')
    ylabel('E_3 (N/mm^2)')
    zlabel('f_{obj} (mm^2)')
    view([1,0.5,0.8])
end
```

Step 5: validation of the inverse analysis procedure based on pseudo-experimental data

The inverse analysis procedure is verified in the following by considering pseudo-experimental results of two simple case studies that consider linear-elastic material properties for the truss system and that aim at the identification of the elastic modulus for one or two truss members.

The pseudo-experimental results are generated based on the geometry specified in Fig. 3.13. Each member is assumed to have a cross-sectional area of 600 mm^2 and the elastic properties are depicted by elastic moduli $E_1 = E_2 = E_3 = E$=200 GPa.

3.3.3 Case study 1: identification of one elastic modulus

In case study 1, we assume to know everything about the truss system except the elastic modulus of member 1, i.e., E_1. By applying the previously introduced inverse analysis approach, we are able to determine the sought material parameter. The identification process works by considering the following residual determined based on the horizontal displacement of the unrestrained truss node:

$$r_u = u^{com} - u^{exp} \tag{3.17}$$

that is included in the objective function whose variation is plotted in Fig. 3.14 to highlight the stability of the solution and the fact that the sought parameter is correctly identified.

The identification process could also be implemented by considering only the vertical component of the displacement of the unrestrained node whose residual can be determined as:

$$r_v = v^{com} - v^{exp} \tag{3.18}$$

that would produce similar results for the objective function to those previously observed in Fig. 3.14.

The various predictions calculated with the inverse analysis before converging to the sought parameter (and matching the pseudo-experimental value for E_1) are presented in Fig. 3.15.

3.3.4 Case study 2: identification of two elastic moduli

The second case study aims at the identification of the elastic moduli possessed by members 1 and 3, i.e., E_1 and E_3, while assuming to know the elastic properties of member 2, i.e., E_2.

The first attempt considered for the inverse analysis solution is to reconsider and implement the procedure previously presented for case study 1. The initial values for E_1 and E_3 adopted in the trials carried out considered variations of $\pm$ 10%. When considering only the residual r_u (Eq. (3.17)), it is not possible to identify both elastic moduli as the objective function does not contain a defined global minimum as depicted in Fig. 3.16. In such situation the inverse analysis problem is classified as ill posed because there is not sufficient experimental information to distinguish and characterize E_1 and E_3.

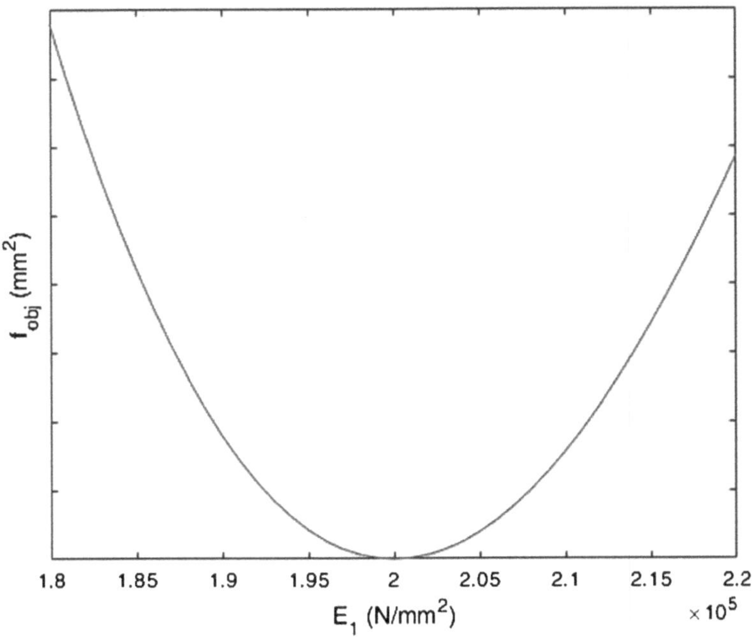

FIGURE 3.14 Overview of the objective function for the identification of one elastic modulus in case study 1.

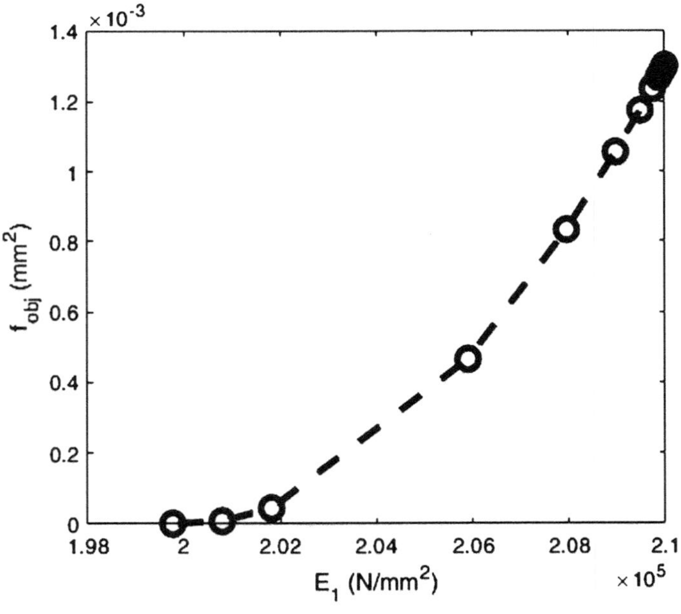

FIGURE 3.15 Example of the convergence process for case study 1.

Similar considerations would apply to the inverse problem implemented with the objective function determined with the residual r_v associated to the vertical displacement v (Eq. (3.18)).

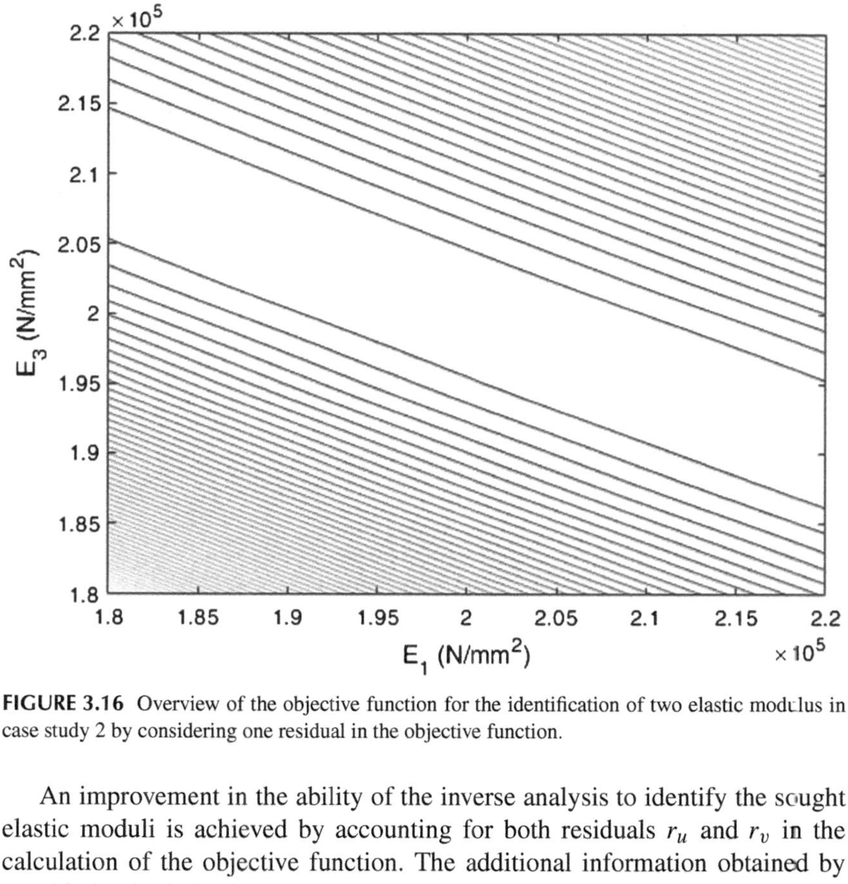

FIGURE 3.16 Overview of the objective function for the identification of two elastic modulus in case study 2 by considering one residual in the objective function.

An improvement in the ability of the inverse analysis to identify the sought elastic moduli is achieved by accounting for both residuals r_u and r_v in the calculation of the objective function. The additional information obtained by considering both displacement components simultaneously enables to move to a well-posed inverse problem as highlighted by the defined minimum in Fig. 3.17 around the correct sought parameters (as values $E_1 = E_3 = 200$ GPa were used to generate the pseudo-experimental data). An example of the different predictions carried out by the inverse analysis procedure when reaching this solution is presented in Fig. 3.18.

3.4 Brief overview of optimization strategies

In the previous sections, we have looked at the overall inverse analysis and applied it to a number of case studies to better illustrates some of its features The identification procedure relies on the use of adequate optimization algorithms as depicted in step 4 of Table 3.1. Depending on the problem considered, the

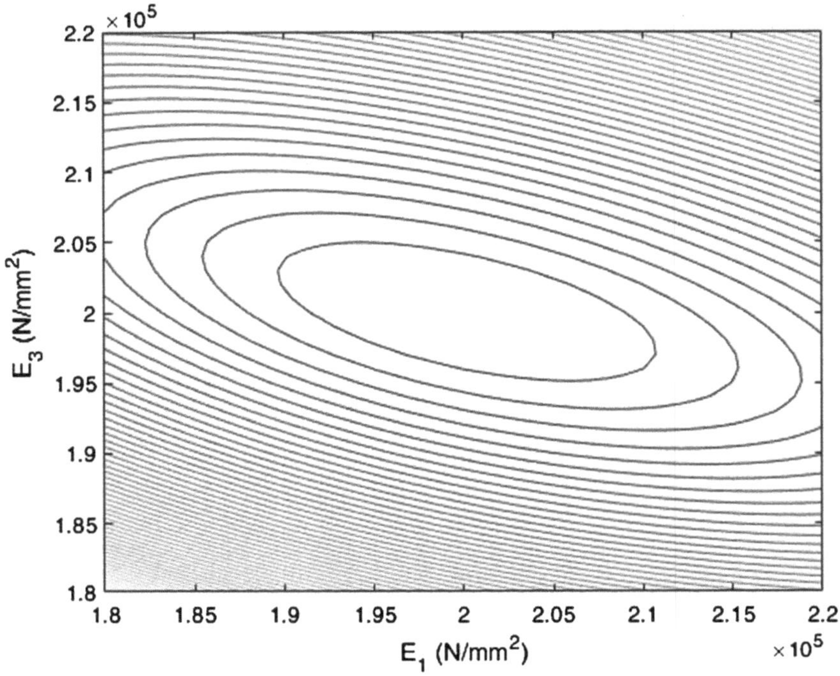

FIGURE 3.17 Overview of the objective function for the identification of two elastic modulus in case study 2 by considering two residuals in the objective function.

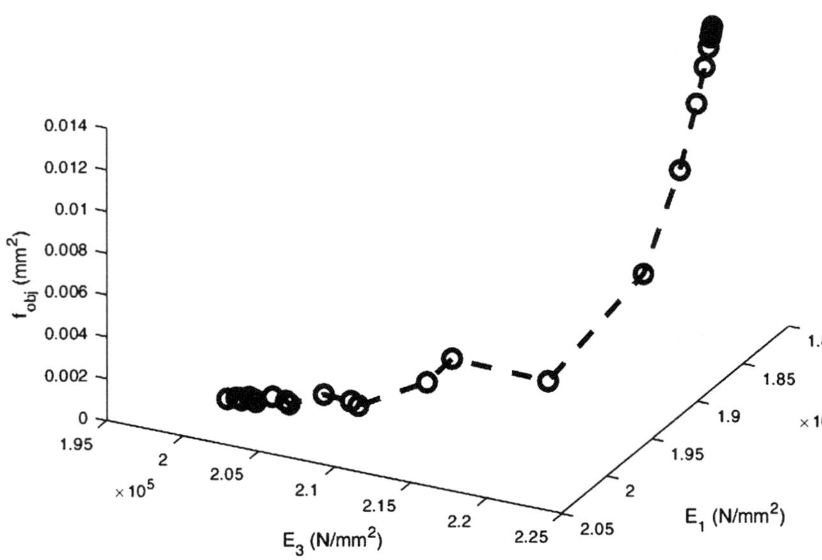

FIGURE 3.18 Example of the convergence process for case study 2 when considering two residuals in the evaluation of the objective function.

choice of a suitable optimization algorithm to perform the typical constrained minimization of the objective function can be fundamental for the development of an effective and robust identification procedure. In this section, we provide a brief overview of some of the available minimization approaches and refer to specialized textbooks for additional details (e.g., [1–4]).

Considering the wide range of available optimization algorithms, it is convenient to group them based on the information that is required to be provided by the objective function in their implementation. From a modeling viewpoint, this is a useful approach as the number of operations needed to be performed on the objective function has an important impact on the computational effort required by the numerical calculations. With this arrangement, we can distinguish between following typologies:

- *zero-order algorithms*—that require the evaluation of only the objective function as input into the optimization process. Approaches that fall under this category include genetic algorithms and the direct search based on the Nelder–Mead algorithm;
- *first-order algorithms*—that are based on the evaluation of the first derivatives of the objective functions, therefore requiring additional computational efforts (in comparison to zero order algorithms) when implemented. Among the others, these include the line search methods and the trust region methods; and
- *second-order algorithms*—that rely on the determination of the second derivatives of the objective functions, as it would be the case, for example, for the line search Newton method.

It is quite common in inverse analysis procedures to perform the required derivatives numerically by means of the finite difference method. With this approach the number of computations increase with the order of the derivatives to be performed. Despite this, in typical material parameter identifications problems, it can be observed that the computational cost produced by first-order algorithms can be smaller than the one associated with zero-order algorithms because the former require fewer simulations in their execution. This consideration should be applied with care as the ability of a first-order algorithm to converge is highly dependent on the shape of the objective function. For example, in the presence of numerous local minima, the effectiveness of first-order algorithms is highly dependent on the initialization values adopted for the sought parameters and the overall procedure can become ineffective computationally because it may require a large number of initialization and the comparison among all estimated minima to establish the global minimum. When dealing with such problems, other strategies, such as those based on the genetic algorithms become attractive because their higher computational demand is counterbalanced by their ability and higher efficiency in dealing with problems with numerous local minima.

Based on the previous considerations, it is recommended to test inverse analysis procedures by employing pseudo-experimental data to clearly distinguish and compare the effectiveness of different optimization algorithms. In this manner, once the most efficient identification procedure is selected, it can be applied to the real-world problem under consideration while addressing the additional noise and variability features expected to occur in real measurements.

3.5 Summary

This chapter provided a brief overview of the inverse analysis procedure and its advantages for the characterization of material parameters required by constitutive models. Conventionally, the identification of material properties is performed by means of destructive testing and the use of inverse analysis techniques represents an attractive alternative to support a wider use of non-destructive approaches and to characterize material parameters that do not have a direct physical meaning.

A key focus of the chapter was devoted to the setting up of an inverse procedure and this process has been articulated in five steps. In the first step, the experiment to be used to collect the measurement data for the characterization process is defined. It is convenient, where possible, to maintain the complexity of the experiment as simple as possible to enable its easy performance while ensuring that sufficient information is collected for a robust identification. The second step involves the establishment of a numerical model to carry out the forward analysis and to predict the response of the experiment. In this case, care must be placed in keeping the numerical model as simple as possible to minimize the computational effort required for the forward analysis calculations. Such an approach can have a significant impact on the duration of the entire inverse analysis procedure, especially when seeking the identification of several material parameters. A sensitivity analysis is performed in the third step to gain insight into the influence of varying the sought parameters on the monitored experimental measurements. This evaluation is useful to determine possible suitable experimental measurements to be included in the analysis. At this stage, it is not possible to draw conclusions on the identifiability of the sought parameters as the latter depends on a number of factors. The fourth step focuses on the selection of a suitable objective function and of an adequate optimization algorithm. Simple arrangements for this choice have been presented to demonstrate their influence on the overall inverse analysis performance. The fifth step consists in the validation of the inverse analysis procedure. As it is very unlikely to have access to a comprehensive set of experimental measurements for a robust calibration of the procedure it has been pointed out that pseudo-experimental results provide a very useful alternative to gain insight into the performance of the procedure. In this case, the variability of the measurements that is observed in real problems can be accounted for by the introduction of different noise patterns in the generation of the pseudo-experimental data to be used in input to the inverse analysis.

The setting up process has been illustrated by considering two simple illustrative examples to provide insight into the details of the numerical solution and its implementation. The identification process performed for these examples has been supported by either analytical expressions or by numerical solutions coded in MATLAB®. Attention has been placed in maintaining the complexity of the examples to a minimum to enable a better interpretation of the impact of different choices made during the setting up of the inverse analysis on its performance.

A brief introduction of available optimization strategies has been presented in the final part of the chapter and grouped into zero-, first-, and second-order algorithms based on the information required in input from the objective function. Different possible identification scenarios have been considered and discussed with the aim of providing an initial indication for the selection of optimization strategies. In general, it has been recommended to make use of pseudo-experimental data to test the efficiency of the adopted optimization algorithms to identify the most suitable one that can then be applied to the problem considered with real experimental measurements.

References

[1] V. Buljak, Inverse Analyses with Model Reduction: Proper Orthogonal Decomposition in Structural Mechanics, Springer, 2012.

[2] C.W. Groetsch, Inverse Problems: Activities for Undergraduates, The Mathematical Association of America, Washington D.C., 1999.

[3] G.R. Liu, X. Han, Computational Inverse Techniques in Nondestructive Evaluation, CRC Press, Boca Raton, 2003.

[4] A. Tarantola, Inverse Problem Theory and Methods for Model Parameter Estimation, Society for Industrial and Applied Mathematics, Philadelphia, 2005.

Chapter 4

Elastic constitutive models

An elastic deformation can be defined as a deformation produced by the application of an action that, upon its removal, it completely vanishes. In reality, all components of engineering structures and machinery can be regarded as deformable bodies. Because of this, they always experience some degree of elastic deformations under working conditions and, on this basis, it is useful to study the elastic material behavior to relate the elastic deformation to the stresses induced in the body. From a constitutive modeling viewpoint, elastic constitutive models are the easiest to formulate since they depict scenarios in which stresses depend solely on the current state of strain and nothing else. The formulation of elastic constitutive models is important as it plays a role also in more complicated material behaviors which involve plastic, or visco-plastic deformation. In the latter models, the developed form of the elastic constitutive models are extended and refined to account for the additional material complexity. In the first part of the chapter, we start by presenting simple material models and then gradually progress toward more complicated ones. At the end of the chapter, we show the generalization of the constitutive formulation that is valid for any type of elastic material.

Routine design usually involves stress analysis of structural components subjected to certain service actions. These components are generally sized so that there are no permanent deformation within the working conditions, therefore ensuring all deformations to remain elastic. For most structural materials, these elastic deformations are fairly small and, therefore, it is usually sufficient to use linear-elastic constitutive models for their analysis.

4.1 Fully symmetric elastic behavior—isotropic materials

A material is denoted as *isotropic* when it exhibits the same mechanical behavior along all directions. This response enables to simplify the material description and, as shown later, to reduce the number of constants describing this behavior to a minimum (when compared to other material models). These elastic constants relate the strain components to the stress components expressed in any arbitrary coordinate system within the material point of a deformable body.

The easiest elastic model that can be formulated and implemented is the isotropic linear-elastic constitutive model. A wide range of materials can be described with this constitutive model. Polycrystalline metals with randomly oriented crystal grains exhibit macroscopic isotropy and could be modeled as

Constitutive Modeling of Engineering Materials. https://doi.org/10.1016/B978-0-12-814696-5.00009-5

an isotropic material when simulating its large scale response of a sample with dimensions of at least two orders of magnitude larger than the typical grain size. This *homogenization* can be applied to materials with random grain orientation, and not to plastically deformed metals that generally introduce preferable grain orientation.

By considering that stresses of an elastic material depend only on strains, the elastic constitutive model can be described by k equations that express each stress component as a function of the strain components as follows:

$$\sigma_{ij} = f_k(\varepsilon_{11}, \varepsilon_{22}, \varepsilon_{33}, \varepsilon_{12}, \varepsilon_{13}, \varepsilon_{23}) \tag{4.1}$$

and, by determining the inverse functions (that should be available), each strain component can be expressed as a function of the stress components as

$$\varepsilon_{ij} = g_k(\sigma_{11}, \sigma_{22}, \sigma_{33}, \sigma_{12}, \sigma_{13}, \sigma_{23}) \tag{4.2}$$

Based on the symmetry of both stress and strain tensors, there are only six independent components in each of them and, therefore, in Eqs. (4.1) and (4.2), index k assumes values $k = 1, .., 6$. These equations are written in the most general case when each stress component is expressed as a function of all strain components. In the case of isotropic materials, these relations can be simplified because, for these materials, the mechanical response is independent on the direction considered. Based on this observation, the material response can be deduced from simple considerations that involve an uniaxial state of stress.

Let us start from the simplest form of elastic behavior—*linear elasticity*. The first proposal of a linear-elastic model is attributed to Robert Hooke, who, when studying the behavior of springs, concluded that the force applied to the spring is proportional to the elongation of the spring. He summarized the results regarding general elastic bodies in his work published in 1678, claiming that: "It is very evident that the Rule or Law of Nature in every springing body is, that the force or power thereof to restore itself to its natural position is always proportionate to the distance or space it is removed therefrom" [1]. Even though the definition of modulus of elasticity, as a constant which relates strain to stress, was formulated only later by Thomas Young (1773–1829), the expression that relates the normalized force (i.e., stress) to the normalized elongation (i.e., strain) is commonly referred to as *Hooke's law* and is written as:

$$\sigma = E \cdot \varepsilon \tag{4.3}$$

Considering that stress and strain are tensors, generalization of the scalar form of Hooke's law given by (4.3) formulates stresses as linear transformation of strains. Stress and strain are second order tensors, i.e., their components have two indexes: σ_{ij}, with $i = x, y$ or z and $j = x, y$ or z, having total of 9 components (i.e., $3^2 = 9$). To formulate the linear transformation, each stress component needs to be written as a linear combination of all the strain components. As an example for one stress component, we can write:

$$\begin{aligned}\sigma_{ij} = D_{ijxx}\varepsilon_{xx} + D_{ijyy}\varepsilon_{yy} + D_{ijzz}\varepsilon_{zz} + D_{ijxz}\varepsilon_{xz} + D_{ijxy}\varepsilon_{xy} \\ + D_{ijyz}\varepsilon_{yz} + D_{ijzx}\varepsilon_{zx} + D_{ijyx}\varepsilon_{yx} + D_{ijzy}\varepsilon_{zy}\end{aligned} \tag{4.4}$$

Clearly, coefficients of this linear transformation need to have four indexes, therefore constitute a fourth-order tensor $\boldsymbol{D}$, called the *material stiffness tensor* (or material elasticity tensor), with 81 (i.e., $3^4 = 81$) components which are not all independent, as it will be demonstrated further on. The generalized Hooke's law in tensorial form can then be written as:

$$\boldsymbol{\sigma} = \boldsymbol{D} : \boldsymbol{\varepsilon} \tag{4.5}$$

Since strain is a symmetric tensor (i.e., $\varepsilon_{ij} = \varepsilon_{ji}$) from Eq. (4.4) it stems that $D_{ijkl} = D_{ijlk}$. Further on, also stress is a symmetric tensor which yields that $D_{ijkl} = D_{jikl}$. This reduces the number of independent components within material stiffness tensor to 36. The reduction allows us to represent it in Voigt notation as a matrix, usually called the *material stiffness matrix* (or material elasticity matrix), to emphasize that within the reduced formulation it loses the tensorial properties and cannot be treated as a second-order tensor. Eq. (4.5) written in matrix notation reads:

$$\begin{bmatrix}\sigma_{xx}\\ \sigma_{yy}\\ \sigma_{zz}\\ \sigma_{yz}\\ \sigma_{xz}\\ \sigma_{xy}\end{bmatrix} = \begin{bmatrix} D_{xxxx} & D_{xxyy} & . & . & . & D_{xxxy}\\ D_{yyxx} & D_{xxyy} & . & . & . & .\\ . & . & . & . & . & .\\ . & . & . & . & . & .\\ . & . & . & . & . & .\\ D_{xyxx} & . & . & . & . & D_{xyxy}\end{bmatrix} \begin{bmatrix}\varepsilon_{xx}\\ \varepsilon_{yy}\\ \varepsilon_{zz}\\ \varepsilon_{yz}\\ \varepsilon_{xz}\\ \varepsilon_{xy}\end{bmatrix} \tag{4.6}$$

It is evident that the last equation represents a matrix expression of six constitutive equations given in the form (4.1) and the generalized Hooke's law represents the simplest linear-elastic constitutive model. For its complete formulation, it is required to quantify the constants entering into matrix $\boldsymbol{D}$. The symmetry of the fourth-order tensor $\boldsymbol{D}$ has already reduced the number of constants from 81 to 36. Based on theory of elasticity, the 36 constants can be further reduced to only 21 independent constants, which essentially means that also matrix $\boldsymbol{D}$ is symmetric (see, e.g., [2] and [3]). More general treatments are provided within continuum mechanics theory which demonstrate what is the minimum amount of independent coefficients required to formulate a tensor $\boldsymbol{D}$ for various cases of material symmetry (see, e.g., [4]). For an isotropic material, we can deduce the entries of the matrix $\boldsymbol{D}$ from considerations on the uniaxial stress state.

Let us look into the portion of the material that is only loaded with a normal stress along one direction (say along the x-axis). In addition to the elongation that takes place along the loading axis, there is also a shortening occurring along the other two perpendicular directions that can be computed based on the material Poisson's ratio (see Fig. 4.1). The three components of deformation are equal to:

$$\varepsilon_{xx} = \frac{\sigma_{xx}}{E}, \quad \varepsilon_{yy} = -\nu\frac{\sigma_{xx}}{E}, \quad \varepsilon_{zz} = -\nu\frac{\sigma_{xx}}{E} \tag{4.7}$$

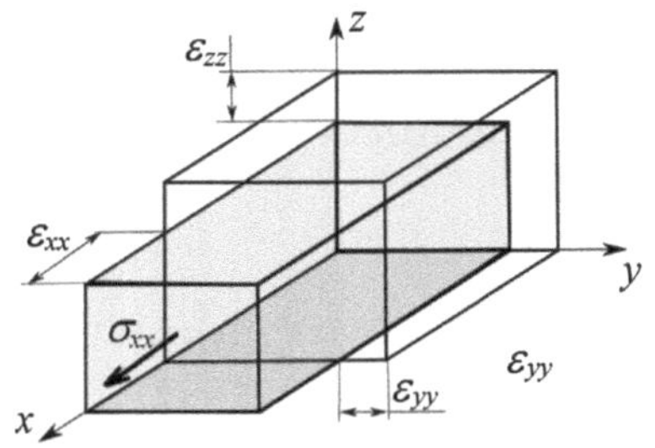

FIGURE 4.1 Small material volume stretched along the x-axis.

Similar expressions can be obtained for the specimen loaded along the y- and z-directions. By superposing the deformation exhibited in each of the three cases, we can write the deformation components for the general case in which all three stresses are acting (i.e., σ_{xx}, σ_{yy}, and σ_{zz}):

$$\begin{aligned} \varepsilon_{xx} &= \frac{\sigma_{xx}}{E} - \nu\frac{\sigma_{yy}}{E} - \nu\frac{\sigma_{zz}}{E} \\ \varepsilon_{yy} &= -\nu\frac{\sigma_{xx}}{E} + \frac{\sigma_{yy}}{E} - \nu\frac{\sigma_{zz}}{E} \\ \varepsilon_{zz} &= -\nu\frac{\sigma_{xx}}{E} - \nu\frac{\sigma_{yy}}{E} + \frac{\sigma_{zz}}{E} \end{aligned} \tag{4.8}$$

In order to emphasis that components of strain and stress in Eq. (4.8) are components of a second-order tensors, the adopted indexes are doubled (i.e., xx is used for the x-direction). Eqs. (4.8) are valid for an isotropic material in which the elastic properties are independent on the direction, hence the same elastic constants (i.e., Young's modulus E and Poisson's ratio ν) can be used for all three directions.

The same two elastic constants can be used to relate shearing stress to shearing strain. This can be demonstrated by considering the response of the rectangular component loaded in compression along the horizontal direction (depicted by axis x' in Fig. 4.2a) with a stress σ, and in tension in the vertical direction (depicted by axis y' in Fig. 4.2a) with a stress of the same intensity. From equilibrium considerations, it can be shown that, by considering a reference system rotated by 45° (depicted by axes $x - y$ in Fig. 4.2), the normal stress vanishes

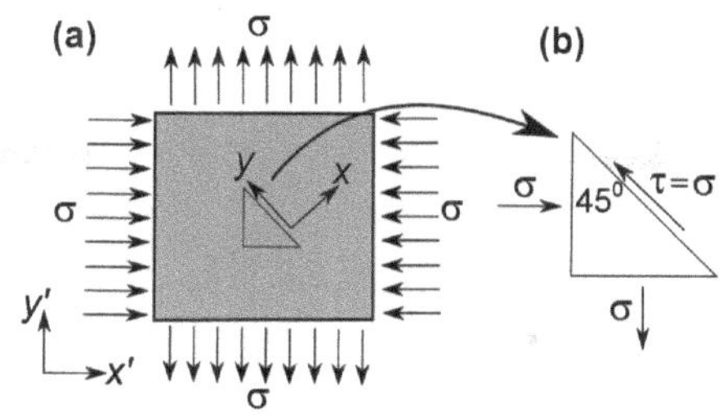

FIGURE 4.2 Pure shear loading of a rectangular component.

and the shearing stress becomes $\tau = \sigma$ (see Fig. 4.2b).[1] Within the reference system $x - y$, the stress is uniaxial as it has only one component different from zero, that is the shear stress component. This stress state is referred to as *pure shear.*

Let us now establish the relationship between shear stress and shear strain by observing the deformation of a square region depicted by *abcd* in Fig. 4.3 and loaded in pure shear. By rotating the $x - y$ system by 45°, the shear stress vanishes and the two normal stresses become equal and of opposite sign. With the adopted notation, the absolute shortening of the diagonal *bd* equals the elongation of the diagonal *ac*. By comparing the deformed and undeformed configurations, we can write the following relationship for the change in angle in the shaded triangle of Fig. 4.3:

$$\tan\left(\frac{\pi}{4} - \frac{\gamma}{2}\right) = \frac{1-\varepsilon}{1+\varepsilon} \tag{4.9}$$

where ε represents the dilation which is the same for the shortened and elongated diagonal (even if opposite in sign). For ease of reference, ε is here written without any index; however, it formally refers to the reference system rotated by 45° depicted by axes $x' - y'$ (such rotated reference system corresponds to the horizontal and vertical axis of Fig. 4.2a). Since the loading in this reference system is represented by two normal stresses of same intensity—stretching in one direction and shortening in the perpendicular one, we can determine the dilation from:

$$\varepsilon = \frac{1}{E} \cdot (1+\nu)\sigma \tag{4.10}$$

For small angles, we can write:

$$\tan\left(\frac{\pi}{4} - \frac{\gamma}{2}\right) = \frac{\tan\frac{\pi}{4} - \tan\frac{\gamma}{2}}{1 + \tan\frac{\pi}{4} \cdot \tan\frac{\gamma}{2}} \approx \frac{1-\frac{\gamma}{2}}{1+\frac{\gamma}{2}} \tag{4.11}$$

---

[1] By summing up the forces acting on triangle *abc*, it can be noted that, from projection to the normal of the side *bc*, the normal stress induced on this side is equal to zero, while the projection along the *bc* side gives: $\tau \cdot dA = \sigma \cdot dA\frac{\sqrt{2}}{2} \cdot \frac{\sqrt{2}}{2} + \sigma \cdot dA\frac{\sqrt{2}}{2} \cdot \frac{\sqrt{2}}{2} \rightarrow \tau = \sigma$.

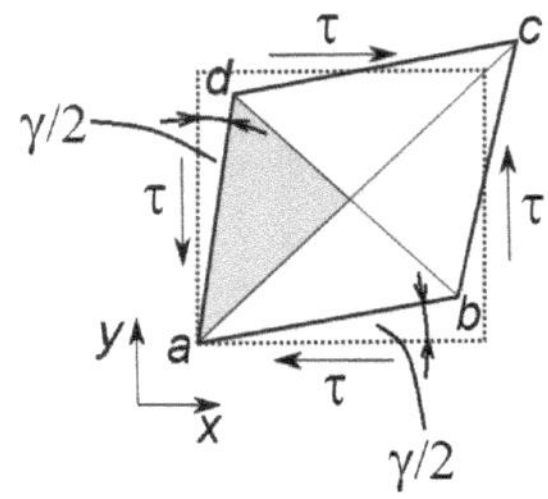

FIGURE 4.3 Deformation in pure shear.

By combining Eq. (4.9) and Eq. (4.11), we obtain the relation between the two components of deformation as:

$$\gamma = 2\varepsilon \tag{4.12}$$

that relates the deformation components determined in different reference systems. This expression is useful because, through Eq. (4.10) in which ε is defined as a function of the applied stresses and by considering that $\tau = \sigma$, we can write:

$$\gamma = \frac{2(1+\nu)}{E}\tau \quad \rightarrow \quad \gamma = \frac{\tau}{G} \tag{4.13}$$

where G is denoted as the *shear modulus* and is equal to:

$$G = \frac{E}{2(1+\nu)} \tag{4.14}$$

The shear modulus is not an additional material elastic constant, as it is computed with Eq. (4.14) as a function of the Young's modulus and the Poisson's ratio. The same geometrical reasoning may be applied for the remaining two planes. If a shear stress acts on all three planes, the resulting shear deformations are independent from each other and can be written as:

$$\gamma_{xy} = \frac{1}{G}\tau_{xy}; \quad \gamma_{xz} = \frac{1}{G}\tau_{xz}; \quad \gamma_{yz} = \frac{1}{G}\tau_{yz} \tag{4.15}$$

Eqs. (4.8) and (4.15) relate all strain components to the stress components and can be written in matrix form:

$$\begin{bmatrix} \varepsilon_{xx} \\ \varepsilon_{yy} \\ \varepsilon_{zz} \\ \gamma_{xy} \\ \gamma_{xz} \\ \gamma_{yz} \end{bmatrix} = \frac{1}{E} \begin{bmatrix} 1 & -\nu & -\nu & 0 & 0 & 0 \\ -\nu & 1 & -\nu & 0 & 0 & 0 \\ -\nu & -\nu & 1 & 0 & 0 & 0 \\ 0 & 0 & 0 & \frac{E}{G} & 0 & 0 \\ 0 & 0 & 0 & 0 & \frac{E}{G} & 0 \\ 0 & 0 & 0 & 0 & 0 & \frac{E}{G} \end{bmatrix} \begin{bmatrix} \sigma_{xx} \\ \sigma_{yy} \\ \sigma_{zz} \\ \sigma_{xy} \\ \sigma_{xz} \\ \sigma_{yz} \end{bmatrix} \rightarrow \boldsymbol{\varepsilon} = \boldsymbol{C} \cdot \boldsymbol{\sigma} \tag{4.16}$$

It is worth pointing out that, in order to keep the matrix $\boldsymbol{C}$ in the above form, the strain tensor in Eq. (4.16) is written as a vector that takes the modified form of $\boldsymbol{\varepsilon} = [\varepsilon_{xx}, \varepsilon_{yy}, \varepsilon_{zz}, 2\varepsilon_{xy}, 2\varepsilon_{xz}, 2\varepsilon_{yz}]^T$, so that the out-of-diagonal terms represent the complete shear deformation. This is a preferred and generally used compact description of stress and strain tensors, since in this form scalar product of stress and strain expressed as vectors; $\boldsymbol{\varepsilon}^T \cdot \boldsymbol{\sigma}$ represents the strain energy. Alternatively, entries C_{44}, C_{55}, and C_{66} of matrix $\boldsymbol{C}$ should be divided by 2.

The compliance matrix $\boldsymbol{C}$ is symmetric, positive-definite, and hence, invertible. Its inverse represents the matrix $\boldsymbol{D}$, required to formulate the isotropic linear-elastic constitutive model in the form of Eqs. (4.5) and (4.6), that can be written as:

$$\boldsymbol{D} = \frac{E}{(1+\nu)(1-2\nu)} \begin{bmatrix} 1-\nu & \nu & \nu & 0 & 0 & 0 \\ \nu & 1-\nu & \nu & 0 & 0 & 0 \\ \nu & \nu & 1-\nu & 0 & 0 & 0 \\ 0 & 0 & 0 & \frac{1-2\nu}{2} & 0 & 0 \\ 0 & 0 & 0 & 0 & \frac{1-2\nu}{2} & 0 \\ 0 & 0 & 0 & 0 & 0 & \frac{1-2\nu}{2} \end{bmatrix} \tag{4.17}$$

4.2 Orthotropic elastic materials

So far we examined isotropic materials, whose mechanical response to external actions is independent of the direction considered. This class of material is characterized by the fact that the entries to matrix $\boldsymbol{D}$ (generally denoted as the *elastic constants*) are independent on the orientation. When these conditions are not satisfied, the material is said, in the most general case, to be *anisotropic*. For this class of materials, the values of entries to the matrix $\boldsymbol{D}$ depend on the orientation of the adopted coordinate system.

In between the two extremes depicted by the isotropic materials, for which any transformation of coordinate system leaves the constitutive equations invariant, and the anisotropic materials, for which any transformation changes the constitutive equations, there are materials which have *some* transformations that are not altering the elastic constants. These particular transformations define the *material symmetry group* or the *isotropy group*.

The isotropy groups are commonly defined through the *elastic symmetry planes*. If the elastic constants at a point do not change for every pair of coordinate systems that are mirror images of each other in a certain plane, this particular plane is denoted as the plane of elastic symmetry. In an isotropic material any plane is a plane of elastic symmetry, thus it denotes the most complete form of material symmetry.

There are many materials which, owing to their microstructure, exhibit different properties when tested and measured along certain specific directions. A typical example is wood that has structural components with generally quite different response when loaded along any of the three perpendicular directions with respect to the grain (see Fig. 4.4a). A similar effect is obtained with man-made heterogeneous materials, such as fiber-reinforced composite materials (see Fig. 4.4b). The two constituents of such materials possess quite different properties: the fibers are usually more rigid but frequently brittle, while the matrix is softer and significantly more ductile. Owing to the presence of the fibers, the fiber reinforced composites have a stiffer response when loaded along the fiber direction than perpendicular to it.

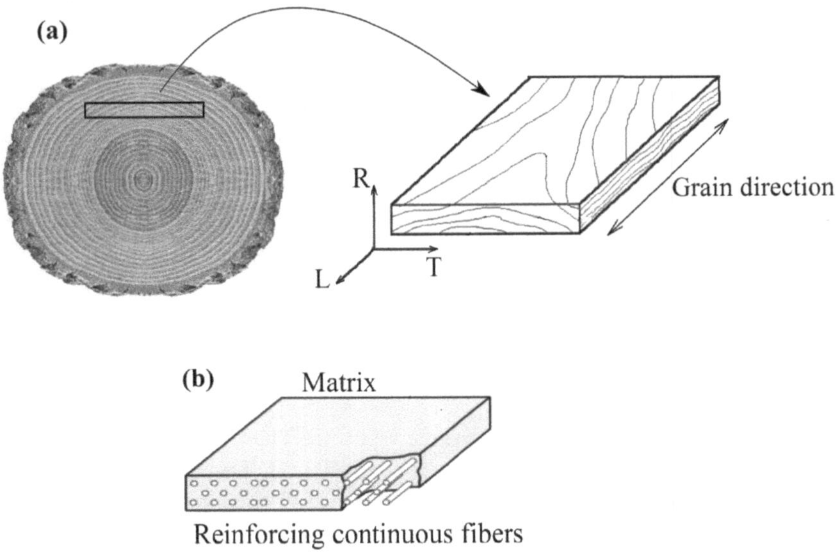

FIGURE 4.4 Examples of orthotropic materials: (a) natural material wood and (b) fiber-reinforced composite.

The different response of these materials is due to their heterogeneity at the constituent level. Such heterogeneity affects also the stresses that, when measured at the micro level, change from one point to the other, depending on the constituent that is occupying the particular material point. From an engineering viewpoint, it is more convenient to express the stresses in averaged terms and, consequently, to formulate effective properties of the equivalent homogeneous material.

In terms of the elastic response, it is usually possible to identify three orthogonal planes of elastic symmetry for these materials that are referred to as *orthotropic materials*. Intuitively, it is expected that the number of elastic constants required to describe the behavior of an orthotropic material is larger than the one required for an isotropic material. As already anticipated with the example of the wood specimen, the exhibited mechanical response is quite different when the material is loaded along the grain direction, the radial direction or the tangential direction (denoted as L, R and T in Fig. 4.5). The elastic re-

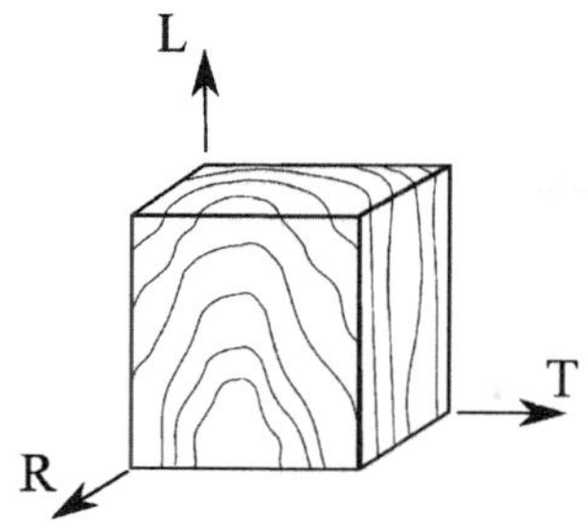

FIGURE 4.5 Principal material coordinates of a wood specimen.

sponse along these three directions is governed by different Young's moduli. The strain components are therefore related to the stress components by the following equation:

$$\begin{bmatrix} \varepsilon_{LL} \\ \varepsilon_{RR} \\ \varepsilon_{TT} \\ \gamma_{LR} \\ \gamma_{LT} \\ \gamma_{TR} \end{bmatrix} = \begin{bmatrix} \frac{1}{E_L} & -\frac{\nu_{RL}}{E_R} & -\frac{\nu_{TL}}{E_T} & 0 & 0 & 0 \\ -\frac{\nu_{LR}}{E_L} & \frac{1}{E_R} & -\frac{\nu_{TR}}{E_T} & 0 & 0 & 0 \\ -\frac{\nu_{LT}}{E_L} & -\frac{\nu_{RT}}{E_R} & \frac{1}{E_T} & 0 & 0 & 0 \\ 0 & 0 & 0 & \frac{1}{G_{LR}} & 0 & 0 \\ 0 & 0 & 0 & 0 & \frac{1}{G_{LT}} & 0 \\ 0 & 0 & 0 & 0 & 0 & \frac{1}{G_{TR}} \end{bmatrix} \begin{bmatrix} \sigma_{LL} \\ \sigma_{RR} \\ \sigma_{TT} \\ \sigma_{LR} \\ \sigma_{LT} \\ \sigma_{TR} \end{bmatrix} \quad (4.18)$$

These three directions define the three orthogonal planes of elastic symmetry. They are also denoted as the *principal material coordinates*. The amount of elastic constants required to define the elastic behavior of an orthotropic material is the smallest if it is done with respect to the principal material coordinates. They are easily identified within the material as they are associated to certain reinforcement directions. By inverting the compliance matrix of Eq. (4.18), the material stiffness matrix of an orthotropic material can be obtained as follows:

$$\boldsymbol{D} = \begin{bmatrix} D_{11} & D_{12} & D_{13} & 0 & 0 & 0 \\ D_{21} & D_{22} & D_{23} & 0 & 0 & 0 \\ D_{31} & D_{32} & D_{33} & 0 & 0 & \\ 0 & 0 & 0 & D_{44} & 0 & 0 \\ 0 & 0 & 0 & 0 & D_{55} & 0 \\ 0 & 0 & 0 & 0 & 0 & D_{66} \end{bmatrix} \quad (4.19)$$

in which the locations of the nil terms is identical to those specified in the isotropic material stiffness matrix. The difference with respect to the isotropic material relies on the fact that there is now the need to know 9 independent elastic constants to define the material stiffness matrix: three Young's moduli and six Poisson's ratios. For orthotropic materials, engineering books (see, e.g.,

[5], [6]) usually provide the relevant elastic constants with reference to material principal coordinates.

Another important form of symmetry can be formulated by considering a particular case of orthotropy. In most composites that are reinforced by long fibers, the arrangement of fibers is random so that the homogenized properties are practically the same for all directions perpendicular to the fibers. Such materials are said to be *transversely isotropic*. In terms of material principal directions (see Fig. 4.6), the properties measured along direction 2 are the same as those evaluated along direction 3. The elastic response along these directions is governed by the same Young's modulus E_t that differs from the one governing the response along the fibers E_f valid for direction 1. In addition to the two elastic moduli, there are three independent Poisson's ratios: ν_{ft} quantifying the deformation in the perpendicular direction caused by a stress induced in the fiber direction; ν_{tf} quantifying the deformation along the fibers caused by a stress resisted in the perpendicular directions, and ν_{tt} for the deformation in the orthogonal direction to the fibers direction for the stresses induced along the other orthogonal direction to the fibers direction. The number of independent elastic constants that are required to formulate the behavior of a transversely isotropic material is now reduced to 5.

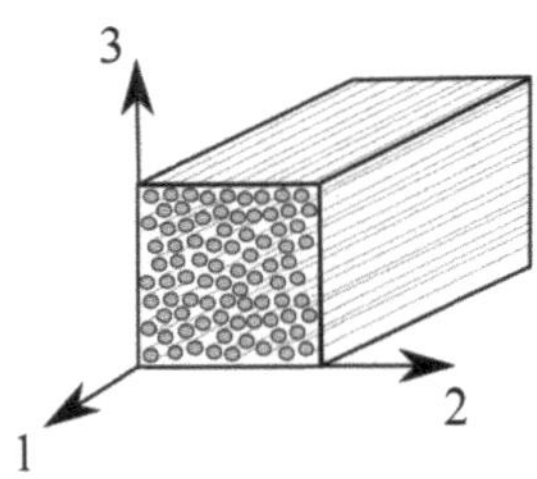

FIGURE 4.6 Composite material with continuous long fibers.

The compliance matrix has the form of the matrix already introduced for the isotropic and orthotropic materials (i.e., it has the same 12 elements which are nonzero) with the only difference that, for a transversely isotropic material, only 5 independent constants are required for its determination:

$$
\boldsymbol{C} = \begin{bmatrix}
\frac{1}{E_f} & -\frac{\nu_{tf}}{E_t} & -\frac{\nu_{tf}}{E_t} & 0 & 0 & 0 \\
-\frac{\nu_{ft}}{E_f} & \frac{1}{E_t} & -\frac{\nu_{tt}}{E_t} & 0 & 0 & 0 \\
-\frac{\nu_{ft}}{E_f} & -\frac{\nu_{tt}}{E_t} & \frac{1}{E_t} & 0 & 0 & 0 \\
0 & 0 & 0 & \frac{1}{G_{ft}} & 0 & 0 \\
0 & 0 & 0 & 0 & \frac{1}{G_{ft}} & 0 \\
0 & 0 & 0 & 0 & 0 & \frac{1}{G_{tt}}
\end{bmatrix} \tag{4.20}
$$

4.3 Anisotropy

An anisotropic material is defined as a material for which there is no plane of elastic symmetry. The material formulation is based on the same relationship between stress and stain as previously introduced in Eq. (4.6), even if in this case the material stiffness matrix $\boldsymbol{D}$ is fully populated (i.e., there are no zero terms in it). In addition, the values of the elastic constants D_{ij} vary when considering different reference systems. The symmetry of matrix $\boldsymbol{D}$ is still preserved based on which it can be concluded that there is a need to quantify 21 independent elastic constants to define this material behavior. It is worth mentioning that completely anisotropic materials are quite rare in nature. In reality, most of the materials, depending on the scale at which the material is observed, usually exhibit at least some form of elastic symmetry, therefore enabling to simplify the material mechanical behavior and to reduce the number of corresponding elastic constants.

A completely populated material stiffness matrix $\boldsymbol{D}$ and its associated compliance matrix $\boldsymbol{C}$ give rise to more complex deformation mechanisms for anisotropic materials when compared to the response of isotropic materials. For example, when considering an uniaxial loading with σ_{xx} being the only nonzero stress component within an isotropic material, the consequent deformations are depicted by three non-zero dilatation components, without any shear deformation. It is evident that this could not be the case for an anisotropic material, since its compliance matrix is fully populated:

$$\begin{bmatrix} \varepsilon_{xx} \\ \varepsilon_{yy} \\ \varepsilon_{zz} \\ \gamma_{xy} \\ \gamma_{xz} \\ \gamma_{yz} \end{bmatrix} = \begin{bmatrix} C_{11} & C_{12} & C_{13} & C_{14} & C_{15} & C_{16} \\ C_{21} & C_{22} & C_{23} & C_{24} & C_{25} & C_{26} \\ C_{31} & C_{32} & C_{33} & C_{34} & C_{35} & C_{36} \\ C_{41} & C_{42} & C_{43} & C_{44} & C_{45} & C_{46} \\ C_{51} & C_{52} & C_{53} & C_{54} & C_{55} & C_{56} \\ C_{61} & C_{62} & C_{63} & C_{64} & C_{65} & C_{66} \end{bmatrix} \begin{bmatrix} \sigma_{xx} \\ \sigma_{yy} \\ \sigma_{zz} \\ \sigma_{xy} \\ \sigma_{xz} \\ \sigma_{yz} \end{bmatrix} \tag{4.21}$$

The uniaxial state of stress leads to shear deformation equal to $\gamma_{xy} = C_{41}\sigma_{xx}$, $\gamma_{xz} = C_{51}\sigma_{xx}$, and $\gamma_{yz} = C_{61}\sigma_{xx}$. In more general terms of stress and strain analysis, it can be noted that, for an anisotropic material, the directions of principal stresses and of principal dilatations do not match. This can be easily observed by considering that, if a reference system $x - y - z$ represents the principal coordinate system for stresses (i.e., $\sigma_{xy} = \sigma_{xz} = \sigma_{yz} = 0$), the resulting shear deformation (e.g., the one measured in $x - y$ plane) is equal to:

$$\gamma_{xy} = C_{41}\sigma_{xx} + C_{42}\sigma_{yy} + C_{43}\sigma_{zz} \tag{4.22}$$

and hence, in the general case it differs from zero.

The mechanical characterization of a fully anisotropic material is generally more difficult. The obvious complexity with respect to materials that possess some form of elastic symmetry is related to the larger number of elastic constants required to be quantified, here equal to 21. Additionally, the constants are dependent on the reference system and change with it. Therefore, any form of symmetry, if present, depending on the required scale for the material mechanical modeling, should be exploited whenever possible. For orthotropic materials, the material stiffness matrix has the reduced form (i.e., the one with 12 entries different from zero) only when defined with respect to the material principal coordinates. For any other reference system, it will be a fully populated matrix like the matrix determined with anisotropy. This should be kept in mind when dealing with an orthotropic material (i.e., to define its mechanical response with respect to material principal coordinates).

4.4 Large elastic deformation—generalization of the concept

By definition, the elastic deformation is the one that completely vanishes upon the removal of the induced action producing it. By describing the elastic deformation in the strain space, a material point that is moved from point A to point B through the application of an externally applied action, is expected to go back to the initial point A upon the removal of the action. The transformation of the deformation state from A to B is therefore a *reversible* transformation. Let us illustrate this concept by considering the two-dimensional strain space spanned by $\varepsilon_{ij} - \varepsilon_{kl}$ (see Fig. 4.7). Formally, the return path from the point B to the point A does not have to be the same as the one initially followed to reach point B, for the transformation $A - B$ to be regarded as reversible.

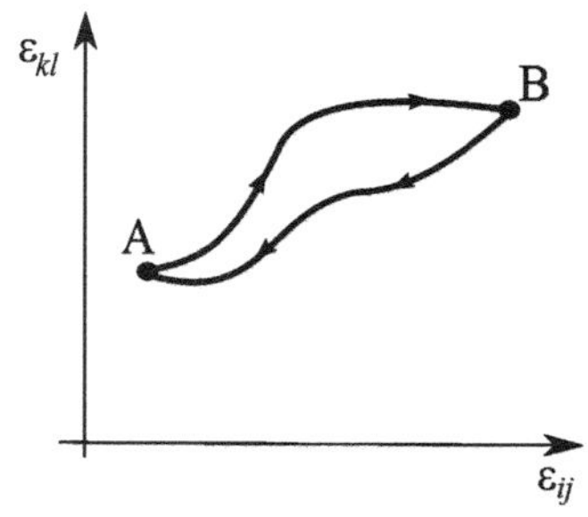

FIGURE 4.7 Schematic representation of a reversable deformation within a two-dimensional strain space.

By considering a deformable body, the position of the point of application of an external force is expected to move as a consequence of deformation. In this process, the external force is expected to conduct some work during the deformation. For the elastic process to be reversible, this work has to be completely stored within the deformed body in the form of *deformation energy*. By subsequent unloading, this energy is completely released and the deformed body is brought exactly to the same starting point A. In an elastic deformation, there is no exchange of energy between the deformable body and the external ambient for one loading-unloading cycle, regardless of the $A-B-A$ path. The variation of the deformation energy (or strain energy) has therefore to be a function *only* of the initial (point A) and final deformation state (point B).

In a particular material point, let us define a *strain energy density* (i.e., the energy per unit volume) as a function of the strain tensor. Within the strain space, this can be defined as a line integral as follows:

$$U(\boldsymbol{\varepsilon}) = \int_0^{\boldsymbol{\varepsilon}^C} dU \tag{4.23}$$

where 0 symbolizes the strain free state of the material and $\boldsymbol{\varepsilon}^C$ depicts the current state of deformation. If the deformation that changes the state from point A to point B is the elastic deformation, the corresponding change in strain energy density depends only on its initial and final value and does not depend on the path. In such a case, we can write:

$$\Delta U(\boldsymbol{\varepsilon}) = \int_{\boldsymbol{\varepsilon}^A}^{\boldsymbol{\varepsilon}^B} dU = U(\boldsymbol{\varepsilon}^B) - U(\boldsymbol{\varepsilon}^A) \tag{4.24}$$

The previous line integral is solved within the strain space that, in the general case, is 6-dimensional. For convenience, we can write the integrand function in the following form:

$$dU = f_1 d\varepsilon_{xx} + f_2 d\varepsilon_{yy} + f_3 d\varepsilon_{zz} + f_4 d\varepsilon_{xy} + f_5 d\varepsilon_{yz} + f_6 d\varepsilon_{xz} \tag{4.25}$$

where f_i are selected functions of strain components.

By definition, the line integral is independent on the path if the integrand function is an exact differential (see [7] for mathematical proof), which means that function $\boldsymbol{F} = [f_1, f_2, f_3, f_4, f_5, f_6]$ must represent a gradient of a single-valued (i.e., scalar) function U. Based on this, we can write the integrand function as follows:

$$dU = \frac{\partial U}{\partial \varepsilon_{xx}} d\varepsilon_{xx} + \frac{\partial U}{\partial \varepsilon_{yy}} d\varepsilon_{yy} + \frac{\partial U}{\partial \varepsilon_{zz}} d\varepsilon_{zz} + \frac{\partial U}{\partial \varepsilon_{xy}} d\varepsilon_{xy} + \frac{\partial U}{\partial \varepsilon_{yz}} d\varepsilon_{yz} + \frac{\partial U}{\partial \varepsilon_{xz}} d\varepsilon_{xz} \tag{4.26}$$

For a small change in the deformation measured by $d\varepsilon_{xx}$, $d\varepsilon_{yy}$, $d\varepsilon_{zz}$, $d\varepsilon_{xy}$, $d\varepsilon_{yz}$, and $d\varepsilon_{xz}$ it may be assumed that the corresponding values of the stress have not changed during the deformation. The specific work done by the internal forces (i.e., work per unit volume) along this deformation is then equal to:

$$dW = \sigma_{xx} \cdot d\varepsilon_{xx} + \sigma_{yy} \cdot d\varepsilon_{yy} + \sigma_{zz} \cdot d\varepsilon_{zz} + \sigma_{xy} \cdot d\varepsilon_{xy} + \sigma_{xz} \cdot d\varepsilon_{xz} \tag{4.27}$$

If the deformation is completely elastic, the work done by the internal forces is exactly equal to the change in the strain energy density, thus we may equate

(4.26) and (4.27). Based on this, we can deduce the following relation:

$$\sigma_{ij} = \frac{\partial U}{\partial \varepsilon_{ij}} \tag{4.28}$$

that represents a direct expression of the elastic constitutive model where for the formulation of such constitutive model, it is required to specify the strain density function. This form of constitutive relation is general enough and thus valid for an arbitrary elastic behavior. For the uniaxial case, Fig. 4.8 provides a clear graphical interpretation, therefore highlighting how the relationship is valid also for a nonlinear elastic deformation.

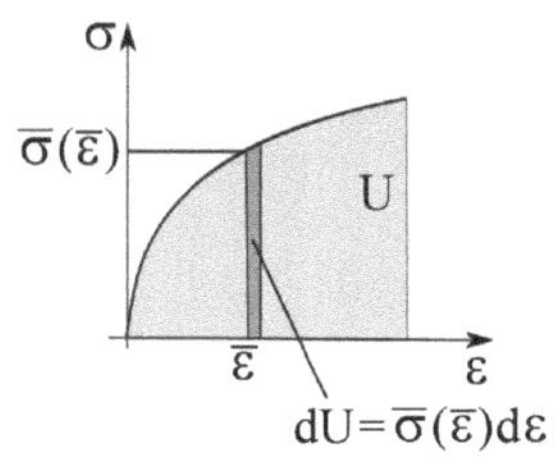

FIGURE 4.8 Graphical representation of the strain energy density for an uniaxial state of stress.

In mathematical terminology when a vector (or tensor) field is obtained as the gradient of a scalar function, the latter scalar function is referred to as the *potential* of the corresponding vector field [7]. In the direct formulation of an elastic constitutive model, given by Eq. (4.28), the stress field is obtained as a gradient of the elastic energy density, which can then be denoted as the *elastic potential.*

Green [8], [9] was the first to suggest the definition of an elastic material as one for which a strain energy density function must exist. From previous discussions, we saw that this function plays a role of potential of the stress field. The formulation of an elastic constitutive model in the form of Eq. (4.28) has general validity and is not restricted to linear-elastic behavior. Most materials, when deformed by small deformation, show a linear dependency of stresses on strains. Certain materials, such as rubbers, have the possibility to develop large elastic deformations. For this type of materials, the relationship between stress and strain is not linear. A typical stress–strain curve for rubber-like materials with the capability of large elastic deformations is plotted in Fig. 4.9. Also for these materials, there is a linear relationship between stress and strain for small deformations, so the Young's modulus can be defined, but valid only for a very limited range of deformation. In order to describe the mechanical behavior of these materials, through a wider range of deformation, the constitutive model is generally expressed by means of the strain energy density function.

Since the formulation of the constitutive model through the elastic potential is typical for materials that are expected to exhibit large elastic deformations

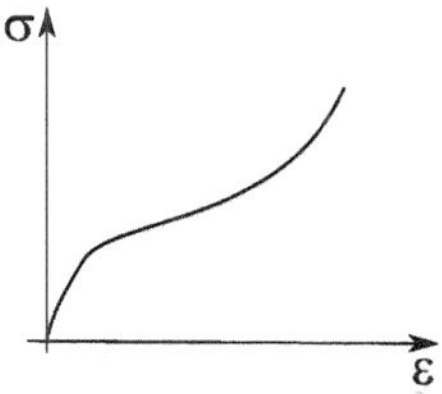

FIGURE 4.9 Typical stress-strain curve for a hyperelastic rubber-like material.

and such materials are referred to as hyperelastic or Green-elastic materials (see, e.g., [10]). The existence of the strain energy density function, which plays a role of the potential of the stress field, is typical for all elastic materials, and it will be further demonstrated that the generalized Hooke's law is just a particularization of it.

In linear elasticity, the stresses are linearly related to strains. Thus, by considering Eq. (4.28), we may conclude that the strain energy density function for linear elasticity can only contain terms up to the quadratic order of strains:

$$\begin{aligned} U = D_0 + D_1\varepsilon_{xx} + D_2\varepsilon_{yy} + D_3\varepsilon_{zz} + D_4\varepsilon_{xy} + D_5\varepsilon_{xz} + D_6\varepsilon_{yz} + \\ \frac{1}{2}D_{11}\varepsilon_{xx}^2 + D_{12}\varepsilon_{xx}\varepsilon_{yy} + D_{13}\varepsilon_{xx}\varepsilon_{zz} + D_{14}\varepsilon_{xx}\varepsilon_{xy} + D_{15}\varepsilon_{xx}\varepsilon_{xz} + D_{16}\varepsilon_{xx}\varepsilon_{yz} + \\ \frac{1}{2}D_{22}\varepsilon_{yy}^2 + D_{23}\varepsilon_{yy}\varepsilon_{zz} + D_{24}\varepsilon_{yy}\varepsilon_{xy} + D_{25}\varepsilon_{yy}\varepsilon_{xz} + D_{26}\varepsilon_{yy}\varepsilon_{yz} + \\ \frac{1}{2}D_{33}\varepsilon_{zz}^2 + D_{34}\varepsilon_{zz}\varepsilon_{xy} + D_{35}\varepsilon_{zz}\varepsilon_{xz} + D_{36}\varepsilon_{zz}\varepsilon_{yz} + \\ \frac{1}{2}D_{44}\varepsilon_{xy}^2 + D_{45}\varepsilon_{xy}\varepsilon_{xz} + D_{46}\varepsilon_{xy}\varepsilon_{yz} + \\ \frac{1}{2}D_{55}\varepsilon_{xz}^2 + D_{56}\varepsilon_{xz}\varepsilon_{yz} + \\ \frac{1}{2}D_{66}\varepsilon_{yz}^2 \end{aligned} \tag{4.29}$$

with D_i and D_{ij} being arbitrary constants. As the constants are arbitrary, the generality is preserved even by writing the coefficient $\frac{1}{2}$ in front of all D_{ii} constants. Written in the present form, these constants are directly casted into entries of material stiffness matrix $\boldsymbol{D}$, as shown later.

By requiring that U vanishes for the unstrained state (i.e., all the strain components are zero) constant D_0 has to equal zero. Further on, by applying Eq. (4.28) to obtain the stress component σ_{xx}, we can write:

$$\sigma_{xx} = \frac{\partial U}{\partial \varepsilon_{xx}} = D_1 + D_{11}\varepsilon_{xx} + D_{12}\varepsilon_{yy} + D_{13}\varepsilon_{zz} + D_{14}\varepsilon_{xy} + D_{15}\varepsilon_{xz} + D_{16}\varepsilon_{yz} \tag{4.30}$$

Here, we exclude coupling effects (e.g., thermo-mechanical or hygro-mechanical coupling) as these effects may introduce certain preexisting stresses even without strain, and require that the stress must be zero for the unstrained state. Therefore, the constant D_1 must also be equal to zero.

Through the same reasoning applied to the other five stress components, also constants D_2 to D_6 must equal zero. It therefore results, that the elastic potential for the linear elastic body is a homogeneous quadratic function of strains (i.e., the quadratic form), which can be written in matrix form as:

$$U = \frac{1}{2}\boldsymbol{\varepsilon}^T \cdot \boldsymbol{D} \cdot \boldsymbol{\varepsilon} \tag{4.31}$$

with $\boldsymbol{D}$ being a symmetric matrix of constants, and $\boldsymbol{\varepsilon}$ being a vector collecting the six components of the strain tensor.

By applying Eq. (4.28) to the elastic potential given by Eq. (4.31), we can determine the stress components as follows:

$$\begin{aligned}
\sigma_{xx} &= \frac{\partial U}{\partial \varepsilon_{xx}} = D_{11}\varepsilon_{xx} + D_{12}\varepsilon_{yy} + D_{13}\varepsilon_{zz} + D_{14}\varepsilon_{xy} + D_{15}\varepsilon_{xz} + D_{16}\varepsilon_{yz} \\
\sigma_{yy} &= \frac{\partial U}{\partial \varepsilon_{yy}} = D_{12}\varepsilon_{xx} + D_{22}\varepsilon_{yy} + D_{23}\varepsilon_{zz} + D_{24}\varepsilon_{xy} + D_{25}\varepsilon_{xz} + D_{26}\varepsilon_{yz} \\
\sigma_{zz} &= \frac{\partial U}{\partial \varepsilon_{zz}} = D_{13}\varepsilon_{xx} + D_{23}\varepsilon_{yy} + D_{33}\varepsilon_{zz} + D_{34}\varepsilon_{xy} + D_{35}\varepsilon_{xz} + D_{36}\varepsilon_{yz} \\
\sigma_{xy} &= \frac{\partial U}{\partial \varepsilon_{xy}} = D_{14}\varepsilon_{xx} + D_{24}\varepsilon_{yy} + D_{34}\varepsilon_{zz} + D_{44}\varepsilon_{xy} + D_{45}\varepsilon_{xz} + D_{46}\varepsilon_{yz} \\
\sigma_{xz} &= \frac{\partial U}{\partial \varepsilon_{xy}} = D_{15}\varepsilon_{xx} + D_{25}\varepsilon_{yy} + D_{35}\varepsilon_{zz} + D_{45}\varepsilon_{xy} + D_{55}\varepsilon_{xz} + D_{56}\varepsilon_{yz} \\
\sigma_{yz} &= \frac{\partial U}{\partial \varepsilon_{xy}} = D_{16}\varepsilon_{xx} + D_{26}\varepsilon_{yy} + D_{36}\varepsilon_{zz} + D_{46}\varepsilon_{xy} + D_{56}\varepsilon_{xz} + D_{66}\varepsilon_{yz}
\end{aligned} \tag{4.32}$$

that can be written in matrix form as:

$$\boldsymbol{\sigma} = \boldsymbol{D} \cdot \boldsymbol{\varepsilon} \tag{4.33}$$

that demonstrates that matrix $\boldsymbol{D}$ is the material stiffness matrix for a general anisotropic linear-elastic material describing the generalized Hooke's law. This derivation also proves the symmetry of the material stiffness matrix, that implies that for an anisotropic material there are 21 independent elastic constants.

The symmetry of matrix $\boldsymbol{D}$ is easily evidenced if we notice from (4.32) that the components of the fourth order tensor $\boldsymbol{D}$ (there represented as matrix according to Voigt reduced notation) can be expressed as:

$$D_{ijkl} = \frac{\partial^2 U}{\partial \varepsilon_{ij}\varepsilon_{kl}} = \frac{\partial^2 U}{\partial \varepsilon_{kl}\varepsilon_{ij}} = D_{klij} \tag{4.34}$$

as the order of differentiation can be inverted without affecting the final result.

With this formulation, it is easy to show that for an isotropic linear-elastic material there can be only two independent elastic constants required to define the mechanical behavior. For such materials, there is no preferable direction and, therefore, the constitutive model is independent of the reference system. The elastic potential should then be defined as a function of the strain invariants:

$$U = U(I_1, I_2, I_3) \tag{4.35}$$

where I_1, I_2 and I_3 are the well-known three invariants of the strain tensor (see, e.g., [2]). By relying on the fact that the elastic potential has to be of quadratic form, and by recalling that the three invariants are linear (I_1), quadratic (I_2) and cubic (I_3) functions, we can conclude that the elastic potential has to be written as a function of the following type:

$$U = aI_1^2 + bI_2 \tag{4.36}$$

where a and b are material constants. Indeed, for a linear-elastic isotropic material, there are only two independent elastic constants. The constitutive model is frequently written with three elastic constants, as shown in Eq. (4.16), however only two of these three constants are independent, as earlier shown in Eq. (4.14).

4.5 Thermoelastic constitutive equations

Most material properties depend on temperature. The same holds also for the elastic constants specified within the elastic constitutive models. When a structural problem is analyzed within a constant temperature, the dependency is merely reduced to the selection of the appropriate values of the elastic constants that are valid for the temperature of interest. If it is required to study the response of a deformable body over a wider range of temperatures, the problem to be analyzed becomes more complex and some modifications need to be applied to the governing equations introduced so far.

The first and obvious consequence of the material elastic constants being a function of temperature is that it affects the validity of the adopted assumption of material homogeneity. In most phenomenological models the internal material structural variation (i.e., presence of inclusions, voids, etc.) is usually excluded and, on larger scales, the material is studied as a continuum with the same properties at all points. This clearly does no longer hold in a nonuniformly heated body if the elastic constants become function of temperature. Therefore, to study such structural problem, it is required to know also the temperature field over the deformable body.

The second consequence is to enhance the complexity of the analysis to account for the well-known feature that a temperature change in an unrestrained elastic body produces deformation. If constrains are prohibiting this deformation, internal stresses are expected to develop within the body only due to the temperature change without additional external actions.

Based on these considerations, if the structural analysis problem is to be solved over a wider range of temperatures, an appropriate modification of the governing equations needs to be carried out. In this case, the only required change relates to the stress-strain relations, now provided by thermo-elastic constitutive equations.

Let us start performing this modification by formulating that the total strain is decomposed into a mechanical and a thermal component. The thermal strain in an unrestrained body can be written as:

$$\varepsilon_{ij}^{T} = \alpha_{ij}(T - T_0) \tag{4.37}$$

where T_0 represents a reference temperature, and T is an arbitrary temperature, from which it can be observed that only the *temperature difference* can induce a change in strain. α_{ij} is a material property that relates a temperature difference (i.e., a scalar) to a strain (that is a second-order tensor). It follows that α_{ij} needs to be a component of a second-order thermal expansion tensor. It is worth pointing out that, for certain materials, the thermal expansion coefficient is itself a function of temperature, therefore the relationship between temperature change and strain is a nonlinear one, while Eq. (4.37) represents a linearizion of this relationship, valid over a certain range of temperatures.

For an elastically isotropic material, the thermal expansion tensor must be an isotropic tensor defined by one independent material constant α. The resulting three dilations induced by a temperature change are therefore equal to:

$$\varepsilon_{ij} = \alpha(T - T_0) \cdot \delta_{ij} \tag{4.38}$$

where δ_{ij} is the Kronecker delta.

For an isotropic material, no shear strain is generated by a temperature change. We can therefore combine the first three components of mechanical strain (i.e. the dilations) with thermal strain to obtain:

$$\begin{aligned}
\varepsilon_{xx} &= \frac{\sigma_{xx}}{E} - \nu\frac{\sigma_{yy}}{E} - \nu\frac{\sigma_{zz}}{E} + \alpha(T - T_0) \\
\varepsilon_{yy} &= -\nu\frac{\sigma_{xx}}{E} + \frac{\sigma_{yy}}{E} - \nu\frac{\sigma_{zz}}{E} + \alpha(T - T_0) \\
\varepsilon_{zz} &= -\nu\frac{\sigma_{xx}}{E} - \nu\frac{\sigma_{yy}}{E} + \frac{\sigma_{zz}}{E} + \alpha(T - T_0)
\end{aligned} \tag{4.39}$$

while the remaining three strain components are unchanged and given by Eq. (4.15).

The thermoelastic constitutive relationship that expresses stress in terms of strain can be obtained by solving the system of Eqs. (4.39) for the stress components as follows:

$$\sigma_{xx} = \frac{E}{(1-2\nu)(\nu+1)}\left[(1-\nu)\varepsilon_{xx} + \nu\varepsilon_{yy} + \nu\varepsilon_{zz}\right] - \frac{E}{1-2\nu}\alpha(T-T_0)$$
$$\sigma_{yy} = \frac{E}{(1-2\nu)(\nu+1)}\left[(1-\nu)\varepsilon_{yy} + \nu\varepsilon_{xx} + \nu\varepsilon_{zz}\right] - \frac{E}{1-2\nu}\alpha(T-T_0)$$
$$\sigma_{zz} = \frac{E}{(1-2\nu)(\nu+1)}\left[(1-\nu)\varepsilon_{zz} + \nu\varepsilon_{yy} + \nu\varepsilon_{xx}\right] - \frac{E}{1-2\nu}\alpha(T-T_0) \quad (4.40)$$

The last result is referred to in the literature as the *Duhamel–Neumann thermo-elastic constitutive law*.

Developed thermo-elastic constitutive laws provide a framework to study combined thermo-mechanical problems. There are many engineering problems in which structures are subjected to both mechanical and thermal actions. Thermal actions can be particularly important in the presence of constrains, or for multicomponent materials. For example, heating up a fully restrained steel bar through a temperature difference of 100 degrees gives rise to thermal stresses that can be in the order of yielding for certain steels. Similarly, the internal constrains present in multicomponent materials can cause stresses that, for brittle materials, could lead to internal cracks and/or structural failure. In general, any nonuniform temperature distribution through a body can create internal stresses.

If the variation of temperature is significant, the structural analysis should account also for thermal effects. In such a case, the analyses fall within the theory of thermoelasticity (i.e., elasticity with thermal effects) and should provide, in addition to displacements, and stress and strain fields, also the temperature field. These fields, depending on the type of interaction, can be solved simultaneously or separately. In order to include thermal variables (i.e., temperature and heat flux), the energy balance has to be incorporated within the formulation of the problem.

The heat transfer within the solid body is denoted as the heat conduction and is governed by the Fourier's law (for details, see [11]). According to this law, the flow of heat, measured by the heat flux, is proportional to the temperature gradient and is directed from a region of higher temperature to a region of lower temperature. The vector of heat flux $\boldsymbol{q}$ is thus equal to:

$$\boldsymbol{q} = -\boldsymbol{K}\cdot\nabla T \quad (4.41)$$

where $\boldsymbol{K}$ represents a material property and the sign minus reflects the fact that heat flows in the direction of temperature drop. Since both $\boldsymbol{q}$ and ∇T are vectors, $\boldsymbol{K}$ has to be a second-order tensor and is denoted as the *thermal conductivity tensor*. Thermal conductivity is a symmetric tensor that, for isotropic material, has only one independent material parameter defined as follows:

$$\boldsymbol{K} = k\cdot\boldsymbol{I} \quad (4.42)$$

where k is simply referred to as the thermal conductivity and $\boldsymbol{I}$ is the identity matrix.

In most engineering problems, the heat produced by the deformation is very small and, therefore, can be neglected. The change in temperature exhibited within an elastic body, except scenarios depicting thermal shocks, is usually relatively slow and, as a consequence, the corresponding thermal strains propagate slowly. These cases fall within quasi-static thermo-elasticity, in which the temperature distribution in the body can be determined independently of the state of stress and strain. The solution strategy in such analysis problems consists, firstly, in finding the temperature distribution on the basis of the heat energy equation and then, based on the thermo-elastic constitutive equation, in determining the relevant stresses and strains.

The heat energy equation can be derived on the basis of energy balance considerations for a unit volume element (for details, see [11] [12]). According to the energy balance, the sum of the rate of the heat flux that comes into the control volume (given by writing the rate form of Fourier's law) and of the rate of the generated heat inside the control volume must equal the sum of the rate of the heat flux transmitted outside of the volume and the rate of heat energy stored inside the control volume. By assuming that there is a reference uniform temperature distribution T_0 for the body under consideration for which no stresses exist inside the body in the absence of mechanical loads, we can write the heat energy balance for the temperature increment with respect to the stress-free temperature:

$$c\rho \frac{\partial T}{\partial t} = k\nabla^2 T + S \tag{4.43}$$

in which S represents the strength of an internal heat source or sink distribution term, ρ is the material density and c is the specific heat capacity at constant volume (i.e., the amount of heat necessary to rise the temperature of a unit mass by one degree), and k is the thermal conductivity, here written for an isotropic material. Alternatively, the latter should be expressed as a second-order tensor. For a detailed derivation of the energy balance equation it is required to incorporate thermodynamic considerations that are not discussed here in details and reference should be made to dedicated textbooks, e.g., [11] [12] [13].

Within thermo-elasticity, it is common to deal with analysis problems that do not include an internal heat source (i.e., S=0) and the energy equation can be reduced to a parabolic partial differential equation as follows:

$$c\rho \frac{\partial T}{\partial t} = k\nabla^2 T \tag{4.44}$$

It is common for engineering problems to be only interested in steady-state solutions, for which the energy equation reduces to the well-known Laplace equation:

$$\nabla^2 T = \frac{\partial^2 T}{\partial x^2} + \frac{\partial^2 T}{\partial y^2} + \frac{\partial^2 T}{\partial z^2} = 0 \tag{4.45}$$

The Laplace equation is an elliptic partial differential equation, thus constituting a boundary value problem, and it can be solved with the same methods adopted for the displacement boundary value problem. Both problems discussed so far, namely the transient given by Eq. (4.44) and the steady-state given by Eq. (4.45), represent an uncoupled formulation of the problem because the temperature field can be determined independently of the stress-field calculation, by solving the uncoupled heat transfer problem first, to get the temperature distribution. Once the temperature is obtained, the elastic stress analysis is performed on the basis of the thermo-elastic constitutive model to complete the solution. This represents what is called a sequentially-coupled thermo-mechanical problem, since the temperature field is solved independently, but its result influences the deformation field that is subsequently solved for.

A fully coupled problem represents the case in which the mechanical and thermal solutions affect strongly each other and, therefore, they must be solved simultaneously. This occurs, for example, in metal forming analysis where significant heat due to inelastic deformation can be generated. If the process is performed rapidly, the heat is expected not to spread within the solid and to create an inhomogeneous thermal field throughout the body. In general, within a transient thermal analysis, whenever the rate of change of thermal sources or the boundary conditions are comparable with the structural vibration characteristics, thermal stress waves are expected to occur. Under these circumstances, the fields of the stress and temperature must be obtained simultaneously through a fully coupled analysis.

Governing equations of the fully coupled thermo-mechanical problem are clearly more complicated. The generalization of the thermo-elasticity equations presented so far starts from the modification of the energy balance equation given by Eq. (4.43). This equation should now include also the rate of change of the elastic energy due to the deformation by adding the term $\sigma\dot{\varepsilon}$. Considering that the stresses within thermo-elasticity are given by Eq. (4.40) and are functions of both the strain components and temperature, this modification will obviously introduce a mixed (i.e., coupled) term that contains both strain and temperature. The detailed derivation of the governing equations for this coupled problem involves thermodynamics and is omitted here. Reference could be made to [12] and [13] for details. Only the final form of the energy equation is presented here in index notation[2]:

$$c\rho\dot{T} = kT_{,ii} + S + \frac{E}{2\nu - 1}\alpha T_0 \dot{\varepsilon}_{ii} \tag{4.46}$$

This equation differs from the one describing the uncoupled analysis problem given by Eq. (4.43) by the presence of the last term that involves both thermal and mechanical variables, therefore representing the *coupling term*.

[2] For the consistency in writing, the index notation is now adopted also for the nabla operator, namely: $\nabla^2 T = T_{,ii}$.

It is worth mentioning that Eq. (4.46) is a linearized one, as it is derived by employing a linear-elastic constitutive model given by (4.39) or (4.40). As an alternative and more general formulation of the coupled problem, it may be stated that the elastic potential U, required for the direct formulation of elastic constitutive models (see Eq. (4.28)), depends on the strain components and also on the temperature. This formulation is usually provided in thermoelasticity books (see, e.g., [12] and [13]).

4.6 Summary

Elastic constitutive models represent the simplest material models as the stress solely depends on the current state of deformation. These models are important because they are applicable to a wide range of scientific and engineering applications. All materials have certain ranges of actions for which only the elastic deformation is introduced. Elastic models are also relevant when dealing with more complex elasto-plastic constitutive models because the latter are built as an extension of the elastic constitutive model by splitting the deformation into an elastic and a plastic part, as shown in the next chapter.

Routine design usually involves stress analysis of structural components subjected to certain service loads, with the sizing of the components being generally decided to ensure no permanent deformation occur at working conditions. Therefore, all the resulting deformations should be elastic. For such a structural analysis, it is recommended to use an appropriate elastic constitutive model, without the need to include additional complexities, e.g., related to plastic deformation, fracture, or damage of the material.

The easiest elastic model that can be formulated and implemented consists of the isotropic linear-elastic constitutive model. It can be applied to a wide range of materials. For example, polycrystalline metals with randomly oriented crystal grains exhibit macroscopic isotropy, and can be modeled as an isotropic material when simulating their response at a sufficiently large scale and considering a sufficiently large sample. This *homogenization* can be applied to materials with random grain orientation, and should not be used for plastically deformed metals that generally introduce preferable grain orientations. For most structural materials, the elastic deformation is fairly small, therefore it usually suffice to use linear-elastic constitutive models for such analysis.

When dealing with hyperelastic materials, the linearized Hooke's law cannot be used since the stresses over the wider range of deformation are not defined as a linear function of the strains. For this purpose, a direct formulation of the constitutive model relies on the use of the elastic potential. The formulation of the constitutive behavior of a hyperelastic materials is by far a more difficult task than the one required for a linear-elastic material. Even though the elastic potential exists for all elastic materials, and thus all elastic constitutive models can be defined through it, it is common practice to use this methodology only for hyperelastic materials when analyses with large deformation are performed.

In all other situations, it is more convenient to adopt the linearized version of the constitutive model as it is expressed by the generalized Hooke's law.

References

[1] R. Hooke, De Potentia Restitutiva, or of Spring. Explaining the Power of Springing Bodies, John Martyn, London, 1678.
[2] Stephen Timoshenko, James Goodier, Theory of Elasticity, McGraw-Hill, New York, 1951.
[3] Martin Sadd, Elasticity: Theory, Applications and Numerics, Academic Press, 2014.
[4] Lawrence Malvern, Introduction to the Mechanics of a Continuous Medium, Prentice-Hall, New Jersey, 1969.
[5] Maurizio Piazza, Roberto Tomasi, Roberto Modena, Strutture in legno, Hoepli, Milano, 2005.
[6] Ronald Gibson, Principles of Composite Material Mechanics, McGraw Hill, New York, 1994.
[7] Erwin Krezszig, Advanced Engineering Mathematics, second edition, John Willie and Sons, New York, 1967.
[8] George Green, On the laws of laws of reflection and refraction of light at the common surface of two non-crystallised media, Transactions of the Cambridge Philosophical Society 7 (1839) 1–24.
[9] George Green, On the propagation of light in crystallized media, Transactions of the Cambridge Philosophical Society 7 (1841) 121–140.
[10] C. Truesdell, W. Noll, The non-linear field theories of mechanics, in: S. Flugge (Ed.), Encyclopedia of Physics, vol. 3(3), Springer-Verlag, Berlin, 1965.
[11] F. Kreith, M. Bohn, Principles of Heat Transfer, sixth edition, Brooks / Cole - Thomson Learning, Pacific Grove, CA, USA, 2001.
[12] Heinz Parkus, Thermoelasticity, second revised and enlarged edition, Springer Verlag, Vienna, 1976.
[13] Richard Hetnarski, M. Reza Eslami, Thermal Stresses - Advanced Theory and Application, Springer Verlag, Berlin, 2009.

Chapter 5

Constitutive models for inelastic deformation

There are number of engineering problems in which a material exhausts its elastic capacity and undergoes inelastic deformation. In such cases, a suitable modeling of the material response requires the material to deform beyond its elastic limit. To obtain reliable results, it is important to model correctly the mechanisms which are at the origin of the inelastic deformation by employing appropriate constitutive models. Even though most of the mechanisms that provoke the inelastic deformation take place at the atomistic, or crystalline lattice scale, they can be quite well observed at the macroscopic scale. This scale is of the main interest for the phenomenological models that treat the material as an homogeneous continuum. Such models can be successfully applied to simulate rather complicated deformation scenarios, even though they do not actually model physical phenomena that are taking place at the smaller scales.

Within the framework of the phenomenological models, the main variables consist of the macroscopic stress and the macroscopic strain. For a particular material, the type of constitutive model that should be adopted depends on the type of deformation that is macroscopically observed in the stress–strain curve recorded during a loading/unloading cycle. If the deformation induced during the loading is completely recovered upon unloading, and subsequent loading/unloading cycles follow the same path (during the loading and unloading process), the deformation can be classified as elastic. This behavior could be described by means of a suitable (linear or nonlinear) elastic model. In the case an irreversible deformation remains upon complete unloading, an elasto-plastic constitutive model should be adopted. Alternatively, if the unloading reveals the complete recovery of deformation, but the material unloads following a different path (generally with a reduced inclination in the stress-strain plane), a damage model that accounts for the reduction of the elastic constants has to be adopted. In the case the material is strained to the same amount of deformation and resists different levels of stress depending on the loading duration and velocity, a visco-(elastic or plastic) model should be considered, depending whether the strain rate dependence is observed within the elastic or plastic range of deformation.

When dealing with an elastic deformation the stress depends only on the current value of strain. Because of this, from a modeling viewpoint an elastic constitutive model should express stresses as a function of strains. This is consistent with the definition of stress at any point of a Cauchy elastic body.

Constitutive Modeling of Engineering Materials. https://doi.org/10.1016/B978-0-12-814696-5.00010-1

For situations in which a material exhibits some degree of inelastic deformation the stress depends on the current value of strain as well as on "something else." An introduction to the formulation of this "something else" is the focus of this chapter. Such a dependency was already outlined in Chapter 1 with example 1.2, considering an elasto-plastic response, in which the stress depended not only on the deformation but also on the previous deformation history. Such a dependency on the history of previous deformation is typical for most inelastic models and not only for plasticity models.

When quantifying the current state of stress and strain with an inelastic model at a particular point in time, it is required to know the deformation state at all preceding times. While this requirement leads to marginal theoretical difficulties, it imposes practical limitations from a computational viewpoint because requiring that the full history of the previous deformation is stored for each computational point. A convenient way to overcome this difficulty (as discussed in Chapter 1) is through the use of internal variables. The way in which internal variables are defined and incorporated into the governing equations distinguishes one inelastic constitutive model from the other. In this chapter, we provide an overview of the main features of different groups of inelastic constitutive models with particular attention devoted to the formulation of the governing equations.

The first set of inelastic constitutive models considered in the following consists of the visco-elastic models. These are followed by plasticity models and damage models. In the final part of the chapter, models dealing with crack initiation and propagation are introduced.

5.1 Visco-elastic models

The constitutive modeling developed for rheological models[1] provides a useful framework for the representation of deformation that can occur in materials that exhibit rate sensitivity. In this section, we provide an introduction to a particular group of materials that fall within this category and that is known as visco-elastic materials.

The rate sensitivity in materials is generally manifested through larger deformation produced by slow stressing than the one produced by rapid stressing with the same amount of load. The phenomenon is called *creep*, and is pronounced quite notably for some materials (e.g., like for certain polymers). The rate sensitivity however tends to increase with temperature, so materials that behave linear elastically at room temperature, for the same amount of stress may become markedly nonlinear at an elevated temperature due to creep. Such behavior is typical for metals, so the phenomenon cannot be ignored when considering high temperature applications.

Alternative manifestation of the same phenomenon is observed by interchanging the roles of stress and strain. By considering strain as independent

[1] Rheology is the study of the flow of matters in liquid or gas state.

variable, due to strain rate sensitivity the stress generally decreases with time at constant strain. The behavior is typically called *relaxation.*

We now illustrate the visco-elastic response by considering two limit loading conditions in which the material is either subjected to a constant stress or to a constant deformation. In particular, in the case a constant stress σ_0 is applied to the material, the strain tends to increase over time after the occurrence of an initial elastic deformation ε_0 (see Fig. 5.1). This response is usually referred to as *creep*. If the load is removed at a time t_0, the strain decreases by the same time-independent amount, followed by further slow decrease to the null value. The second limit loading condition represents the case in which the material is subjected to a constant deformation ε_0. The stress, induced in the material at the time of the initial elastic deformation, tends to reduce over time. This behavior is commonly denoted as *relaxation.*

In the following, we consider only the linear visco-elastic behavior, to outline the main modeling strategies in formulating the constitutive models.

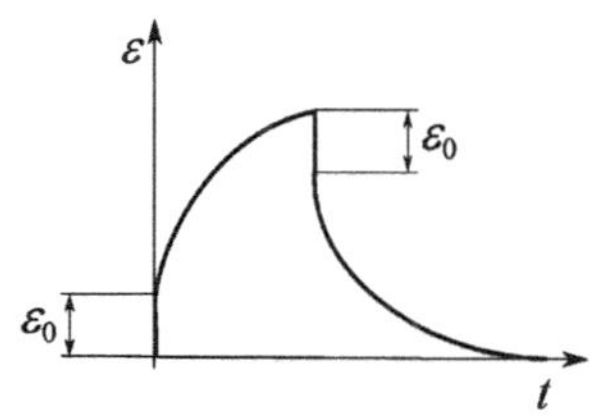

FIGURE 5.1 Time dependent response of strain.

5.1.1 Linear visco-elastic models

Within the previous description, we observed that the strain for a visco-elastic behavior consists of two parts: an elastic one, and a viscous one, or more generally termed the inelastic strain. This represents one important feature which is the core of practically all the inelastic models: the strain is additively decomposed into the elastic and inelastic parts, that is:

$$\varepsilon^{tot} = \varepsilon^{el} + \varepsilon^{ie} \tag{5.1}$$

Commonly, as also adopted in this book, the superscript *tot* is avoided, so within the context of inelastic constitutive models, ε denotes the total strain.

From a modeling viewpoint, the visco-elastic response is usually represented by combining a spring and a dashpot. For simple one-dimensional models, the force resisted by the elements represents stress and the elongation depicts deformation. The two elements, i.e., spring and dashpot, are commonly assumed to exhibit a linear response, relating the spring force to the elongation and the damping force to the strain rate (see Figs. 5.2 and 5.3). These two elements can be connected in different ways to produce different mechanical responses. We will now look at some of the constitutive representations that can be used to describe the visco-elastic response.

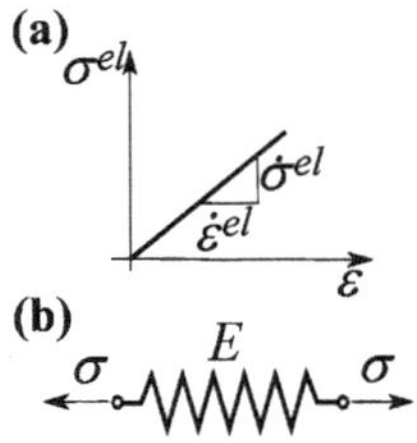

FIGURE 5.2 (a) Linear-elastic behavior and (b) schematic response.

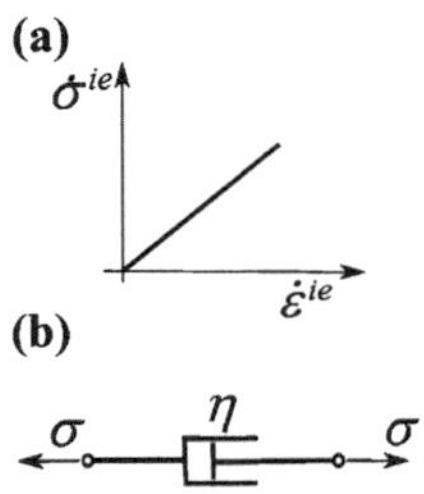

FIGURE 5.3 (a) Linear viscosity and (b) schematic representation.

One of the simplest one-dimensional models is obtained by considering a mechanical system consisting of spring and dashpot connected in series (see Fig. 5.4). Such visco-elastic model is called the *Maxwell model*, for which:

$$\varepsilon^{el} = \frac{\sigma}{E} \qquad \dot{\varepsilon}^{ie} = \frac{\sigma}{\eta} \tag{5.2}$$

Linear-elastic relationship remains unaltered also when expressed in terms of rates (see Fig. 5.2), so when expressing Eq. (5.1) in rate form and by combining it with Eq. (5.2) the total stain rate can be expressed as:

$$\dot{\varepsilon} = \frac{\dot{\sigma}}{E} + \frac{\sigma}{\eta} \tag{5.3}$$

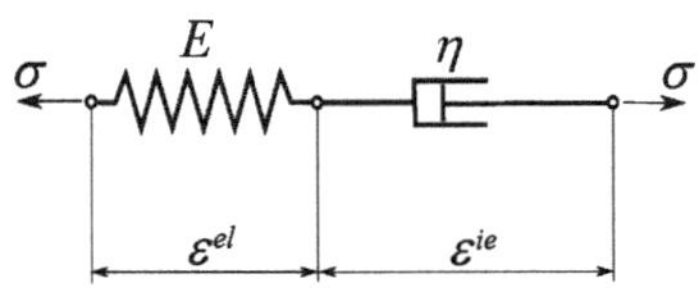

FIGURE 5.4 Mechanical system for the visco-elastic (Maxwell) model.

Let us now use the Maxwell model to predict the response of a creep test (i.e., constant sustained stress producing time-dependent deformation). By denoting the constant sustained stress applied at time $t_0 = 0$ as σ_0, we can express the stress at any time t (≥ 0) as:

$$\sigma(t) = \sigma_0 = const \tag{5.4}$$

Considering that the stress is constant ($\dot{\sigma} = 0$), we can write based on Eq. (5.3):

$$\dot{\varepsilon} = \frac{\sigma_0}{\eta} \tag{5.5}$$

which, after integration, results in a linear increase in the deformation with time as:

$$\varepsilon(t) = \frac{\sigma_0}{\eta} \cdot t + C \tag{5.6}$$

with C being a constant of integration to be determined based on the initial conditions. In this case, we consider the condition that, at time t_0 (here taken as $t_0 = 0$), the deformation induced in the material is equal to the instantaneous elastic deformation σ_0/E:

$$\varepsilon(t) = \frac{\sigma_0}{E} + \frac{\sigma_0}{\eta} \cdot t \tag{5.7}$$

that can be rewritten as follows by introducing the relaxation time $\tau = \eta/E$:

$$\varepsilon(t) = \frac{\sigma_0}{E}(1 + \frac{t}{\tau}) \tag{5.8}$$

Upon removing load σ_0 at time t_1, the elastic part of deformation vanishes immediately, while the creep deformation remains as an irreversible deformation (see Fig. 5.5a).

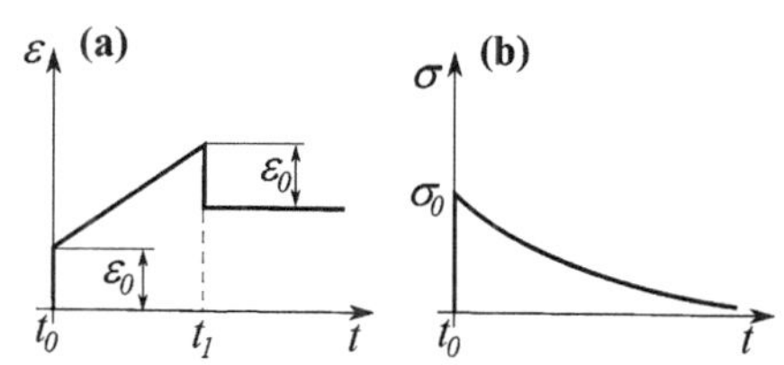

FIGURE 5.5 Visco-elastic response with the Maxwell model in: (a) creep test and (b) relaxation test.

Let us now consider the use of the Maxwell model to describe the response of the material when a strain is applied to the material at time $t_0 = 0$ and maintained constant over time. The initial (elastic) deformation ε_0 is assumed to be produced by a stress σ_0. From Eq. (5.3) and by setting $\dot{\varepsilon}$ equal to zero (condition of constant deformation), the following differential equation is obtained:

$$\frac{\dot{\sigma}}{E} + \frac{\sigma}{\eta} = 0 \tag{5.9}$$

that represents a homogeneous ordinary differential equation expressed as a function of σ. The solution is found in the form[2]:

$$\sigma(t) = Ce^{-\frac{E}{\eta}t} \tag{5.10}$$

The integration constant C is found from the initial conditions, i.e., stress σ_0 applied at time t_0, and the solution for the relaxation test is given by:

$$\sigma(t) = \sigma_0 e^{-\frac{t}{\tau}} \tag{5.11}$$

where τ represents, previously introduced relaxation time.

Stress relaxation modeled by Maxwell model is depicted in Fig. 5.5b that highlights how the stress tends to relax to zero as time approaches infinity, which is not physical for any solid. In the material characterization of the model parameters, it is common to use the instantaneous deformation to identify the value of E and to rely on the stress value at $t = \tau$ to evaluate the parameter η. This behavior is a consequence of the connection in series between the spring and the dashpot, in which the deformation in the dashpot over time allows for the complete relaxation of the spring (i.e., the elastic deformation), and thus also the reduction in stress.

Even if time-dependent responses predicted by the Maxwell model for the creep and relaxation test (Fig. 5.5) are not representative of the behavior exhibited by real solids, its formulation is useful from a modeling viewpoint and can find wide applicability when combined in different arrangements with other elastic and/or dashpot components.

A possible improvement of the Maxwell model could consist in the addition of a spring in a parallel connection as shown in Fig. 5.6. Such mechanical model represents a limit condition of what is usually referred to as the *Generalized Maxwell model* where different Maxwell models (Fig. 5.4) are combined in parallel with the addition of a separate spring. This simplified model is sometimes referred to as the standard linear solid based on the Maxwell model. In Fig. 5.6, we denote the stiffness coefficient of the elastic spring as E^{∞} to reflect the fact that it governs the response as time tends to infinity.

In this case, the strain ε resisted by the two parallel branches is equal, i.e., the deformation of the branch with the Maxwell model ε_M equals the strain induced in the elastic spring branch ε_E:

$$\varepsilon = \varepsilon_M = \varepsilon_E \tag{5.12}$$

and the stress of the generalized Maxwell model can be determined as the sum of the stresses resisted by the two branches as follows:

$$\sigma = \sigma_M + \sigma_E \tag{5.13}$$

---

[2] The differential equation in implicit form is of the type given by $\frac{dy}{dx} + ky = 0$ and the solution is found in the form $y = e^{Cx}$, considering that $\frac{d}{dx}\left(e^{Cx}\right) = Ce^{Cx} = Cy$, with C being the constant of integration. [1]

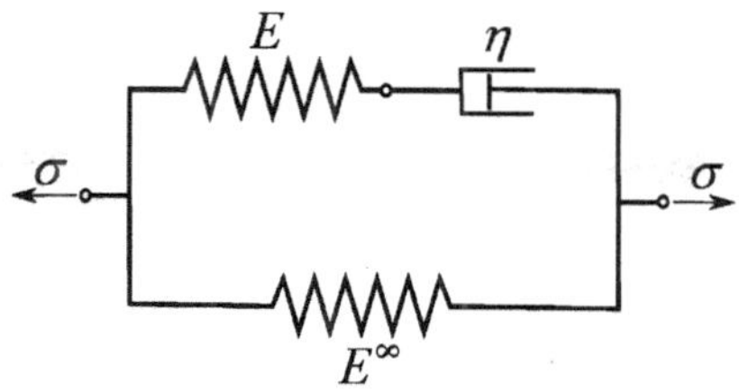

FIGURE 5.6 Mechanical system for the generalized Maxwell model.

where σ_M denotes the stress resisted by the Maxwell branch over time and σ_E depicts the stress induced in the elastic branch. Both components can vary over time.

In addition to Eq. (5.13), the governing equations of the generalized Maxwell model are obtained when Eq. (5.3) is written separately for the Maxwell branch resulting in:

$$\dot{\varepsilon} = \frac{\dot{\sigma}_M(t)}{E} + \frac{\sigma_M(t)}{\eta} \tag{5.14}$$

with the stress in the elastic branch being equal to:

$$\sigma_E = E^{\infty} \cdot \varepsilon \tag{5.15}$$

The strain component in Eq. (5.15) and its rate in Eq. (5.14) correspond to two different branches but represent the same component since the branches are connected in parallel.

In the generalized Maxwell mode, the case of a sustained stress applied at time t_0 (i.e., creep test) is described by an immediate elastic deformation and by a time-dependent creep deformation that occurs over time. In the case of the relaxation test (with an initial deformation applied at time t_0 and maintained constant over time), the stress resisted by the dashpot tends to zero due to the relaxation of the Maxwell model branch while the stress in the parallel elastic branch remains, as it is only resisted by means of the elastic spring. As a consequence, the time-dependent stress tends to a limit value at infinity, depicted by σ^{∞} in Fig. 5.7b.

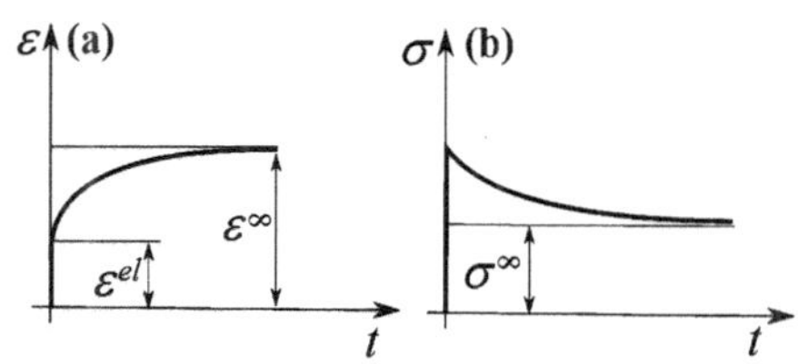

FIGURE 5.7 Response of generalized Maxwell model to: (a) creep test and (b) relaxation test.

Another model, simpler than the generalized Maxwell one, can be obtained by connecting an elastic spring and a dashpot in parallel as shown in Fig. 5.8.

This representation is usually denoted as the *Kelvin–Voigt model* (or simply the *Kelvin model*). With this model the elastic and viscous (i.e., inelastic) strains are identical:

$$\varepsilon = \varepsilon^{el} = \varepsilon^{ie} \tag{5.16}$$

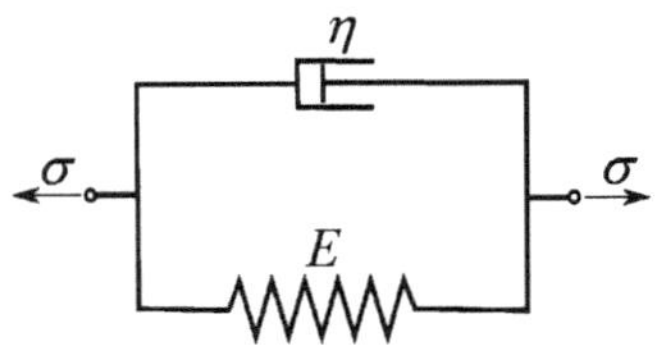

FIGURE 5.8 Mechanical system for the Kelvin–Voigt model.

The overall stress resisted by the Kelvin–Voigt model is determined by summing the stresses induced in the elastic (σ^{el}) and viscous (σ^{v}) branches as follows:

$$\sigma = \sigma^{el} + \sigma^{v} = E\varepsilon + \eta\dot{\varepsilon} = E\left(\varepsilon + \tau\dot{\varepsilon}\right) \tag{5.17}$$

where τ was previously defined as the relaxation time.

By applying load very slowly, the deformation rate is negligibly small (i.e., $\dot{\varepsilon} \approx 0$), so the material would respond elastically, which results from Eq. (5.17), by setting $\dot{\varepsilon}$ to zero:

$$\sigma = E\varepsilon \tag{5.18}$$

If an instantaneous load of σ_0 is applied, the integration of the rate equation (5.17) is required to establish the response of the system. This represents a linear ordinary differential equation, which is now inhomogeneous in a view of the presence of another function of time, here represented by $\sigma(t)$, or simply σ, (aside of the unknown $\varepsilon(t)$, here simply denoted as ε, for which the differential equation has to be solved).

From a modeling viewpoint, function σ is considered as *input* and, therefore, considered as a known function, while ε represents an *output*, or response of the system, and mathematically is a quantity for which the differential equation has to be solved. Different loading conditions correspond to different differential equations. In the following, we consider only one case, corresponding to the creep test.

By considering the creep test, an instantaneous initial stress σ_0 is applied at time $t_0 = 0$ and maintained constant over time. As a consequence, the left-hand side of Eq. (5.17) is considered to be constant (and equal to σ_0). Eq. (5.17) is then solved to determine the expression for ε. In particular, the homogeneous part of the solution of Eq. (5.17)) can be written as:

$$\varepsilon_h(t) = Ce^{-\frac{t}{\tau}} \tag{5.19}$$

with C being the integration constant and the particular part of the solution can be determined based on:

$$\varepsilon_p = \frac{\sigma_0}{E} \tag{5.20}$$

By combining the homogeneous and the particular parts of the solution, and by setting the initial condition related to the deformation to satisfy $\varepsilon(0) = 0$, it is possible to evaluate the integration constant and the complete solution can be expressed in the following form:

$$\varepsilon(t) = \frac{\sigma_0}{E}\left(1 - e^{-\frac{t}{\tau}}\right) \tag{5.21}$$

Let us now assume that the Kelvin–Voigt model is then unloaded at a point in time t_1 as shown in Fig. 5.9. Up to the instant in time $t = t_1$, the system response is governed by the inhomogeneous differential equation since there is a stress different from zero. The solution of this equation is given by Eq. (5.21), and it provides the deformation at time t_1 as follows:

$$\varepsilon(t_1) = \frac{\sigma_0}{E}\left(1 - e^{-\frac{t_1}{\tau}}\right) \tag{5.22}$$

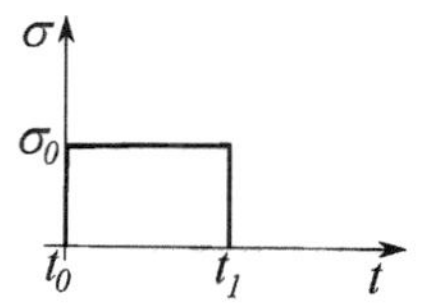

FIGURE 5.9 Example of the induced stress history modeled with the Kelvin–Voigt model.

The material does not resist any stress after time t_1 and the solution of the problem is then described by the homogeneous part of the solution previously derived in Eq. (5.19) that can be rewritten (to account for the removal of the stress at time t_1) as:

$$\varepsilon(t) = Ce^{-\frac{t-t_1}{\tau}} \qquad t > t_1 \tag{5.23}$$

and the constant of integration C can be obtained from the initial conditions related to the known strain at time $t = t_1$ (given by Eq. (5.22)). Based on this, we can write:

$$C = \frac{\sigma_0}{E}\left(1 - e^{-\frac{t_1}{\tau}}\right) \tag{5.24}$$

By substituting Eq. (5.24) into Eq. (5.23), the solution at times $t > t_1$ can be expressed as:

$$\begin{aligned}\varepsilon(t) = \frac{\sigma_0}{E}\left(1 - e^{-\frac{t_1}{\tau}}\right)e^{-\frac{t-t_1}{\tau}} &= \frac{\sigma_0}{E}\left(e^{\frac{t_1}{\tau}} - 1\right)e^{-\frac{t_1}{\tau}} \cdot e^{-\frac{t-t_1}{\tau}} \\ &= \frac{\sigma_0}{E}\left(e^{\frac{t_1}{\tau}} - 1\right)e^{-\frac{t}{\tau}}\end{aligned} \tag{5.25}$$

whose response is depicted in Fig. 5.10.

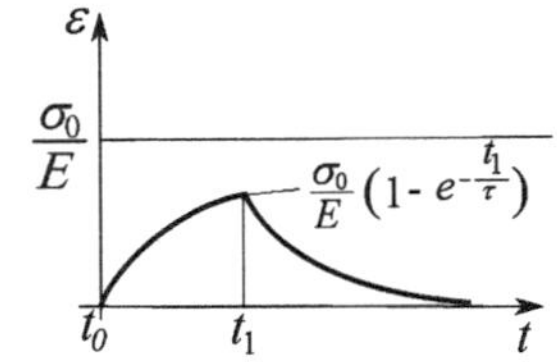

FIGURE 5.10 Time-dependent response of the Kelvin–Voigt model subjected to the stress history of Fig. 5.9.

An improvement of the Kelvin–Voigt model can be obtained by combining a Kelvin–Voigt model in series with an elastic spring as shown in Fig. 5.11. This representation is usually denoted as the *standard solid model* based on the Kelvin–Voigt model. In this model, two springs are contributing to the modeling of the elastic response, but the presence of a branch with a single spring assures that within the relaxation test the stresses will not relax to a zero value. Also for this model there is a clear distinction between the elastic strain ε^{el} (the elongation of left spring) and the inelastic strain ε^{ie} (the displacement of the parallel connection of the spring and the dashpot), that can be written as:

$$\varepsilon = \varepsilon^{el} + \varepsilon^{ie} = \frac{\sigma}{E_0} + \varepsilon^{ie} \tag{5.26}$$

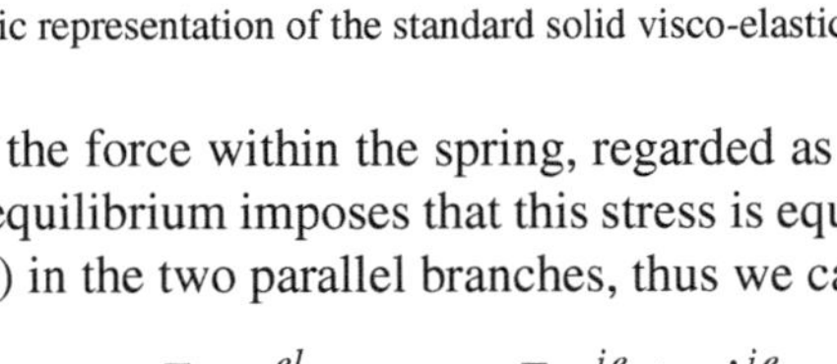

FIGURE 5.11 Schematic representation of the standard solid visco-elastic model.

where σ represents the force within the spring, regarded as a stress like in previous models. The equilibrium imposes that this stress is equal to the sum of the stresses (i.e., forces) in the two parallel branches, thus we can write:

$$\sigma = E_0 \cdot \varepsilon^{el} \qquad \sigma = E_1 \varepsilon^{ie} + \eta \dot{\varepsilon}^{ie} \tag{5.27}$$

The differential equation of the inelastic strain in explicit form results directly from the last expression and it reads:

$$\dot{\varepsilon}^{ie} = \frac{1}{\eta}\sigma - \frac{E_1}{\eta}\varepsilon^{ie} \tag{5.28}$$

which together with Eq. (5.26) provides the governing equation for the standard solid constitutive model giving the relationship between total strain and stress, and the way the inelastic strain evolves.

Additional complexities may be added to the visco-elastic constitutive models by combining the spring and dashpot elements with different characteristics, i.e., different E and η values, and with different arrangements in which the elements act in parallel or in series. More complex alternatives can be found in [2] with generalization for multicomponent state of stress given, for example, in [3] and [4].

The linear visco-elastic models outlined in this section are regarded as simple and they were introduced to highlight some of the key features in which viscous deformation can be formulated in a constitutive model because it can be treated as an inelastic deformation. In the proposed formulations, the material parameters have been assumed to be non-aging in nature to simplify the notation and expressions of the solution. Under this assumption, the solution of the problem depends on the duration of loading and not on the time of loading as it would occur when dealing with aging materials. Such assumption can be released and the problem expressed to account for material aging effects.

An important point raised in this section regarded the additive decomposition of strain into an elastic and a viscous (inelastic) part. Such a decomposition is generally present in most inelastic constitutive models.

For any inelastic deformation the stress does not depend only on the current state of stress, but also on "something else." It was evidenced that the dependency on the latter was introduced through the use of internal variables. Within the visco-elastic constitutive models previously considered, the viscous strain represents the internal variable. From a mathematical viewpoint, it may be stated that the main goal of an inelastic constitutive model is to provide governing equations to establish the dependency of stress on the current value of strain and on the current value of the internal variables, together with the rate equation quantifying how the internal variables vary with the deformation. As an example, for the standard solid model the rate equation may be expressed with Eq. (5.28).

We also noted that visco-elastic models can be formulated with governing equations in a rate or differential form. As all visco-elastic phenomena are a transient phenomena, it is crucial to account for the time dependency within the governing equations considered in the rate form. In the following, it is shown that it is convenient to introduce the rate form of equations also for plasticity models, even if these problems are time independent. The use of the rate form in the definition of constitutive models is common.

The time dependency of the visco-elastic phenomena results in the existence of nonequilibrium states. It was shown that, even for the simple one-dimensional examples considered in this section, the problem must be solved as a transient one and suitable numerical solutions need to be implemented when dealing with these problems.

5.2 Elasto-plastic constitutive models

Plastic deformation represents, by definition, an irreversible deformation that remains upon the removal of stress. The word "plastic" is derived from the classical Greek word *plastikos* and it means "capable of shaping or modeling." It therefore refers to the property of certain materials, like metals, to have the capability of retaining a new (deformed) shape, even after the stress applied to achieve the deformation is removed. Classical theory of plasticity has been primarily developed to describe the behavior of ductile metals. It is therefore useful to consider some specific features of metals, which are at the origin of the plastic deformation.

Metals possess a small amount of electrons on their outer shell, and hence, need a large number of electrons to fill this shell. Atoms within metals are grouped in a way that several neighboring atoms can share electrons on their outer shells to form stable structure. This type of inter-atomic bond is denoted the *metallic bond*. The structure is possible if the atoms are arranged in a regular, well-ordered arrangement usually referred to as *crystals*.

A crystal can be considered as an arrangement of atoms that has a regular, periodic structure that repeats itself. For example, a simple cubic crystal structure is depicted in Fig. 5.12. This structure is formed by repeating one unit cell containing 8 atoms arranged at the corners of a cube. There are overall 14 different possible unit cell arrangements of atoms where the lattice looks the same from each point. These cell arrangements are denoted as *Bravais lattices*, named after their discoverer Auguste Bravais (see [5]).

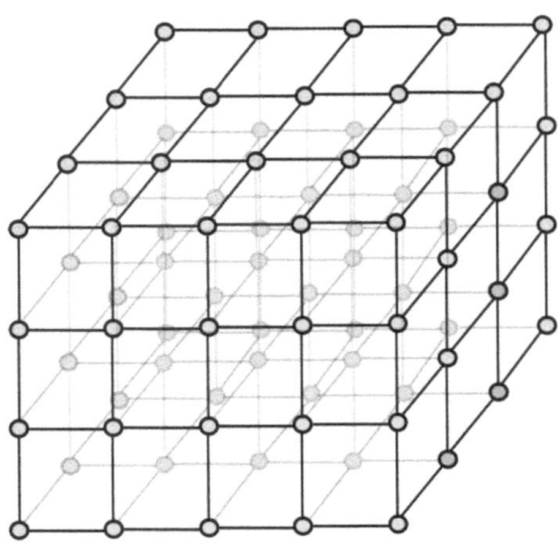

FIGURE 5.12 Schematic representation of a cubic lattice.

The atoms oscillate about their fixed positions within the crystalline lattice, but generally do not move away from them as they are kept in equilibrium by the presence of other neighboring atoms. Stretching and compressing these interatomic bonds can be caused by an external action that can lead to the accumulation of internal energy and development of internal stress. This behavior corresponds to a macroscopic elastic deformation, as the atoms would go back to their equilibrium position upon the removal of the external actions.

When a block of atoms within a crystal is subjected to an external force that is represented macroscopically by an applied shear stress τ, it exhibits the occurrence of a slide, symbolically denoted by x in Fig. 5.13, of one block of atoms over another that stretches the corresponding connections. If the shear stress is sufficiently large to displace the block of atoms so that x reaches the value of $\frac{d}{2}$ (where d depicts the distance between the centers of adjacent atoms), an unstable equilibrium configuration occurs. The instability is due to the fact that the configuration is now equidistant from the two stable configurations: the original one before the application of stress, as shown in Fig. 5.13a, and the one depicted in Fig. 5.13d after one block of atoms *slipped* over an adjacent one. This slip represents the origin of plastic deformation within metals and it takes place over specific crystallographic planes when subjected to shear stress along these planes.

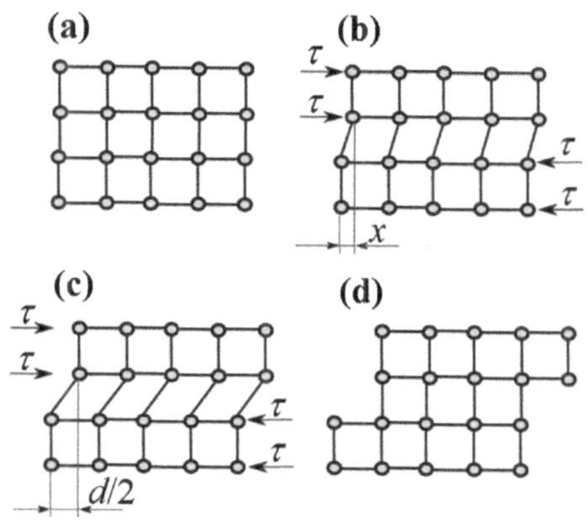

FIGURE 5.13 Plastic slip induced by the application of a shear stress.

The metallic bonds over several atoms are formed so that they share electrons. The electrons are thus distributed over a larger region and a small force is sufficient to displace single atoms. Macroscopically, this behavior is at the basis of the ductile response of metals and of their capability to deform plastically.

This elementary reasoning revealed several important features of plastic deformation. While the elastic deformation involves only the stretching of interatomic bonds, slip requires the breaking and re-establishing of these bonds and the consequent motion of one plane of atoms over another. The points listed below intend to summarize this behavior and are generally valid for metal plasticity:

- The atomic structure remains locally unchanged, apart from its extremes where the change in shape can be observed. Therefore plastic slip does not lead to volume change. This is usually referred to as the incompressibility condition of plasticity.

- The plastic slip is a shearing process and, therefore, the hydrostatic stress at macro level can often be assumed not to influence the onset of plastic deformation.
- Unlike the elastic deformation where the work done by the external forces is stored within the deformed body, the plastic deformation requires the breakage of atomic bonds. For this reason, during plastic deformation energy is dissipated.
- Plastic deformation does not generate stress in materials that can change only shape (not volume).

Shearing of a single crystal may be used to estimate what is called the *theoretical shear strength.* It is computed as the shear stress required so that the bonds between the atoms are stretched elastically until they can switch to the next atom. The calculation leads to a result which estimates shear strength at about $\frac{G}{6}$, with more refined calculations resulting in $\frac{G}{30}$, with G being the elastic shear modulus of the material (for details of this calculation see [2] or [5]). Experimentally measured shear strengths are reported to be smaller, even by a couple of orders of magnitude. The reason for such a large discrepancy is because, in reality, plastic deformation does not occur as a consequence of sliding of a complete block of atoms. Instead, it is related to the specific defects present within a crystalline lattice, denoted as *dislocations*. A detailed discussion of different types of dislocations is outside the scope of this book as crystalline plasticity is not within the focus of constitutive modeling considered here. Here, the goal was merely to present the main elements of microplasticity that are at the origin of the assumptions introduced in continuum plasticity.

5.2.1 Continuum plasticity: formulation of one-dimensional constitutive model

Continuum plasticity studies the overall, homogenized response of a material. The goal of a constitutive model is to provide a mathematical tool to compute stress on the basis of the current value of strain and selected internal variables. The formulation of elasto-plastic constitutive models is first introduced by considering a one-dimensional example, where only one component of stress is computed. For this purpose, we use stress-strain curves obtained from tensile tests as the main source of experimental data, such as those depicted in Fig. 5.14.

Two different material behaviors are depicted in Fig. 5.14: one material that possesses significant ductility and pronounced hardening and the other exhibiting a brittle response and a marginal difference between its ultimate stress and yield stress. For modeling purposes, the latter material is commonly modeled as an elastic–perfectly-plastic material, where the stress remains constant upon exhausting the elastic capacity of the material. The former material exhibits a

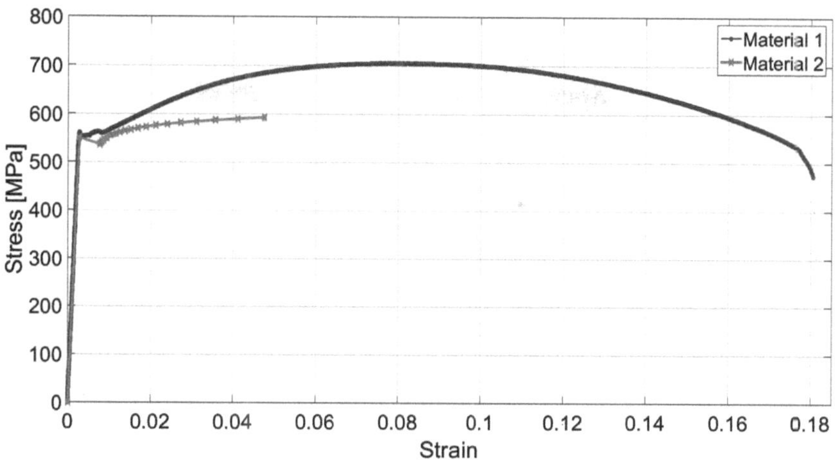

FIGURE 5.14 Typical experimental stress-strain curves obtained for two different materials.

degree of hardening that is usually included in the material model through a linear or exponential law.

An example of an exponential hardening law frequently used is the approximation by Ramberg–Osgood formula that, for sufficiently large plastic deformation, is reduced to a single exponent of hardening (for details, see [2]). According to this approximation, the stress–strain relation reads:

$$\sigma = \begin{cases} E\varepsilon & \varepsilon \leq \frac{\sigma_0}{E} \\ \sigma_0 \left(\frac{E\varepsilon}{\sigma_0}\right)^n & \varepsilon > \frac{\sigma_0}{E} \end{cases} \tag{5.29}$$

where σ_0 is the first yield limit, E is the Young's modulus of elasticity and ε is the total strain.

The measured uniaxial response is usually fitted to selected behavioral models, e.g., with or without hardening. For the materials depicted in Fig. 5.14, the calculated (based on a constitutive model) versus experimental curves are compared in Fig. 5.15.

The softening part visible for material 1 in Fig. 5.14 starts at a fairly large amount of plastic deformation and this trend is usually attributed to the onset of damage. As classical plasticity models do not capture this softening effect, suitable modeling of this behavior should be sought, for example, by coupling plasticity with damage.

The comparisons between the calculated and the experimental curves shown in Fig. 5.15a exhibit increasing discrepancies for strains larger than about 0.09. If the material simulation needs to reach and go beyond this level of deformation in the analysis, it might not be appropriate to model the curve with perfect plasticity model and more complex models, for example, combined with damage, could be considered.

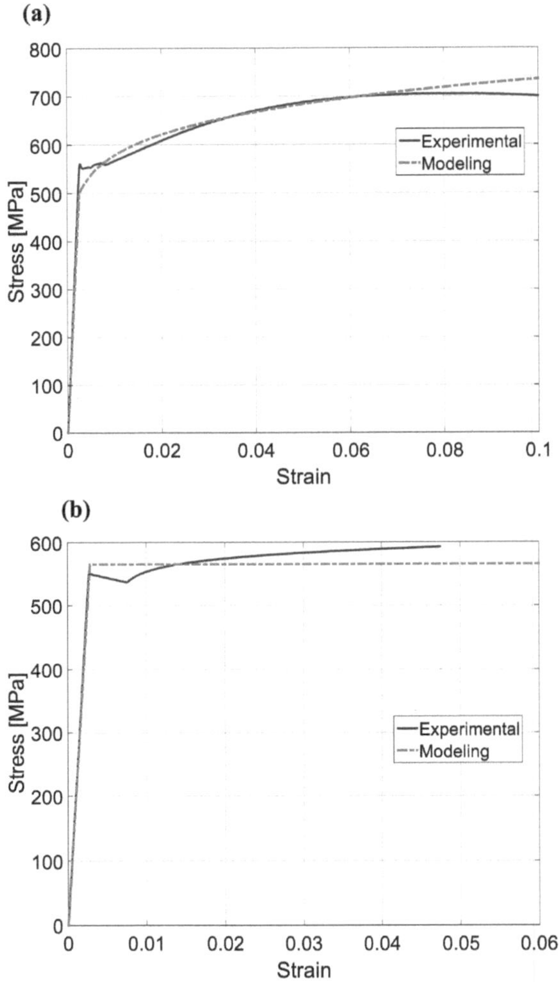

FIGURE 5.15 Calculated vs. experimental curves for: (a) a material with pronounced hardening and (b) an elastic–perfectly–plastic material.

For the second material (shown in Fig. 5.15b), the effect of hardening is limited and the use of a reasonable model that captures an elastic– perfectly-plastic response should be fine. For the convenience of using a simpler model that does not include hardening, a trade off is necessary in overestimating the first yield limit that is then compensated by the use of perfect plasticity (see Fig. 5.15b).

We will now formulate the elasto-plastic constitutive model with the assistance of some model examples. For ease of notation, we start by considering the elastic–perfectly-plastic model and then move to a plastic representation that accounts for strain-hardening. For illustrative purposes, we start by expressing the formulation in absolute (not incremental) values of strain and stress, before introducing the modeling formulated in incremental form, that is inevitable for the elasto-plastic models.

Let us assume that, for the given strain value ε_1, we want to calculate the corresponding stress value. As a first attempt, we assume that the total strain is equal to the elastic deformation from which the corresponding stress value can be evaluated as:

$$\sigma_1 = E \cdot \varepsilon_1 \tag{5.30}$$

We then need to check whether the computed value is plastically admissible by verifying that the following condition is satisfied:

$$\sigma_1 \leq \sigma_0 \tag{5.31}$$

If the last inequality is satisfied, we can accept the value σ_1 as the correct stress value corresponding to the strain value ε_1. If the inequality is not satisfied, the value has to be rejected since plastically non-admissible because the calculated stress exceeds what is the maximal possible stress that the material can resist. This means that not all the strain ε_1 corresponds to an elastic strain and that we need to perform an additive decomposition of the deformation into its elastic and plastic parts as follows:

$$\varepsilon_1 = \varepsilon_1^{el} + \varepsilon_1^{pl} \tag{5.32}$$

where only the elastic part ε_1^{el} should be related to stress through the elastic constant E:

$$\sigma_1 = E \cdot \varepsilon_1^{el} = E \cdot (\varepsilon_1 - \varepsilon_1^{pl}) \tag{5.33}$$

that requires the knowledge of ε_1^{pl} that is now treated as an additional unknown. For the case without hardening, once the plastic deformation starts to develop, the stress remains constant and so does the elastic strain. In this case, the plastic strain can be computed from:

$$E \cdot (\varepsilon_1 - \varepsilon_1^{pl}) = \sigma_0 \quad \rightarrow \quad \varepsilon_1^{pl} = \varepsilon_1 - \frac{\sigma_0}{E} \tag{5.34}$$

In the presence of hardening, the elastic strain (and, therefore, the stress) evolves with the previously accumulated plastic strain based on which the right-hand side of Eq. (5.34) is not constant. In this case, the yield limit is expressed as a function of the plastic strain:

$$E \cdot (\varepsilon_1 - \varepsilon_1^{pl}) = \sigma_0(\varepsilon^{pl}) \tag{5.35}$$

When considering only one single loading cycle, there is no need to distinguish between plastic strain on the left-hand side (ε_1^{pl}) and on the right-hand side (ε^{pl}) of Eq. (5.35) as both terms refer to the current value of plastic deformation within the load cycle. By recalling that plastic deformation hardens the material in such a way that the elastic domain is permanently extended to a new value of stress, the term ε^{pl} on the right-hand side of Eq. (5.35) has to be the function of the plastic strain accumulated during the previous load history, if present. Therefore, in the general case, Eq. (5.34) can be used to calculate the plastic strain, considering σ_0 as a constant for non-hardening case, while for the hardening case the expression for σ_0 can be written as:

$$\sigma_0 = f(\chi) \tag{5.36}$$

where χ is the previously accumulated plastic strain and represents the *internal variable* for this constitutive model.

Let us know revisit the procedure presented above with the assistance of an example based on the widely used relation given by Ramberg–Osgood, previously introduced in Eq. (5.29).

Example 5.1

Let us assume that for the strain increment $\Delta\varepsilon = 0.01$ we want to compute the corresponding stress increment. It is assumed that, at the beginning of the analysis, the material is stress-free without any previous load history (i.e., with no accumulated plastic strain) and that its material parameters are $E = 200{,}000$ MPa, $\sigma_0 = 200$ MPa, and $n = 0.1$.

As the first trial, we compute the stress by assuming a complete elastic response of the material:

$$\sigma_{trial} = E \cdot \Delta\varepsilon = 2000 \text{ MPa} \tag{5.37}$$

By comparing this value with the yield limit, we can see that:

$$\sigma_{trial} = 2000 \text{ MPa} > \sigma_0 = 200 \text{ MPa} \tag{5.38}$$

that implies that not all deformation $\Delta\varepsilon = 0.01$ is elastic because it contains a plastic part as well. Since the initial value of strain was 0, we can treat $\Delta\varepsilon = 0.01$ as the current value of total strain that is separated into the elastic and plastic parts:

$$\varepsilon = \varepsilon^{el} + \varepsilon^{pl} = \Delta\varepsilon \tag{5.39}$$

For the calculation of the sought stress value, we only need to consider the elastic part of the deformation as follows:

$$\sigma_{accepted} = E \cdot (\varepsilon - \varepsilon^{pl}) \tag{5.40}$$

In the last equation, the value of ε^{pl} is unknown. At this point of the analysis, we know that the pure elastic capacity is now exhausted and, therefore, the relationship between total strain and stress can be expressed by Eq. (5.29). By substituting all values of the problem, we can calculate the following stress value:

$$\sigma_{accepted} = \sigma_0 \left(\frac{E\varepsilon}{\sigma_0}\right)^n = 200\left(\frac{200{,}000 \cdot 0.01}{200}\right)^{0.1} = 251.8\ \text{MPa} \qquad (5.41)$$

Since we now know the correct value of the stress that accounts for the elastic-plastic response of the material, we can compute the elastic strain as:

$$\varepsilon^{el} = \frac{\sigma_{accepted}}{E} = 0.00126 \qquad (5.42)$$

and the remaining part of the strain is then attributed to the plastic strain:

$$\varepsilon^{pl} = \varepsilon - \varepsilon^{el} = 0.01 - 0.00126 = 0.00874 \qquad (5.43)$$

The last calculation related a single value of stress to a single value of strain. In elasto-plastic analysis it is usually desirable to trace the solution over the given range of deformation. For this type of solution, the formulation used here is inconvenient as the stress in Eq. (5.29) is given as a function of the total strain. It was already anticipated in Chapter 1 that, for elasto-plastic deformation, the dependency of stress on the history of the previous deformation makes it impossible to establish a one-to-one correspondence between stress and strain. The difficulty that we are facing in using the formulation given by Eq. (5.29) (i.e., relying on the absolute value of strain) is outlined in the following example.

Example 5.2

Let us assume that we are deforming the material that exhibits the exponential hardening uniaxially by the stress that gradually grows up to the value of σ_1, with corresponding total strain equal to ε_1. Since the value is larger than the initial yield limit σ_0 (with the current value of the yield limit taken as $\sigma_Y = \sigma_0$), the material had undergone hardening. In subsequent loading, the material will now behave elastically up to the new yield limit $\sigma_0 = \sigma_1$. If we now start unloading the material up until it reaches a zero stress it undergoes an elastic unloading and there is a permanent plastic deformation that represents the accumulated plastic strain as shown at point 2 in Fig. 5.16. If we further load the material but now with a reversed sign of load, the material is expected to deform elastically up to a certain limit. In the simplest hardening model (denoted as the *isotropic hardening* and explained in more details later in this chapter), we assume that the plastic deformation hardens the material to the same amount regardless of the sign of the stress. Under this assumption, the current loading path is expected to remain linear-elastic until the stress is $\sigma = -\sigma_Y$. By continuing to increase the compressive stress applied to the material, the material will continue to harden and to further enlarge the elastic domain by shifting the new elastic limit σ_Y to higher values, such as σ_3 depicted at point 3 in Fig. 5.16. To compute the new elastic limit, we use the same expression given by Eq. (5.29). In the present case, further hardening occurs even if the current total strain is smaller than ε_1, thus to compute the new value of the elastic limit we cannot use the expression given in terms of the total strain.

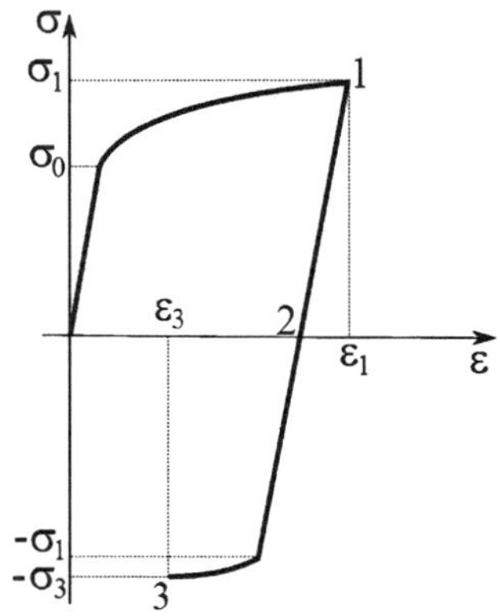

FIGURE 5.16 Example of an uniaxial cyclic elasto-plastic loading.

The illustrated difficulty can be elegantly avoided by expressing the hardening relation in rate form. In general, all the elasto-plastic constitutive equations are defined in a rate form that relate stress rate to strain rate. The rate considered here does not imply the use of derivatives with respect to real time and, instead, it should be regarded as the calculation of a stress increment for a given strain increment.

One procedure to implement and use an incremental formulation is presented below.

At the beginning of the analysis, all variables that are used to update the evolution of the material condition need to be set to their initial values: current stress σ^0, current strain ε^0, and accumulated plastic strain ε^{ap}.

1. Start by defining the current yield stress, here taken at the beginning of the analysis as $\sigma_Y = \sigma_0$, and by determining the current strain increment $\Delta\varepsilon$.
2. Calculate the stress increment while assuming all strain increment to be elastic: $\Delta\sigma_{trial} = E \cdot \Delta\varepsilon$
3. Update the stress to obtain the trial stress increment: $\sigma^{i+1}_{trial} = \sigma^i + \Delta\sigma_{trial}$.
4. Check whether the current trial stress is plastically admissible. It is satisfies the condition: $\sigma^{i+1}_{trial} \leq \sigma_Y$ proceed to step 8, otherwise continue to the next step.
5. Further hardening has occurred in the material and, therefore, the current value for σ_Y needs to be updated. Its new value is computed by

accounting for the current increments of the accumulated plastic strain and elastic strain:
$\sigma_Y = \sigma_0^{1-n} \cdot E^n \left(| \frac{\sigma^i}{E} | + \varepsilon^{ap} + | \Delta\varepsilon | \right)^n$

6. Calculate the current stress increment and trial stress: $\Delta\sigma_{trial} = \sigma_Y - \sigma^i$ and $\sigma_{trial}^{i+1} = \sigma_Y$
7. Evaluate the increment of the accumulated plastic strain:
$\Delta\varepsilon^{ap} = \Delta\varepsilon - \frac{\sigma_{trial}^{i+1}}{E}$
and update the current value of the accumulated plastic strain:
$\varepsilon^{ap(i+1)} = \varepsilon^{ap(i)} + \Delta\varepsilon^{ap}$
8. Update stress and strain:
$\sigma^{i+1} = \sigma^i + \Delta\sigma_{trial}$
$\varepsilon^{i+1} = \varepsilon^i + \Delta\varepsilon$

A few remarks related to the implementation procedure outlined above are presented in the following. The yield limit is stored as a positive value, as there is no need to keep track of its sign (based on the adopted model to distinguish between compression and tensile values). In addition, Eq. (5.29) determines stress values as real numbers only for positive ε, thus negative strain (which would correspond to compression) cannot be specified in the proposed equation. As already pointed out, the current strain value cannot be used as an argument of the hardening function if the simulations include cyclic loads. In such a case, we need to use the summation of current elastic strain and the accumulated plastic strain, where both are taken with positive signs. Based on the previous considerations, the accumulated plastic strain is stored as a non-negative value. For example, let us assume that we are at point A on the hardening curve (see Fig. 5.17) with the strain increment $\Delta\varepsilon$ and that the solution is moving in the direction of further hardening, toward point B. In this case, the initial elastic trial would lead to a plastically non-admissible result and a new, enlarged (i.e., hardened) value of the elastic limit would then need to be evaluated. For this purpose, the argument of the hardening function (5.29) is determined as the summation of the current elastic strain (i.e., corresponding to the converged stress value from the previous increment), the current accumulated plastic strain and the strain increment $\Delta\varepsilon$. In this case, while accumulated plastic strain is always positive, the first and the last members of the summation should be taken as absolute values. When a plastically admissible stress increment is computed, the elastic strain increment can be directly obtained from it. What remains of $\Delta\varepsilon$ is the increment in the accumulated plastic strain, that can be added to the value of the accumulated plastic strain obtained from the converged solution at the end of the previous increment.

The following listing collects the implementation of the above elasto-plastic model with exponential hardening with MATLAB®.

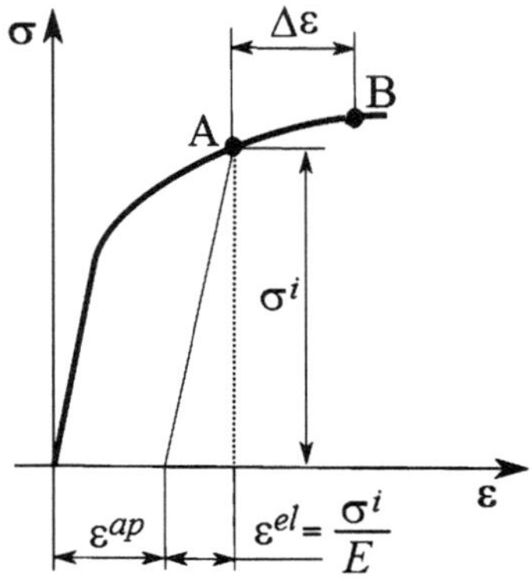

FIGURE 5.17 Diagram illustrating an incremental hardening step.

```
% Implementation of 1-dimensional continuum plasticity model
% ---The incremental formulation---
% ------ Multiple steps ----------
clear
E=200000; Sig0=500; Exp=0.1; % Material parameters
steps=[0,0.03,0.015,0.04]; % Loading steps (in strain values)S
Nstp=size(steps,2); % Number of steps
cYc=0; strain=0; SigY=Sig0; eps_pl=0; % Setting initial values
NMB=1;
SSC(NMB,:)=[0,0];
SigACC=0;
for j=2:Nstp
    D_strain=0.001; % Initial step size
    if steps(j)<strain
        D_strain=-D_strain;
    end
    while abs(strain-steps(j-1))/abs(steps(j)-steps(j-1))<1
        NMB=NMB+1;
        while cYc==0 % Cycle for one increment
          if abs(strain+D_strain-steps(j-1))/abs(steps(j)-steps(j-1))>1
              D_strain=steps(j)-strain;
              strain=steps(j);
          else
              strain=strain+D_strain;
          end
          INC_Str_trial=E*(D_strain); % As if everything is elastic
          Str_trial=SSC(NMB-1,1)+INC_Str_trial; % Absolute value of stress
          if abs(Str_trial)<=SigY
              INC_Str_acc=INC_Str_trial;
          else
              SigY=Sig0^(1-Exp)*E^Exp*(abs(SSC(NMB-1,1))/E+eps_pl+abs(D_strain))^
      Exp;
              if D_strain>0
                  INC_Str_acc=SigY-SigACC;
```

```
34                 else
35                     INC_Str_acc=-SigY-SigACC;
36                 end
37                 eps_pl=eps_pl+abs((D_strain-INC_Str_acc/E));
38             end
39           SigACC=SigACC+INC_Str_acc;
40           cYc=1;
41           end
42       cYc=0; % Updating stress strain curve
43       SSC(NMB,1)=SigACC;
44       SSC(NMB,2)=SSC(NMB-1,2)+D_strain;
45       if abs(SSC(NMB,1)-SSC(NMB-1,1))<20 % Correcting strain increment
46           D_strain=D_strain*2;
47       end
48       end
49   end
```

In the present code, the solution is sought in a multiple step mode that is applicable also to cyclic loads. A prescribed strain history is provided for each loading step. In different zones of the stress–strain curve, diverse step lengths may be required, and for this purpose, the increment size is adjusted if the difference in stress between two increments is less than 20 MPa (see lines 45–47 of the code). The prescribed strain history is specified in code line 6. Typical results are depicted in Fig. 5.18. The incremental solution determined with the code is plotted with circles to identify the discrete number of points obtained throughout the increments. These results are compared against the original stress–strain curve of the adopted hardening behavior. It can be observed that, for the first step (i.e., loading up to the stain value of 0.03), all points of the two curves match. In the second step, after unloading and the application of a reversed load, further hardening takes place when resisting compressive stress. The third step takes the total strain from the value of 0.01 to the final value of 0.04. During this loading phase, the stress undergoes again a sign reversal and produces further hardening in tension. As a consequence of the previous two steps, the points are now above the original stress–strain curve, since the material has hardened and the elastic zone is now enlarged.

The illustrated example demonstrated the need to formulate the elasto-plastic model in an incremental form. In this case, the one-dimensional model was formulated as a function of a single elastic constant E, a yield limit σ_Y and a hardening law given by Eq. (5.29). The internal variable of this constitutive model was the accumulated plastic strain, and the hardening equation provided the rate equation of its evolution. In the following section, we focus on how the constitutive models can be formulated to deal with the general case of continuum plasticity when both stress and strain are considered as second-order tensors.

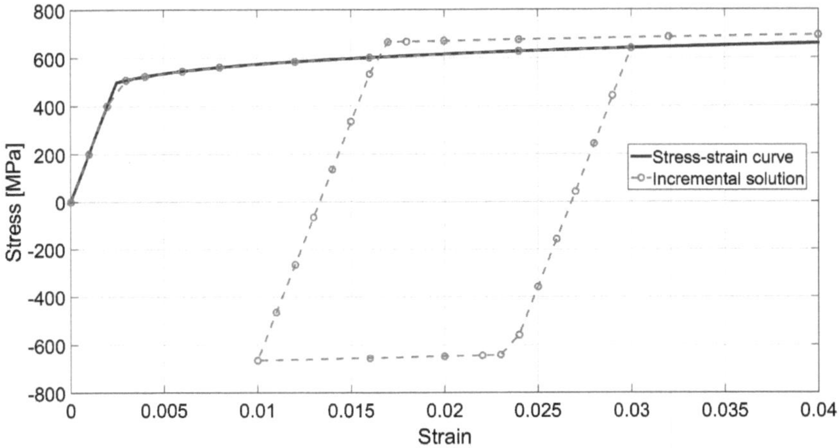

FIGURE 5.18 Stress–strain curve and incremental solution for an uniaxial elasto-plastic loading cycle.

5.2.2 Continuum plasticity for the general three-dimensional state of stress: elastic constants and yield conditions

In the previous section, we illustrated that the main features of an elasto-plastic constitutive model are: additive decomposition of total strain into the elastic and plastic parts; only the elastic part of the strain is related to stress through the elastic constants; it is required to formulate a criterion, usually denoted as the *yield criterion*, to establish whether or not plastic deformation occurs. The concept essentially remains the same also for continuum plasticity, except that both stress and strain are considered as second-order tensors. The strain tensor can be expressed as:

$$\boldsymbol{\varepsilon} = \boldsymbol{\varepsilon}^{el} + \boldsymbol{\varepsilon}^{pl} \tag{5.44}$$

and the stress is linearly related to the elastic strains by:

$$\boldsymbol{\sigma} = \boldsymbol{D} : \boldsymbol{\varepsilon}^{el} \tag{5.45}$$

where $\boldsymbol{D}$ is the material stiffness tensor of fourth order since stress and strain are second-order tensors. In reduced Voigt notation, $\boldsymbol{D}$ can be represented as a matrix and its form depends on the type of symmetry the material possesses (as previously presented in Chapter 4).

In the one-dimensional elasto-plastic constitutive model, the yield criterion was represented by a straightforward comparison between the stress value and the yield limit σ_Y. For a multicomponent state of stress, the yield condition is defined in the form of a function $f(\boldsymbol{\sigma})$, referred to as the *yield function*. For the plastic deformation to occur, the value of this function has to be nil, while for negative values the deformation is elastic. The condition $f(\boldsymbol{\sigma}) > 0$ cannot occur because it represents a plastically non-admissible state of stress. Based on this, we can write:

$$\begin{aligned} f(\boldsymbol{\sigma}) < 0 \quad &\rightarrow \text{Elastic deformation} \\ f(\boldsymbol{\sigma}) = 0 \quad &\rightarrow \text{Plastic flow occurs} \end{aligned} \tag{5.46}$$

Different yield functions can be formulated for different materials to detect the onset of plastic deformation when its value equals zero. As the evaluation of whether or not plastic flow occurs cannot depend on the choice of the reference system, the yield functions are commonly defined as a function of stress invariants (that are independent of the reference system). When principal stresses (σ_1, σ_2, and σ_3) are used to formulate the yield function, it can be visualized in the form of a 3-dimensional surface, referred to as the *yield surface* (see Fig. 5.19). Recalling Eq. (5.46), points within the domain contained in the yield surface depict an elastic state of stress. Points on the surface corresponds to plastic deformation while anything that is outside the surface is plastically non-admissible. In the following, we will briefly present some of the most commonly used yield criteria, present their yield surfaces and discuss their main features.

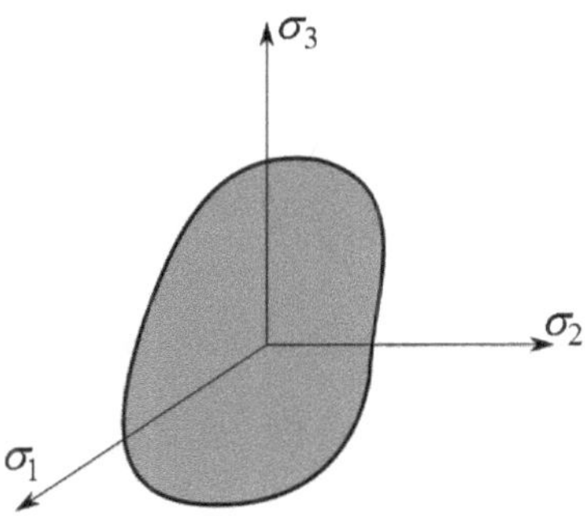

FIGURE 5.19 Yield surface that bounds the elastic domain.

The yield condition represents an important part of a phenomenological constitutive model and its underlying assumptions should be formulated to account for the source of the plastic deformation to adequately capture the global macroscopic behavior. We will exemplify this by considering metal plasticity.

It was previously described that the origin of plastic deformation in metals is due to the occurrence of plastic slips, macroscopically manifested by a change of shape only (leading to plastic incompressibility). The onset of plastic deformation is therefore independent on the volumetric part of the stress tensor. The most popular yield criterion among those formulated as pressure independent is the von Mises criterion that relies on the second invariant of deviatoric stress tensor J_2, responsible for the change of shape. It has the following simple form:

$$\sqrt{J_2} - \tau_0 = 0 \tag{5.47}$$

where τ_0 represents a material property, while the invariant J_2 is equal to the product of the distortion energy and twice the value of shear modulus, and it is therefore a non-negative number (for details, see [6]). Because of this, its square root is always a real number.

Written in terms of the principal stress components, the von Mises yield condition can be expressed as:

$$\sqrt{\frac{1}{6}\left[(\sigma_1-\sigma_2)^2+(\sigma_2-\sigma_3)^2+(\sigma_3-\sigma_1)^2\right]}=\tau_0 \tag{5.48}$$

that, in the principal stress space, represents an inclined cylinder with its axis directed along the diagonal of the first quadrant (i.e., the line defined by $\sigma_1 = \sigma_2 = \sigma_3$), denoted as the *hydrostatic axis* (see Fig. 5.20).[3]

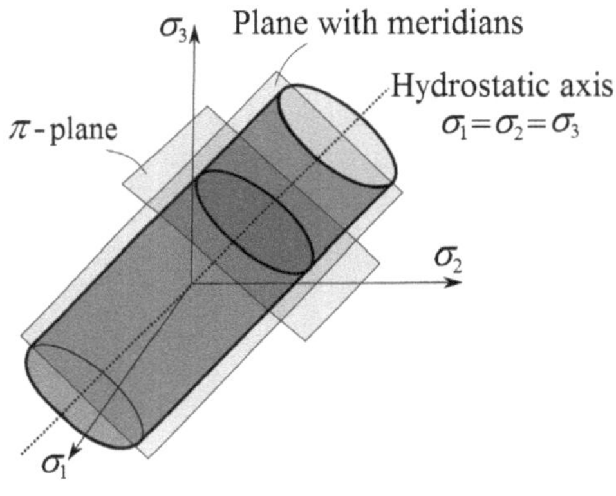

FIGURE 5.20 Von Mises yield surface in principal stress space.

The von Mises cylinder is an infinite cylinder, meaning that the elastic domain for this yield condition expands infinitely along the hydrostatic axis in the direction of both positive and negative pressure. Considering the plastic deformation mechanisms of metallic materials this constitutive behavior is reasonable since, as long as stress state remains on the hydrostatic axis, there are no planes in the material exhibiting shear stress and, therefore, no crystal slip can occur.

The radius of the von Mises cylinder has a magnitude equal to $\sqrt{2}\tau_0$. Let us now see what the material property τ_0 represents. We can write the J_2 invariant as a function of the stress components for an arbitrary xyz reference frame:

$$J_2=\frac{1}{6}\left[(\sigma_{xx}-\sigma_{yy})^2+(\sigma_{yy}-\sigma_{zz})^2+(\sigma_{zz}-\sigma_{xx})^2\right]+\sigma_{xy}^2+\sigma_{xz}^2+\sigma_{yz}^2 \tag{5.49}$$

For a pure shear loading, we have:

$$\sigma_{xx}=\sigma_{yy}=\sigma_{zz}=\sigma_{xz}=\sigma_{yz}=0, \quad \sigma_{xy}\neq 0 \quad \rightarrow J_2=\sigma_{xy}^2 \tag{5.50}$$

[3] Equation of a cylinder with radius r that passes through the diagonal of the first quadrant has the form: $(x-y)^2+(x-z)^2+(y-z)^2=3r^2$.

in which case the von Mises yield condition now becomes:

$$\sigma_{xy} = \tau_0 \tag{5.51}$$

from which the constant τ_0 depicts the magnitude of shear stress at the onset of yielding in pure shear.

For an uniaxial state of stress, we have that $J_2 = \frac{\sigma_{xx}^2}{3}$, so the yield condition becomes:

$$\sigma_{xx}^2 = 3\tau_0^2 \quad \rightarrow \quad \sigma_{xx} = \sqrt{3}\tau_0 \tag{5.52}$$

which implies that, with respect to the case of pure shear, the stress required for yielding with a single normal stress component is $\sqrt{3}$ larger than the stress required for yielding in shear. This difference is reflected, for example, in the evaluation of the resistance of structural bolts whose tangential capacity (i.e., due to shear), by recommendation in engineering design manuals, is considered to be 0.6 times the capacity resisted when subjected to an axial force. The scaling coefficient approximately amounts to $\frac{1}{\sqrt{3}}$ which is the difference between the shear yield stress and the normal yield stress.

Based on the relation between the normal and the shear yield stress, the von Mises criterion given in Eq. (5.47) can be reexpressed in terms of the normal stress yield limit σ_0 as:

$$\sqrt{3J_2} - \sigma_0 = 0 \tag{5.53}$$

that is frequently used within linear-elastic analysis to evaluate an arbitrary three-dimensional state of stress. According to this equation, if the value $\sqrt{3J_2}$ is smaller than σ_0 the state of stress is elastic. The quantity $\sqrt{3J_2}$ is interpreted as a von Mises *equivalent stress* as it is equivalent to the yield stress σ_0 of an uniaxially loaded sample.

Under plane stress conditions (i.e., $\sigma_3 = 0$), the second deviatoric invariant is equal to $J_2 = \frac{\sigma_1^2 - \sigma_1\sigma_2 + \sigma_2^2}{3}$, so the yield condition of Eq. (5.47) takes the following form:

$$\sigma_1^2 - \sigma_1\sigma_2 + \sigma_2^2 = 3\tau_0^2 \tag{5.54}$$

Considering the relationship given by Eq. (5.52), Eq. (5.54) can be rewritten as:

$$\sigma_1^2 - \sigma_1\sigma_2 + \sigma_2^2 = \sigma_0^2 \tag{5.55}$$

where σ_0 is the yield limit under uniaxial stress. The elastic domain for the plane stress is visualized in Fig. 5.21(a). Graphically, the intersection between the von Mises cylinder and the plane $\sigma_1 - \sigma_2$ forms an ellipse that intersects the axes at stress values of σ_0 in both tension and compression.

We will now discuss a few more planes of interest within stress analysis and see what is the shape of the von Mises yield surface under those conditions.

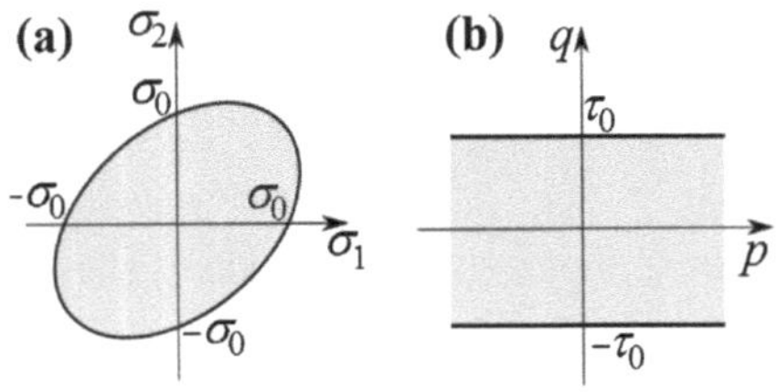

FIGURE 5.21 Von Mises yield surface: (a) in plane stress conditions and (b) in the $p-q$ plane.

Any plane that is perpendicular to the hydrostatic pressure axis represents a deviatoric, or $\pi-$plane. The von Mises yield surface in any deviatoric plane is a circle, and since the criterion is independent on pressure, all the circles are of the same diameter regardless of the position where they intersect the pressure axis. The intersections of the yield surface with any plane that contains the hydrostatic axis are referred to as meridians. For the von Mises yield surface, the meridians are straight lines parallel to the hydrostatic axis. They are commonly represented in the so-called $p-q$ plane, where p depicts a hydrostatic pressure and $q = \pm\sqrt{3J_2}$ is the equivalent von Mises stress (see Fig. 5.21b).

The Tresca criterion is another simple pressure non-sensitive yield criterion. It was formulated by Tresca who was the first to advance the idea of the yield criterion in 1868. According to this criterion, plastic deformation starts to occur when the maximal shear stress reaches a critical value τ_0, that represents a material parameter. Within an arbitrary loaded body, there are six planes with local maxima of shear stress. The Tresca yield condition requires the satisfaction of one of the following six equations:

$$\frac{\sigma_1-\sigma_2}{2} = \pm\tau_0; \quad \frac{\sigma_1-\sigma_3}{2} = \pm\tau_0; \quad \frac{\sigma_2-\sigma_3}{2} = \pm\tau_0 \tag{5.56}$$

These six equations can be packed into a single scalar value yield function so that the yielding occurs when $f(\boldsymbol{\sigma}) = 0$ as described by:

$$f(\boldsymbol{\sigma}) = \left[(\sigma_1-\sigma_2)^2 - 4\tau_0^2\right]\left[(\sigma_1-\sigma_3)^2 - 4\tau_0^2\right]\left[(\sigma_2-\sigma_3)^2 - 4\tau_0^2\right] \tag{5.57}$$

The easiest way to describe the yield surface according to the Tresca condition is to consider the plane stress condition, i.e., $\sigma_3 = 0$. The elastic domain within the $\sigma_1-\sigma_2$ plane is bounded by the following inequalities, resulting from Eqs. (5.56) for the specific case of plane stress:

$$\begin{aligned} -\sigma_0 \leq \sigma_1 - \sigma_2 \leq \sigma_0 \\ -\sigma_0 \leq \sigma_1 \leq \sigma_0 \\ -\sigma_0 \leq \sigma_2 \leq \sigma_0 \end{aligned} \tag{5.58}$$

Within these inequalities $2\tau_0$ is replaced by σ_0 which is the yield limit for an axially loaded sample. This relation results from an elementary stress analysis when an uniaxial load is considered (i.e., $\sigma_1 \neq 0; \sigma_2 = \sigma_3 = 0$). For this case the maximum value of shear stress is equal to $\tau_{max} = \frac{\sigma_0}{2}$.

The first of the double inequalities of Eq. (5.58) results in two inclined lines in Fig. 5.22, requiring σ_2 to remain below the upper one and above the bottom one. The other two double inequalities are simple bounds on the values of σ_1 and σ_2. In principal stress space, the Tresca yield surface is a regular hexagonal prism, inscribed inside a von Mises cylinder (see Fig. 5.23), under the assumption that an uniaxial yield stress σ_0 is the same for both yield conditions.

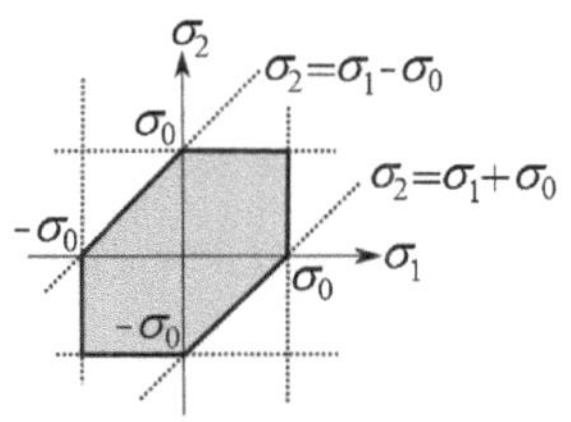

FIGURE 5.22 Tresca yield surface under plane-stress conditions.

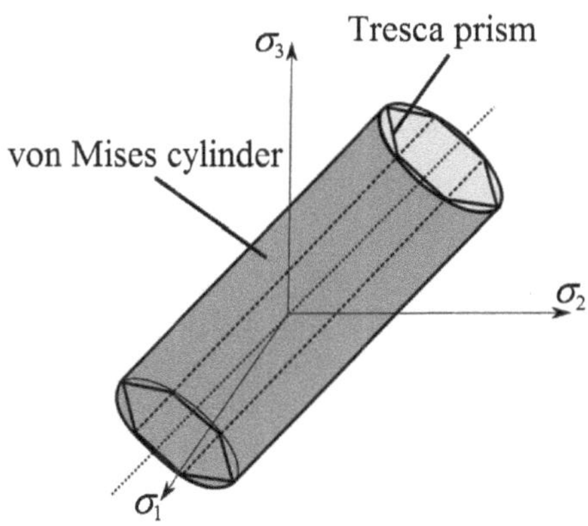

FIGURE 5.23 Tresca yield surface in the principal stress space.

It should be noted that the two yield criteria result in different yield limit τ_0 when loaded by pure shear. For von Mises criterion, it is equal to $\tau_0 = \frac{\sigma_0}{\sqrt{3}}$, while for Tresca it is equal to $\tau_0 = \frac{\sigma_0}{2}$. Alternatively, the same value for τ_0 can be postulated for the two criteria, resulting in different value for corresponding σ_0. In this case, the Tresca yield surface is a hexagonal prism, circumscribed about the von Mises cylinder.

For materials with internal friction, like sand, soil, or rocks, the slip surfaces are rough, so the shear stress needed to activate plastic deformation is influenced by the stress normal to the slip plane. If we imagine that the plastic deformation causes slip of one block over the another (see Fig. 5.24), the force required to provoke the relative movement between the two blocks becomes larger when subjected to a normal force, assuming the contact surface is rough (Fig. 5.24b). A similar effect takes place with materials that exhibit internal friction. These materials are commonly denoted as *pressure sensitive* materials and a yield criterion that aims at describing their response needs to account for the volumetric part of the stress tensor, incorporated through the dependency on the first stress invariant I_1.

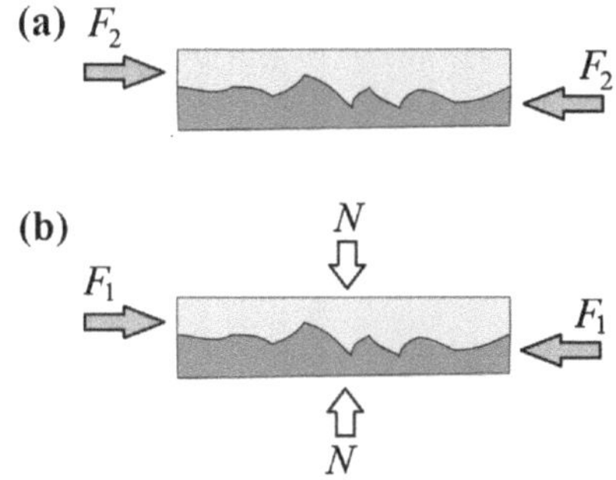

FIGURE 5.24 Rough slip surface and larger tangential force in the presence of normal force.

A popular yield criterion within this family is the Drucker–Prager yield condition. It can be seen as a modified von Mises yield criterion that incorporates the first stress invariant. It is usually formulated as follows:

$$f(I_1, J_2) = \alpha I_1 + \sqrt{J_2} - \tau_0 = 0 \tag{5.59}$$

that includes, in addition to τ_0 (present also in the von Mises yield criterion), another material parameter α that represents the *friction angle*. This formulation highlights the similarities between the Drucker–Prager yield condition and the von Mises one. In particular, the shear stress required for yielding in the Drucker–Prager tends to reduce for increasing values of I_1, i.e., with a drop of the pressure, considering that a positive sign in stress depicts a tensile stress. In the deviatoric planes, the Drucker–Prager yield surface is also a circle but with a variable radius that depends on the position of the plane on the hydrostatic axis. In the principal stress space, it is a circular cone, centered about the hydrostatic axis as depicted in Fig. 5.25.

For constant values of the friction angle, the meridians are straight lines as shown in the $p - q$ plane, while for $\alpha = 0$ the Drucker–Prager is reduced to a von Mises yield criterion as depicted in Fig. 5.26b. For the plane stress condition the intersection of the cone and of the inclined plane forms an ellipse, with the only difference from the von Mises ellipse that it is not symmetric with respect to the origin (see Fig. 5.26a). For this reason, this criterion is sometimes used to evaluate the strength of brittle materials, as they possess a smaller resistance when loaded in tension than when they are subjected to compressive loading.

The two outlined yield conditions represent good examples to demonstrate how different yield criteria are suitable for the modeling of different materials.

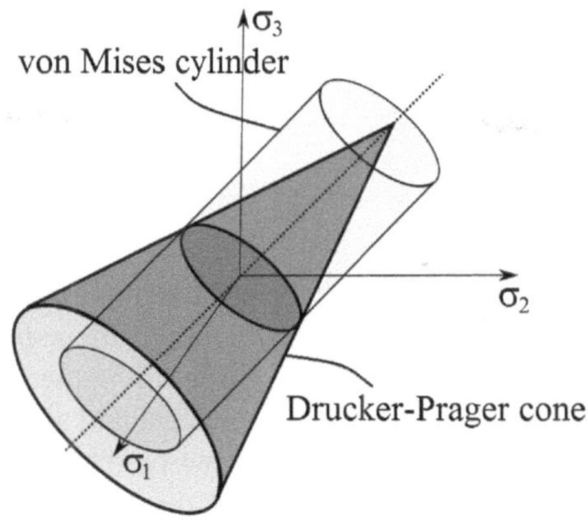

FIGURE 5.25 Comparison between the Drucker–Prager yield surface and the von Mises cylinder.

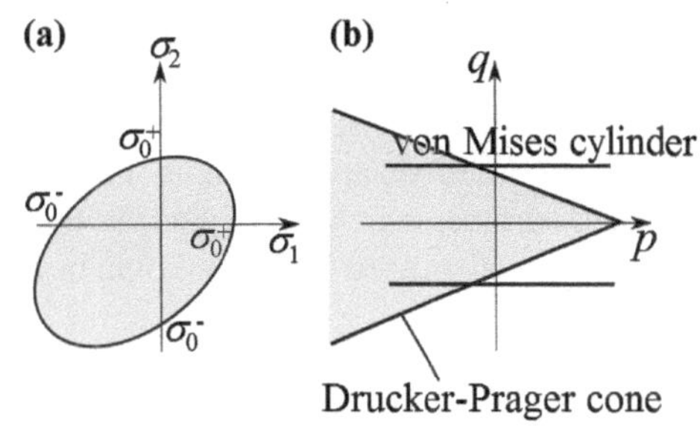

FIGURE 5.26 Drucker–Prager yield surface: (a) in plane-stress condition and (b) in the p-q plane.

There is a wide range of yield criteria developed for specific materials, and the development of novel yield conditions represents a productive branch of mechanics of materials.

The main purpose of a yield criterion is to establish the occurrence of plastic deformation. If a plastic deformation occurs, in order to compute the stress tensor, we need to quantify the plastic part of the strain tensor $\boldsymbol{\varepsilon}^{pl}$ that contains 6 independent unknown components. The yield function provides just one algebraic equation for this purpose and, therefore, additional equations are required. They are provided by the *plasticity flow theory* that defines the evolution of the plastic flow. The complete definition of an elasto-plastic constitutive model requires the yield condition as well as the flow rule to define the direction of the plastic strain.

5.2.3 Flow rule for continuum plasticity

The flow rule for a given material is generally deduced from experiments. However, for most materials it is accustomed to work with flow rules that preserve the validity of the *postulate of maximum plastic dissipation*. This postulate was formulated by Richard von Mises in 1928 and it states that: *for any given plastic strain rate $\dot{\boldsymbol{\varepsilon}}^{pl}$, among all plastically admissible stresses, the actual stress is the one that maximizes the "power"*:

$$D(\dot{\boldsymbol{\varepsilon}}^{pl}) = \boldsymbol{\sigma} : \dot{\boldsymbol{\varepsilon}}^{pl} = \max_{f(\sigma^* \leq 0)} (\boldsymbol{\sigma}^* : \dot{\boldsymbol{\varepsilon}}^{pl}) \tag{5.60}$$

in which the term *power* is placed under the quotation marks to emphasize that here we do not talk in terms of rate as a differential with respect to real time, but rather as the small change in the quantity. By recalling that the double contraction sign of a second-order tensor has the same meaning as the scalar product of vectors formed from symmetric second-order tensor in Voigt notation, Eq. (5.60) can be interpreted as the condition that the projection of the admissible stress state onto the direction of the plastic strain rate is maximized by the actual stress state.

For a convex yield surface, the above requirement has a graphical interpretation: if the vector of increment of plastic strain is brought into the stress space, it would have the direction normal to the yield surface at the point of current stress state (see Fig. 5.27). These conditions are known as the *convexity* and *normality* requirements of a yield surface, which is satisfied for the so-called *standard materials*. This expression was first used by Radenkovic in 1961 [7].

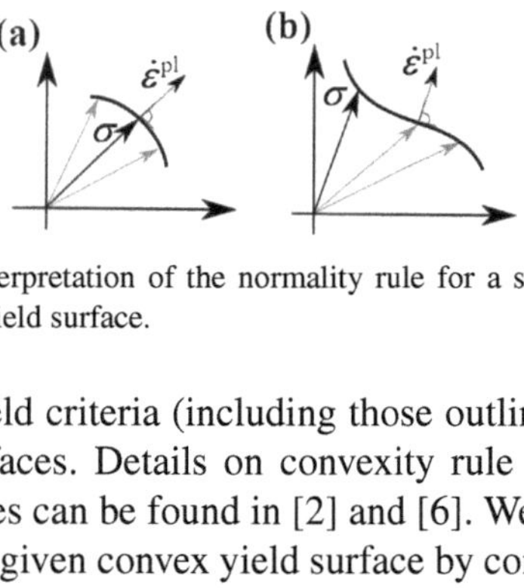

FIGURE 5.27 Graphical interpretation of the normality rule for a segment of: (a) convex yield surface and (b) non-convex yield surface.

Commonly used yield criteria (including those outlined previously) are described by convex surfaces. Details on convexity rule and related proofs for certain specific examples can be found in [2] and [6]. We will here demonstrate the normality rule for a given convex yield surface by considering a simple case of a structural frame element subjected to combined bending and axial load. Within the elasto-plastic analysis of frames, it is common to express the analysis in terms of generalized stress and generalized strain. Under this assumption, the sectional actions, that consist of the bending moment and axial force, can be regarded as generalized stress components. The corresponding generalized strain components are selected in such a way that the product of the generalized stress component and the generalized strain component has the meaning of work. The generalized strain components that correspond to the bending moment and axial force consist of the rotation angle and elongation, respectively.

For this example, we consider an elastic–perfectly-plastic material response (i.e., no hardening) and assume to evaluate the behavior of the beam while loading it gradually in bending with respect to the horizontal axis of symmetry (assuming a rectangular cross-section). From slender beam theory, we can determine the stress distribution to be linear over the cross-section that divides the section into a shortened and a stretched part subjected to compressive and tensile stresses, respectively, as shown in Fig. 5.28a. By increasing the applied moment, the initial linear distribution will continue to grow until the stress in the outermost points of the section (loaded in compression and tension) reach the yield limit σ_0. From this point, the cross-section can resist increasing values of applied moment since there is a large portion of cross-section that is still within the elastic domain. To have plastically admissible stress at every point of the cross-section, the stress distribution needs to account for the maximum yield stress that the material can resist. The distribution will assume the form as shown in Fig. 5.28b. The bending moment can grow up to the moment at which there is (theoretically) no portion of the cross-section in the elastic range and this can be described by the stress distribution shown in Fig. 5.28c. At this stage, we say that a *plastic hinge* has formed, as this section cannot take any further increments in moment.

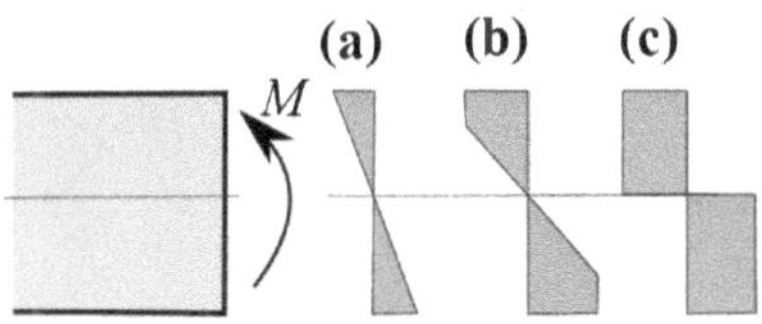

FIGURE 5.28 Frame element subjected to bending—stress distribution of the cross-section under: (a) elastic conditions, (b) elasto-plastic conditions, and (c) completely plastic conditions.

In the case the frame element is loaded only by an axial load, the stress distribution over the cross-section is uniform. By gradually increasing the load the stress at each point of the cross-section grows until the yield limit value σ_0 is reached.

Let us consider now an element that is simultaneously subjected to an axial force and a bending moment. The stress distribution that can occur in this case is illustrated in Fig. 5.29a. The position of neutral axis (i.e., location at which the stress equals 0) depends on the intensity of generalized stress components, i.e., M and N. By increasing the values of both moment and axial force, the stress distribution will change shape until the elastic capacity of the cross-section is exhausted (see Fig. 5.29b and 5.29c). At this point, we say that a *generalized plastic hinge* is formed. A pair of bending moment M and axial force N corresponds to this particular stress distribution (as N and M can be obtained by integration of the stress distribution).

By considering a wide range of different sets of applied N and M it is possible to develop an envelope of all $M - N$ pairs corresponding to the limit cases of generalized plastic hinges in $M - N$ space. The envelope is usually referred to as the *elastic limit envelope*. By recalling that M and N represent the general-

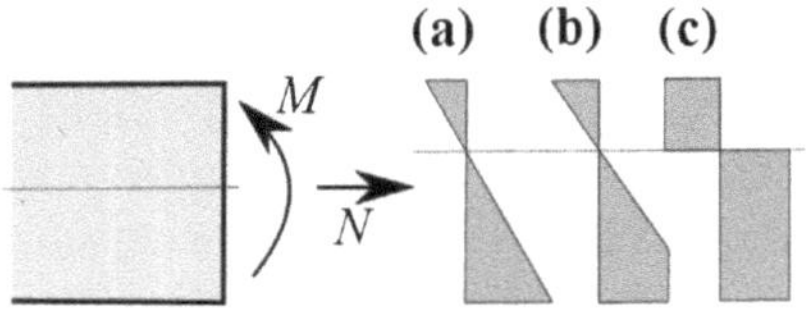

FIGURE 5.29 Frame element subjected to combined bending and axial load—stress distribution of the cross-section under: (a) elastic conditions, (b) elasto-plastic conditions, and (c) completely plastic conditions.

ized stress components of this cross-sectional analysis, the $M - N$ envelope can be treated as a yield surface, since it defines the (generalized) stress state that depicts the plastic condition. The shape of such domain depends on the shape of cross-section and, qualitatively, it can take the form given in Fig. 5.30.

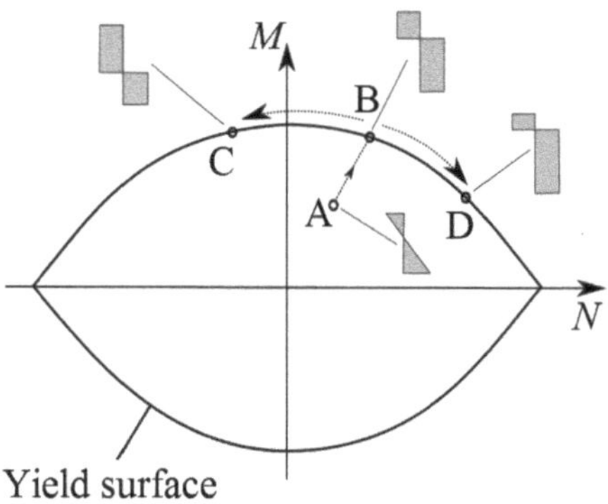

FIGURE 5.30 Envelope of yield surface of the generalized stresses N and M obtained with a cross-sectional analysis.

If the pair of generalized stress components M and N lies inside the yield surface, the deformation can be either elastic or elasto-plastic and still grow until the yield surface is reached as, for example, can happen when moving from point A to point B (see Fig. 5.30). Once the state of stress is on yield surface (e.g., point B), the generalized stress components cannot grow anymore, but they still can change their values in such a way that the point remains on the yield surface, e.g., by moving toward point C or point D.

As expected, a change in the generalized stress components (i.e., M and N) shifts the position of the neutral axis to satisfy equilibrium. The shift can be quantified by assuming that, for a particular cross-section for which a generalized plastic hinge has already formed, the neutral axis moves by dz_n in the positive direction of the z-axis (see Fig. 5.31). This leads to a change in the cross-sectional area that is subjected to compressive and tensile stresses (assumed to occur at the top and bottom parts of the cross-section, respectively) that can be quantified as:

$$\begin{aligned} dA^+ &= -dz_n \cdot b_n \\ dA^- &= dz_n \cdot b_n \end{aligned} \tag{5.61}$$

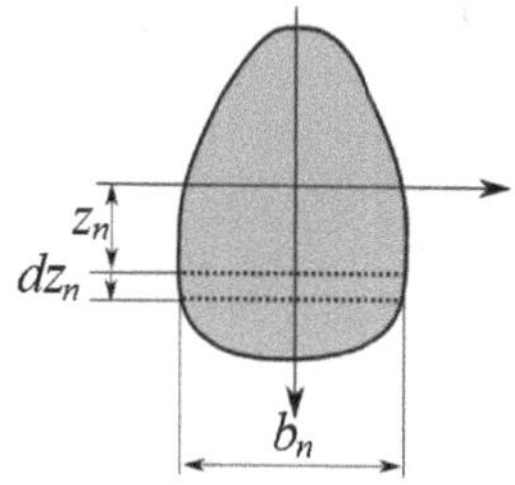

FIGURE 5.31 Neutral axis shift due to change in the values of the (generalized) stress components.

Since it is assumed that a generalized plastic hinge has already formed the stress in all points is equal to σ_0 or $-\sigma_0$ for the tensile and compressive parts of the section, respectively, and the above change leads to the following change in axial force resisted by the cross-section:

$$dN = \sigma_0 \left(dA^+ - dA^-\right) = -2\sigma_0 b_n dz_n \tag{5.62}$$

The movement of the neutral axis changes the first moments of area under tension and compression with respect to the centroidal axis by the amount:

$$\begin{aligned} dS^+ &= z_n dA^+ = -b_n z_n dz_n \\ dS^- &= z_n dA^- = b_n z_n dz_n \end{aligned} \tag{5.63}$$

The associated change in bending moment can be determined as:

$$dM = \sigma_0 \left(z_n dA^+ - z_n dA^-\right) = -2\sigma_0 b_n z_n dz_n = z_n dN \tag{5.64}$$

By dividing Eq. (5.64) by Eq. (5.62), we obtain:

$$\frac{dM}{dN} = z_n \tag{5.65}$$

that implies that the slope of the plastic envelope in the $M - N$ space is equal to the distance of the neutral axis from the centroid.

The generalized strain components consist of the rotation angle and the elongation. Even though the plastic deformation generally takes place also in the neighboring area of a cross-section, it is common to assume (in slender beam theory) that the whole rotation can be lumped at one cross-section (i.e., the one where we consider that the generalized plastic hinge has formed). In this manner, we can write the expression for the plastic elongation at a point defined by the coordinate z at the cross-section as (see Fig. 5.32):

$$\bar{e}^{pl}(z) = e_p + \theta z \tag{5.66}$$

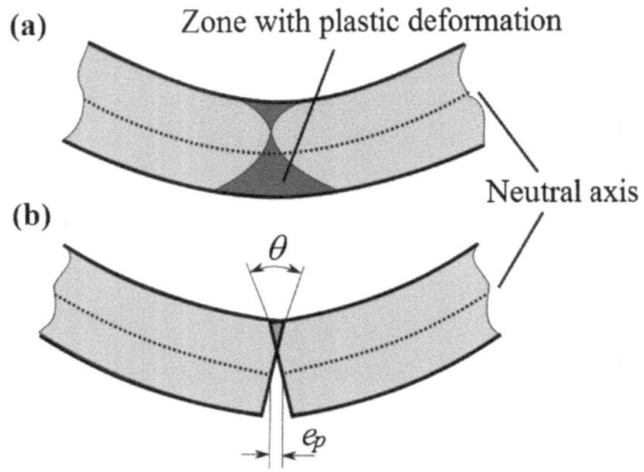

FIGURE 5.32 Formation of a generalized plastic hinge: (a) real plastic zone and (b) lumped model for computation.

The position of the neutral axis is defined as the location where the plastic deformation equals 0:

$$\bar{e}^{pl}(z_n) = e_p + \theta z_n = 0 \tag{5.67}$$

and this can also be written in a rate form (i.e., in terms of small changes in the generalized strain components):

$$\dot{e}_p + \dot{\theta} z_n = 0 \tag{5.68}$$

By combining the last equation with Eq. (5.65) and by considering the change in the generalized stress components M and N in rate form, we can write:

$$\dot{e}_p \dot{N} + \dot{\theta} \dot{M} = 0 \Longrightarrow \dot{\boldsymbol{E}}_p^T \cdot \dot{\boldsymbol{S}} = 0 \tag{5.69}$$

where the generalized strain and generalized stress components are collected into vectors:

$$\boldsymbol{E}_p = \begin{bmatrix} e_p \\ \theta \end{bmatrix}; \quad \boldsymbol{S} = \begin{bmatrix} N \\ M \end{bmatrix} \tag{5.70}$$

Eq. (5.69) represents the orthogonality condition of the plastic strain rate and stress rate vectors, since it states that the scalar product of the two vectors is equal to 0. For a point that lies on the yield surface the direction of stress rate (i.e., stress change) to keep the stress on the yield surface has to be tangential. As a consequence, the plastic strain rate has the direction normal to the yield surface at that point.

Material models for which the direction of plastic flow coincides with the normal to the yield surface belong to the group of *associative plasticity* models. The flow rule provides additional condition required to quantify the plastic part of strain and it is generally written in the form:

$$\dot{\boldsymbol{\varepsilon}}^{pl} = \dot{\lambda}\frac{\partial f}{\partial \boldsymbol{\sigma}} \tag{5.71}$$

The gradient to the yield surface provides only the direction of the plastic strain rate, while its full intensity is determined by an additional scalar multiplier, denoted by $\dot{\lambda}$ in Eq. (5.71), and called the *plastic multiplier*. The condition given by Eq. (5.71) provides 6 algebraic equations that, together with the yield condition, are sufficient to solve for the 6 unknown components of the plastic strain rate plus the additional unknown plastic multiplier λ.

For pressure sensitive materials, experiments have shown that the associative flow rule is unrealistic and that it needs to be replaced by a more general non-associative rule through an additional function:

$$\dot{\boldsymbol{\varepsilon}}^{pl} = \dot{\lambda}\frac{\partial g}{\partial \boldsymbol{\sigma}} \tag{5.72}$$

where function g is called the *plastic potential*.

All the governing equation required to formulate the elasto-plastic constitutive model for the general three-dimensional state of stress are summarized below.

1. For a given strain increment $\Delta\boldsymbol{\varepsilon}$, the trial stress is computed as if all the strain increment is elastic through equation:

$$\boldsymbol{\sigma} = \boldsymbol{D} : \boldsymbol{\varepsilon} \tag{5.73}$$

2. With the updated stress, the yield condition is verified by establishing whether the stress state is plastically admissible: if $f(\boldsymbol{\sigma}) < 0$ the step is elastic and the state of stress can be accepted.
3. If $f(\boldsymbol{\sigma}) > 0$ the state of stress cannot be accepted since it is outside the yield surface. The additive decomposition of the strain is adopted ($\boldsymbol{\varepsilon} = \boldsymbol{\varepsilon}^{pl} + \boldsymbol{\varepsilon}^{el}$) and the plastic strain is introduced to "relax" the stress to get it back to the yield surface.
4. The direction of the plastic strain rate is quantified by the flow rule through the plastic potential function:

$$\dot{\boldsymbol{\varepsilon}}^{pl} = \dot{\lambda}\frac{\partial g}{\partial \boldsymbol{\sigma}} \tag{5.74}$$

For associative plasticity the role of the plastic potential function g is played by the yield function f, while for the non-associative plasticity an additional function is required.

5. The intensity of the plastic strain rate is governed by the plastic multiplier λ to bring the stress back to the yield surface (i.e., to satisfy the equation $f(\boldsymbol{\sigma}) = 0$).

From the above summary, we can highlight that, in order to formulate an elasto-plastic constitutive model, it is required to quantify the elastic response and to adopt a suitable yield condition and accompanying flow rule. The yield condition determines whether the state of stress is elastic (i.e. $f(\boldsymbol{\sigma} < 0)$), in which case there is no plastic deformation and $\dot{\lambda} = 0$. If the plastic deformation occurs the plastic multiplier will assume a value greater than zero ($\dot{\lambda} > 0$) and, in this case, the stress has to remain on the yield surface, so $f(\boldsymbol{\sigma}) = 0$. In order to avoid the inconvenience of having the inequalities within the governing equation, this condition is cumulatively formulated as:

$$f(\boldsymbol{\sigma}) \cdot \dot{\lambda} = 0 \tag{5.75}$$

The proposed solution scheme is valid for the case of perfect plasticity, namely where the accumulated plastic strain is not modifying the yield limit. In the following, we briefly present how to account for the hardening when dealing with the general three-dimensional state of stress.

5.2.4 Introducing hardening for a three-dimensional state of stress

In the one-dimensional example previously considered (Example 5.2), we introduced hardening by postulating the yield limit not to be constant but to be a function of an internal variable. In the example the internal variable was represented by an accumulated plastic strain monitored during the entire load history. The generalization of such concept to a three-dimensional state of stress requires to formulate the dependency of the yield function on the internal variables. Graphically, these changes result in a yield surface that is changing shape, size, orientation, and/or position within the stress space as the material conditions evolve. The complexity of the hardening model influences the number of specific changes (e.g., changes in size or position) that is incorporated in the model.

The basic principles of formulating hardening in a three-dimensional continuum model is here presented for the von Mises yield condition. In the one-dimensional model representation the elastic domain is bounded by a single scalar σ_0, while the domain expands equally in compression and tension with plastic flow. Applying the von Mises yield condition to a three-dimensional stress state the elastic domain is bounded by a cylinder, while the generalization of the previous hardening concept is formulated through a cylinder with increasing radius due to hardening (see Fig. 5.33). The modification is simply introduced by expressing the von Mises yield function so that σ_0 is a function of the selected internal variable. By writing the equivalent von Mises stress q as a function of the principal stresses the yield criterion with dependency on the internal variable ξ assumes the following form:

$$f(\boldsymbol{\sigma}, \xi) = q - \sigma_0(\xi) = \frac{1}{\sqrt{2}}\sqrt{\left[(\sigma_1 - \sigma_2)^2 + (\sigma_2 - \sigma_3)^2 + (\sigma_1 - \sigma_3)^2\right]} - \sigma_0(\xi) \tag{5.76}$$

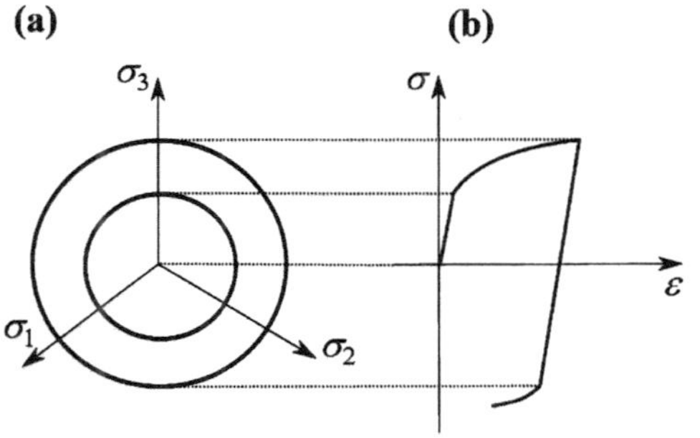

FIGURE 5.33 Isotropic hardening of the von Mises yield surface presented: (a) in π-plane and (b) in uniaxial stress-strain space.

This introduced change affects only the size of the yield surface, by modifying the cylinder radius, without any translation or change in the shape. Such hardening is usually denoted as *isotropic hardening*. For the constitutive description of the isotropic hardening, a single scalar-value internal variable ξ is usually adopted. The internal variable has obviously to be frame invariant and, for this purpose, there are two popular approaches: within a *strain hardening* approach the variable is selected to depend on the invariants of the plastic strain tensor; within a *work hardening* approach it is represented based on the dissipated plastic work.

Commonly adopted plastic strain invariant is the so-called von Mises *effective plastic strain* or *equivalent plastic strain*. It is a scalar-value invariant that represents the work conjugate to the equivalent von Mises stress. Written as a function of stress tensor components, the equivalent von Mises stress can be expressed as:

$$q = \sqrt{3J_2} = \sqrt{\frac{1}{2}\left[(\sigma_{xx} - \sigma_{yy})^2 + (\sigma_{yy} - \sigma_{zz})^2 + (\sigma_{zz} - \sigma_{xx})^2\right] + 3(\sigma_{yz}^2 + \sigma_{zx}^2 + \sigma_{xy}^2)} \tag{5.77}$$

In terms of the deviatoric stress components (that corresponds to the part of stress responsible for the shape change and, therefore, directly related to plastic deformation) the equivalent von Mises stress takes the following form[4]:

$$\sqrt{3J_2} = \sqrt{\frac{3}{2}\boldsymbol{S} : \boldsymbol{S}} \tag{5.78}$$

where $\boldsymbol{S}$ is the deviatoric part of the stress tensor.

From the last expression, it directly stems that the work conjugate of the equivalent von Mises stress is:

$$\varepsilon^{ep} = \sqrt{\frac{2}{3}\boldsymbol{\varepsilon}^{pl} : \boldsymbol{\varepsilon}^{pl}} \tag{5.79}$$

The equivalent plastic strain given by Eq. (5.79) is used in the same way as previously discussed for the one-dimensional case when the accumulated plastic strain was used. It therefore represents the internal variable ξ that modifies the yield surface.

The starting point in the definition of the suitable hardening model is the experimental stress-strain curve of the material. From this curve an idealized hardening curve is constructed usually in linear or exponential form (i.e., linear or exponential dependency of yield stress on plastic strain; see Fig. 5.34). Based on a general stress analysis, the equivalent plastic strain is computed by means of Eq. (5.79). Through the hardening curve, this calculated value is related to the current yield limit $\sigma_0(\xi)$ that is further used to model the isotropic hardening within the yield condition given by Eq. (5.76).

When dealing with work hardening models, the internal variable ξ governing the isotropic growth of the yield surface depicts the work dissipated during the plastic deformation. Within an uniaxial test, this work is represented by the shaded area under the diagram that, together with the elastic part of the work, constitutes the full work done during the deformation of the body (see Fig. 5.35). The part of the energy that is dissipated is defined by the following integral:

$$\xi = W^{pl} = \int_0^t \boldsymbol{\sigma} : \dot{\boldsymbol{\varepsilon}}^{pl} dt \tag{5.80}$$

where $\dot{\boldsymbol{\varepsilon}}^{pl}$ represents the rate of the plastic strain, expressed in terms of a strain change rather than a derivative with respect to the real time, thus the obtained scalar quantity has the units of work (not power). The integrand function is written in the form where both stress and plastic strain are expressed as second-order tensors. In Voigt notation (i.e., symmetric tensors are represented by six-component vectors), the integrand function is replaced by a scalar product of the two vectors.

[4] Since $\boldsymbol{S} : \boldsymbol{S} = S_{11}S_{11} + S_{22}S_{22} + S_{33}S_{33} + S_{12}S_{12} + S_{13}S_{13} + S_{21}S_{21} + S_{23}S_{23} + S_{31}S_{31} + S_{32}S_{32}$, by further expressing the deviatoric stress components S_{ij} in terms of the stress components σ_{ij} the form given in Eq. (5.77) is obtained.

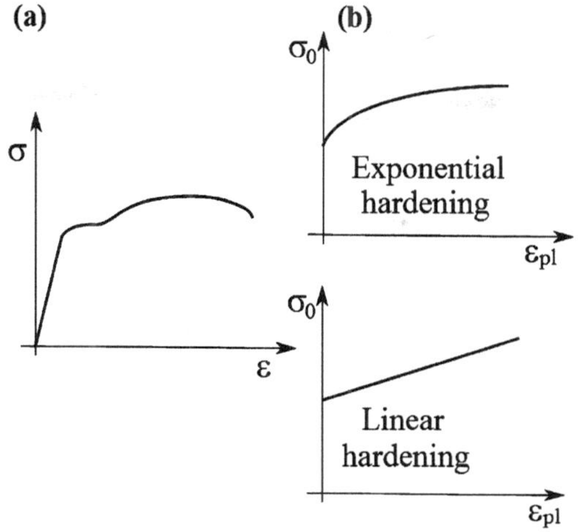

FIGURE 5.34 (a) Typical experimental stress-strain curve and (b) modeling exponential and linear hardening curves.

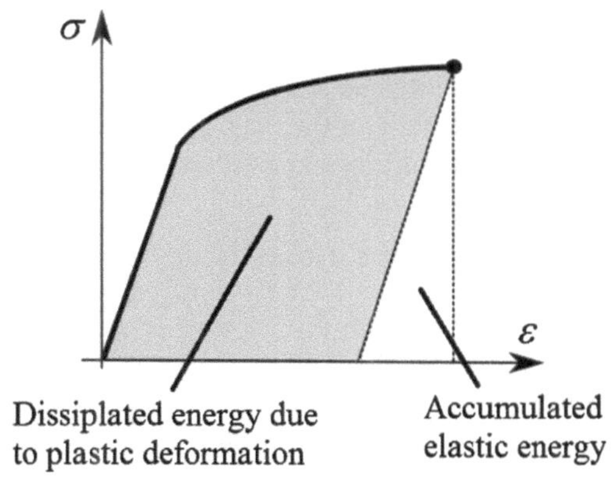

FIGURE 5.35 Qualitative representation of the accumulated elastic energy and dissipated plastic work during an elasto-plastic deformation.

The hardening curve here should provide the dependency of the yield stress (i.e., von Mises cylinder radius) on the dissipated plastic work. The remaining part of the implementation is the same as for strain hardening. Both strategies can be equally used to model the same material response. For the proof of the equivalence between the two approaches, reference should be made to [8].

There are materials for which it has been experimentally observed that, after being loaded and hardened in one direction, they experience a decay of elastic limit when loaded in the opposite direction. The phenomenon is known as the *Bauschinger effect* and can be modeled by a movable yield surface. Such hardening is usually denoted as *kinematic hardening* (see Fig. 5.36). It represents a different type of hardening with respect to the isotropic hardening and it requires an additional internal variable to govern the translation of the yield surface.

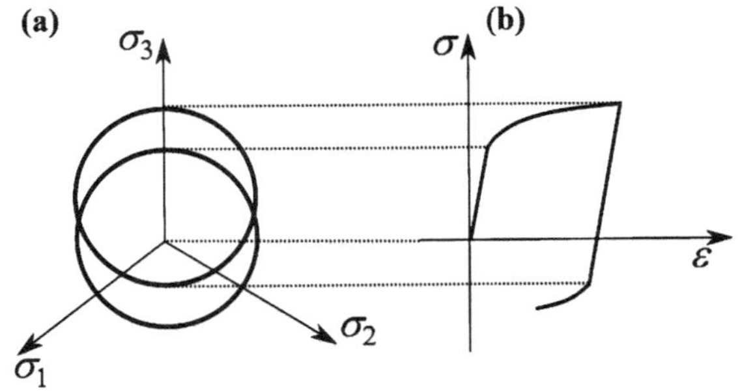

FIGURE 5.36 Kinematic hardening for the von Mises yield surface: (a) in π-plane and (b) in stress–strain space.

Real life materials often experience both isotropic and kinematic hardening. Because of this, more realistic plasticity models are obtained if a mixed isotropic/kinematic hardening is used. This clearly enhances the complexity of the model and makes the material characterization more demanding. It is therefore reasonable to adopt simpler models if they are working reasonably well for the problem considered. In general, if the material is to be modeled for a cyclic load application, the inclusion of kinematic hardening may become essential.

5.3 Constitutive models for inelastic deformation due to material damage

Classical theory of plasticity does not modify the elastic response of the material. This is consistent with the physics of the processes involved in what macroscopically is observed as plastic deformation, mainly due to the plastic slip of one block of atoms over another one that occurs without any destruction of interatomic bonds. Since stretching and compressing of these bonds governs the elastic response of a material, it is clear that the plastic deformation cannot alter the elastic properties.

A material subjected to certain levels of load can exhibit the occurrence of cracks and if the load is taken to ultimate conditions, it can lead to failure due to fracture. Brittle materials, like glass or ceramics, upon exhausting the elastic capacity commonly form one major crack that propagates fast through the material and results in the separation of the body into two or more parts. The inelastic deformation within brittle materials prior to fracture is completely absent, or rather small. On the other hand, for certain materials, usually denoted as *quasi-brittle*, instead of the formation of one or few large cracks, the fracture failure is preceded by an elevated number of microcracks, or generally irreversible changes at the microstructural level, spread over relatively wide zones of the material. Such mechanism provides a certain level of ductility within what is macroscopically observed to be an inelastic deformation. Within this inelastic response, the stress generally increases beyond the level at the elastic limit and a certain amount of energy is dissipated during the loading process. Due to changes in the microstructure, this deformation alters the elastic response of a continuum body. Such behavior therefore cannot be modeled by a plasticity constitutive model. The branch of continuum mechanics that deals with the modeling of this behavior is called *damage mechanics*.

Within the framework of damage mechanics, constitutive models are formulated in such a way that the effects of microstructural changes are incorporated within the continuum model by means of internal variables. These internal variables, in general, can be either scalar or tensor-valued. The concept of a damage model can be illustrated by considering a system composed of N parallel bars (see Fig. 5.37). Each bar is assume to behave in an elastic–perfectly-brittle manner so that, upon exhausting its elastic capacity, it immediately breaks (see Fig. 5.37b). All the bars have the same modulus of elasticity E, but different cross-sectional area A_i (despite the different cross-sectional areas, it is here assumed to maintain the example as simple as possible that all bars elongate or shorten by the same amount). In order to investigate the elastic response of the system, we can consider a linear relationship between the applied force F and the consequent displacement u:

$$F = k \cdot u \tag{5.81}$$

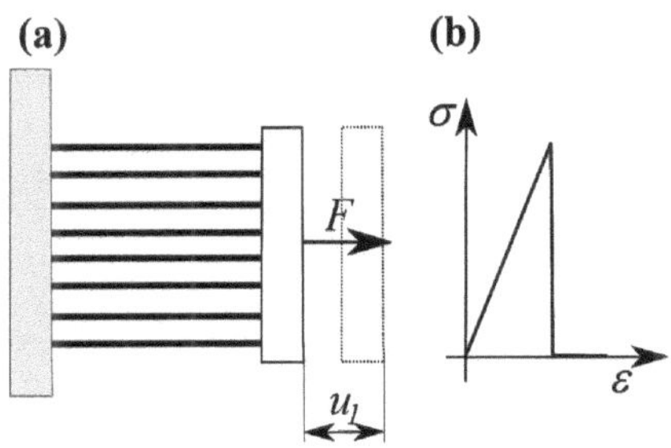

FIGURE 5.37 (a) System of N bars and (b) elastic-perfectly brittle behavior.

From equilibrium we can then write:

$$F = \sum_{1-i}^{N} F_i \tag{5.82}$$

where F_i is the force in the ith bar. Relying on Hooke's law for a single bar we can write:

$$\sigma = E \cdot \varepsilon \Rightarrow \frac{F_i}{A_i} = E\frac{u}{l} \Rightarrow F_i = \frac{EA_i}{l}u \tag{5.83}$$

By inserting the last equation into Eq. (5.82), we get:

$$F = \frac{EA_0}{l} \cdot u \tag{5.84}$$

where $A_0 = \sum_{1-i}^{N} A_i$ depicts the sum of the cross-sectional areas of all N bars.

If we compare Eq. (5.84) with Eq. (5.81), we can calculate the coefficient k depicting the linear elastic response of the whole system as:

$$k = \frac{EA_0}{l} \tag{5.85}$$

In the following, we consider constitutive modeling in the context of continuum damage mechanics. In view of that approach the response of the system is represented by the load F that is resisted by the entire system for a particular level of displacement. By averaging this load over the area of all the bars, we obtain an averaged property that can be observed macroscopically:

$$F = \frac{EA_0}{l} \cdot u \Rightarrow \sigma = \frac{F}{A_0} = E \cdot \varepsilon \tag{5.86}$$

and the continuum damage model can account for this stress quantity when formulating its constitutive model. For example, let us assume that, for a certain displacement u_1, M bars are broken. By following the same reasoning we can write that, in this case, the coefficient that governs the linear-elastic response becomes:

$$k_d = \frac{E\left(A_0 - \sum_1^M A_i\right)}{l} = \frac{E\left(A_0 - A_b\right)}{l} \tag{5.87}$$

which is clearly smaller than the value given by Eq. (5.85) since the area considered here accounts for the area of all broken bars $A_b = \sum_1^M A_i$ (with A_0 being the sum of the areas of all bars).

The macroscopically observable stress as a global property can still be expressed with respect to the initial, intact area A_0:

$$\sigma = \frac{F}{A_0} = \frac{E\left(A_0 - A_b\right)}{A_0} \cdot \varepsilon = E\left(1 - \frac{A_b}{A_0}\right)\varepsilon \tag{5.88}$$

where the ratio $\frac{A_b}{A_0}$ can assume only discrete values that depend on the number of bars. In reality, only the unbroken bars are capable of transmitting load and, therefore, we can introduce an *effective stress* to account for this:

$$\sigma_{ef} = \frac{F}{\left(A_0 - A_b\right)} = \frac{F}{A_0} \cdot \frac{1}{\left(1 - \frac{A_b}{A_0}\right)} = \frac{\sigma}{\left(1 - \frac{A_b}{A_0}\right)} \tag{5.89}$$

By making the transition to a continuous model (i.e., the number of bars tends to infinity while their cross-sectional area tends to zero), the ratio $\frac{A_b}{A_0}$ is replaced by the continuous variable ω within the range [0, 1]:

$$\sigma = E(1 - \omega)\varepsilon \tag{5.90}$$

where ω is denoted as the *damage variable* and it represents the internal variable of the model that quantifies the state of the system (i.e., it quantifies how much of the system is damaged). The value of the damage variable monotonically increases with loading from 0 (i.e., intact state) to 1 (fully broken state), or remains constant within unloading. Eq. (5.90) constitutes a simple one-dimensional nonlinear damage model.

Similarly, the relationship between macroscopically observable stress and effective stress for a continuous model is given by:

$$\sigma = (1 - \omega)\sigma_{ef} \tag{5.91}$$

The formulation of the damage constitutive model involves a nonlinear relation between the total strain and the corresponding stress value. To provide the solution strategy, it is required to express it in rate form (the expression adequate for computing a stress increment for a given strain increment). The following is obtained by differentiating Eq. (5.90):

$$\dot{\sigma} = E(1 - \omega)\dot{\varepsilon} - \frac{d\omega}{d\varepsilon}\dot{\varepsilon}E\varepsilon = E(1 - \omega - \frac{d\omega}{d\varepsilon}\varepsilon)\dot{\varepsilon} \tag{5.92}$$

where ω is treated as a function of time through deformation ε. The solution of the problem expressed in the rate form of the damage model constitutive equation requires the knowledge of the evolution of the internal damage variable ω with the deformation ε. Based on the introduction remarks and simple problem of Fig. 5.37, it can be noted that both the initiation of damage and its evolution depend on the displacement (or deformation).

We now consider a simple numerical example that deals with a one-dimensional continuum damage model for which the initiation criterion is specified as the value of the stress at which damage starts to occur, while the maximum deformation ε_m at which the material is completely damaged is required to describe the damage evolution. It is also assumed that the damage variable ω changes as a linear function of deformation ε. The following MATLAB® code listing provides the numerical implementation of the model for this example.

```
% Implementation of 1-dimensional continuum damage model
% ------ Multiple steps -----------
clear
E=10000; Sig_0=100; % Young's modulus and stress at which damage starts
eps_max=0.05; % Max total strain
epsilonD0=Sig_0/E; % Strain at which damage starts
epsilonD=epsilonD0; % Deformation for damaging
```

```
8  steps=[0,0.03,0.02,0.04,0.01]; % The first one must be zero
9  NoI=20; % Number of increments
10 NoS=size(steps,2); epsilon=0;
11 ss=[0,0]; NMB=1; D=0;
12 for i=1:NoS-1 % Cycle over steps
13     eps_inc=(steps(i+1)-steps(i))/NoI;
14     for j=1:NoI;
15         NMB=NMB+1;
16         if eps_inc>0 % Loading step
17             epsilon=epsilon+eps_inc;
18             if epsilon > epsilonD
19                 D=(epsilon-epsilonD0)/(eps_max-epsilonD0);
20                 if D>1
21                     D=1;
22                 end
23                 epsilonD=epsilon;
24             end
25             stress=E*(1-D)*epsilon;
26         else % Unloading step
27             epsilon=epsilon+eps_inc;
28             stress=E*(1-D)*epsilon;
29         end
30         ss(NMB,:)=[epsilon,stress];
31     end
32 end
```

This implementation is formulated in terms of absolute values (not incremental) of stress and strain and, as such, it is not the most appropriate for a general strategy to the solution of nonlinear problems. Despite this, in this simplified form it can serve the purpose to demonstrate some of the main features of a damage model. Another remark regarding the implementation relates to the fact that it implements the damage *a posteriori* where, on the basis of current stress, the damage variable is modified for the subsequent step increment. This approach can lead to a solution dependent on the size of the stress increments within the calculation. Such formulation of a damage model, without the iterative convergence for the equilibrium state, like the one presented for the plasticity models, is clearly numerically easier to implement. The solution strategy may require small stress increments to provide a reliable response. Due to this limitation, there is a wide range of damage models available in commercial FEM codes that can be specified only with explicit dynamic analyses, where the time steps are generally small.

Within the present listing, the stress is computed for a hypothetical multiple step analysis driven by different values of strain (quantified in line 8 of the code). It consists of an initial loading up to the strain value of 0.03, then a partial unloading to a strain of 0.02, with a subsequent reloading to 0.04, and a final unloading up to a strain value of 0.01. The overall stress–strain response is shown in Fig. 5.38.

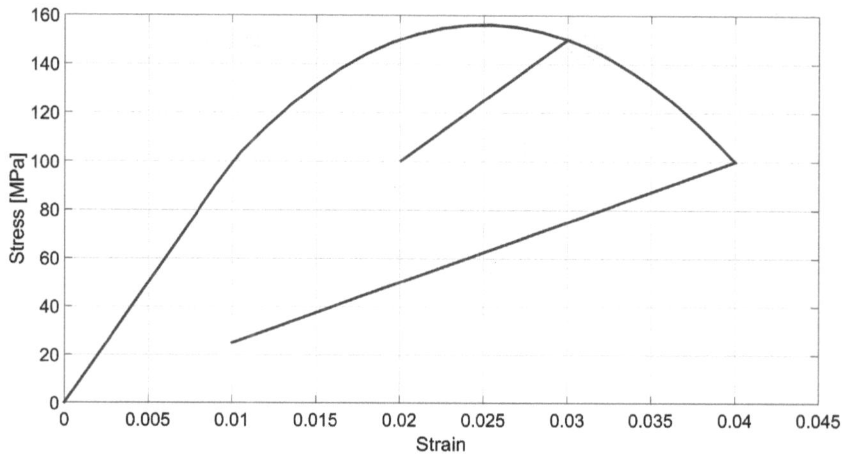

FIGURE 5.38 Stress–strain curve for a simple 1-D damage model with loading and unloading branches.

As it can be noted from the calculated response the inclination of the two partial unloadings changes with respect to the initial linear response corresponding to the undamaged material. Subsequent load steps follow the same line with a reduced inclination since the damage variable reduces the elastic properties of the material. Further damage can occur only if the strain reaches the maximum strain value measured in the previous load history.

This simple example highlights that there are two essential parts in the formulation of damage model: a damage initiation criterion and a damage evolution criterion. For the purpose of the previous example the damage initiation criterion was set as the value of the stress at which damage starts to develop. Considering that the damage model in the form given by Eq. (5.90) relies on the total strain, upon reaching the specified stress threshold the damage variable starts to increase, leading to a nonlinear stress–strain relationship. At this point the stress will still continue to grow for increasing values of strain, even if at a decreasing rate. Once the material reaches its peak stress (at some moderate value of the damage variable ω) the stress will start to decrease and follow its softening range. This behavior results from a specific formulation of the model, but it leads to the response in which the evolution of damage variable governs the peak stress value.

By using the above implementation, it can be shown that, while keeping the damage initiation stress at the same level, different responses are obtained in terms of peak stress for different values of the strain at complete material degradation (i.e., the damage evolution governing parameter). Fig. 5.39 illustrates the stress-strain curves for three different cases with $\varepsilon_{ul} = 0.04$, $\varepsilon_{ul} = 0.05$, and $\varepsilon_{ul} = 0.06$. In all the three cases, the stress-strain curves depart from the linear part at the same point while they differ for the remaining of the curves. Such a behavior is quite obvious considering that the model is formulated in terms of the total strain and, therefore, the value of the stress for a specific strain depends on the current value of the damage variable. This may be inconvenient from a modeling viewpoint because the peak stress of a material represents an important property for engineering applications. It is therefore desirable to have such value as a direct parameter within the model rather than to determine it as a quantity evaluated based on other constitutive parameters. A partial remedy to this aspect can be achieved if the damage evolution law is modeled by means of a bilinear or multiple-linear segments (see Fig. 5.40) based on a tabular formulation of the damage evolution. In such a way, it is possible, for example, to bring the peak stress closer to the damage initiation value that could be followed by a first rapid degradation while preserving the required energy dissipation for subsequent degradation at higher deformation levels.

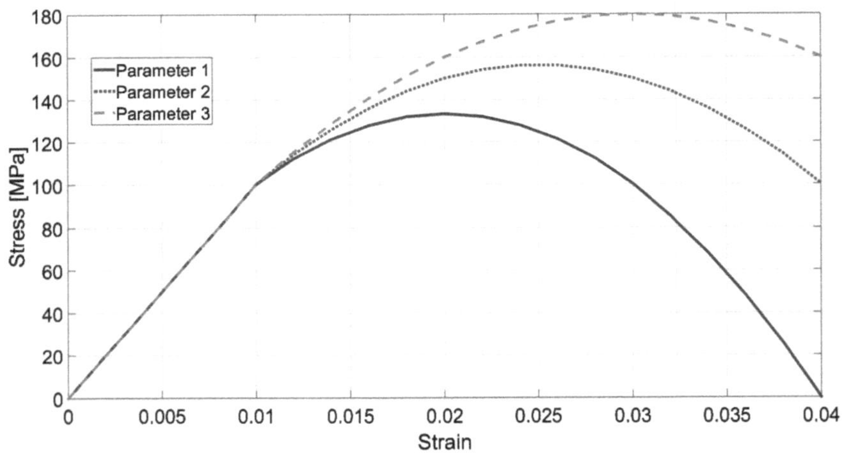

FIGURE 5.39 Stress-strain curves for a simple 1-D damage model with three different damage evolution parameters.

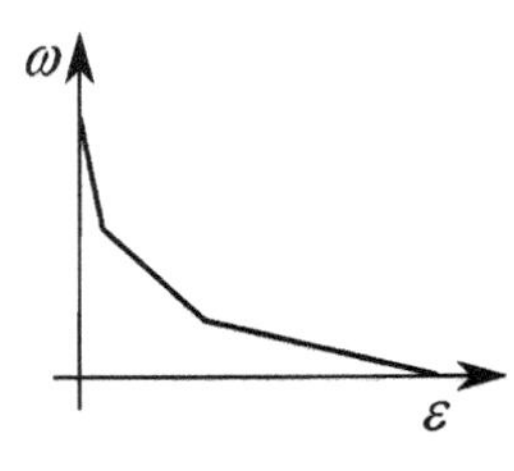

FIGURE 5.40 Damage evolution defined by multiple linear segments.

An alternative way of formulating damage evolution is by specifying the softening part through a direct dependency of stress on displacement instead of on strain. The approach is frequently used for crack modeling within the so-called *traction-separation law*, that is outlined later in the chapter. When used for the damage evolution definition, this strategy offers a direct way of controlling the peak stress value. The use of the displacement does also provide certain advantages over the use of strain when dealing with softening formulations.

The second aspect highlighted with this simple model is related to the formulation of the damage evolution that is expressed as a function of strain. It is generally common in continuum mechanics to express constitutive models in terms of stress–strain relations. However, when the material exhibits strain-softening behaviors, a formulation expressed in terms of strain measures leads to an ill-posed problem without guaranteeing the uniqueness of the solution. This ill-posedness (in a numerical sense) leads to a strain localization and a strong mesh dependency. In the following, we introduce a couple of examples to better understand the nature of the problem.

Let us consider a straight bar of length L with a constant cross-sectional area subjected to the uniaxial tensile load. The material is assumed to behave linear-elastically up to the peak stress σ_m, followed by a linear softening to the strain value ε_f at which the stress transmitted by the bar is reduced to zero (see Fig. 5.41). Let us now assume that the bar is loaded in tension by applying a displacement at its free end equal to u and that the response remains linear-elastic up to a tip displacement of $u_0 = L\varepsilon_0$. At this point, the force transmitted by the bar reaches its maximum value. The material model specifies that, if the strain increases beyond this point, the material enters into a softening regime that leads to a gradually decreasing force that it can transmit. By complying with the constitutive description, at each material point the stress can decrease either at an increased strain (following the softening branch) or at a decreased strain (following the elastic unloading branch). The equation of equilibrium implies that the stress should be uniform along the bar. The strain, however, can assume either of the two values corresponding to the same stress value. Therefore, any piecewise constant distribution of strain that jumps between the two strain values represents a valid solution as it satisfies simultaneously equilibrium and the constitutive model (see Fig. 5.42).

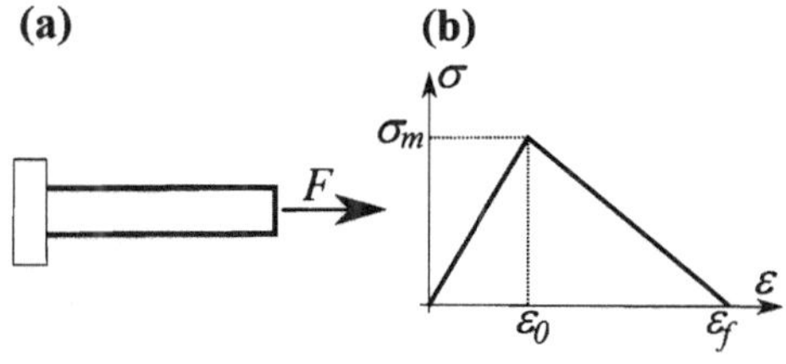

FIGURE 5.41 Axially loaded bar made of a material with a stress-softening behavior.

Once the stress in the bar goes back to zero, the strain in the bar segments that exhibited unloading is equal to zero, while, in segments that experienced softening, it is equal to $\varepsilon = \varepsilon_f$. If we denote by L_u the overall length of the segments that undergo unloading and by L_s the part that softened, we may write

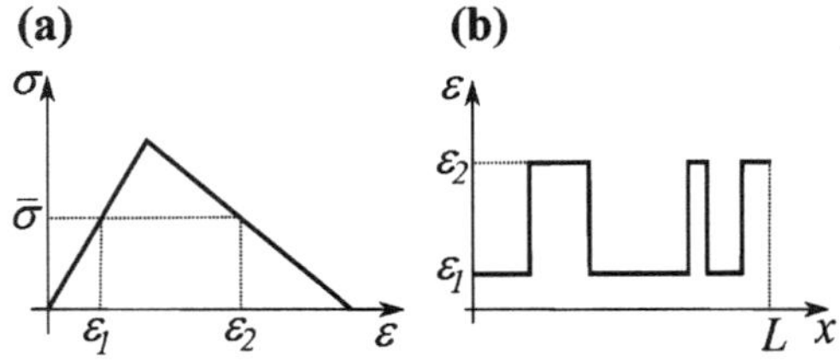

FIGURE 5.42 (a) Two strain values corresponding to the same stress value and (b) distribution of strain over the bar that satisfies both the equilibrium and constitutive equations.

the following expression for the total elongation of the bar:

$$u_b = L_s \varepsilon_f + L_u \varepsilon_u \tag{5.93}$$

There is an infinite number of combinations of the two lengths L_s and L_u that can satisfy both the equilibrium conditions and the constitutive model, and hence represents a valid solution. This means that, for a given softening behavior of material, the elongation of the bar for considered loading case remains indeterminate. Since any piecewise constant strain distribution of the type of Fig. 5.42 represents the solution in terms of force transmitted by the bar vs. elongation of the bar, the post-peak behavior can be represented by a straight line that falls between the line corresponding to the unloading branch and the one corresponding to the case of complete softening (see Fig. 5.43).

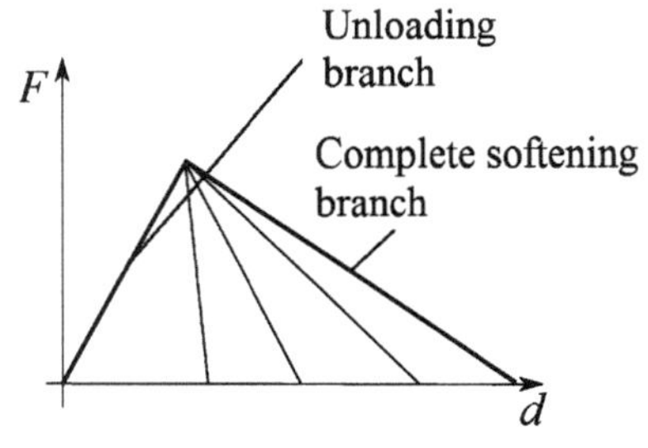

FIGURE 5.43 Possible solution for the load-displacement curve for the bar with strain-softening behavior.

From a numerical viewpoint, the discussed ill-posed nature of the problem is translated into a strong mesh dependency. The difficulty can be illustrate if we assume that the axially loaded bar is modeled by continuum plane-stress finite elements. Since the state of stress is uniform over the bar for the elastic solution, no mesh refinement is required and the exact solution for the linear elastic part of the response is obtained even with a single element. When the problem is modeled by multiple finite elements, the strain should nominally be the same but, due to round-off errors, it can be somewhat larger in one element, sufficient to precede the entering into the softening branch before the neighboring elements. When this happens, the damage model is activated in that particular element resulting in a reduced elastic constant (i.e., due to increased damage variable) for the subsequent increment. This however leads to the relaxation of the neighboring elements that start to follow the unloading branch. Due to the evolution of the damage variable in one element, the deformation produced by following load increments will concentrate in that element only, until failure. The softening branch is thus localized within usually one element, and the phenomenon is known in the literature as *strain localization*.

A consequence of strain localization is the severe mesh-dependency of the global solution. The smaller the element, the faster it approaches to the failure strain when subjected to the same displacement. Because of this, the mesh refinement leads to a steeper post-peak response observed on the load-displacement curve and the energy dissipated decreases with the mesh refinement (see Fig. 5.44).

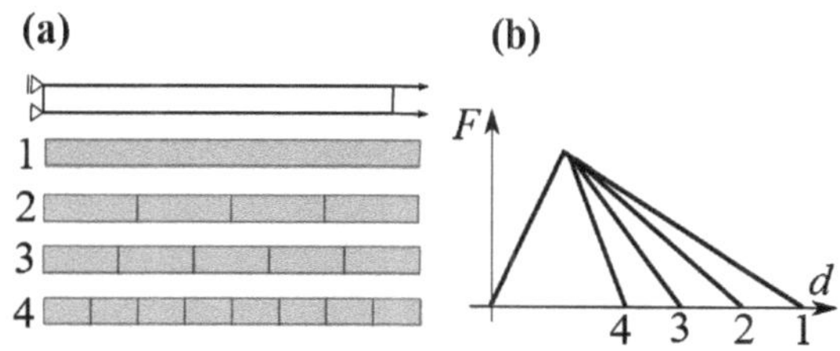

FIGURE 5.44 (a) Bar loaded in pure tension with four different mesh discretization and (b) resulting force-displacement curve corresponding to the four different mesh discretization.

Damage constitutive models that suffer the above outlined problem are denoted as *local damage models*. A possible remedy to this feature is to use, when evaluating the growth of the damage variable, a spatial averaged quantity instead of a state variable (here taken as the strain value) at a single integration point. This approach leads to the formulation of *non-local damage models* [9].

The previously discussed examples have evidenced mesh dependency when dealing with strain-softening materials. This dependency is generally related to the lack of uniqueness of the solution if the material model is formulated in terms of a stress–strain relation. To alleviate this undesired effect, softening models are commonly formulated in terms of stress-displacement, while, for the general integration within finite element context, the displacement is related to strain through a characteristic length that is either specified as a parameter of constitutive model or it is automatically computed by the software on the basis of the element sizes.

When a stress-displacement formulation of the softening part is adopted, the integral of the softening curve represents the dissipated energy per unit area (see Fig. 5.45). The softening behavior is commonly specified through this specific energy as a material parameter that, in fracture mechanics, is denoted as the *energy release rate* as it represents the energy required to create the unitary surface area through a fracturing process. With this formulation the softening part within the damage model can be defined through the stress σ_0 at which the damage starts, and the energy release rate G_f. In the formulation of the final part of the material degradation curve (i.e., the softening part of the stress-displacement curve), it is important to specify the way in which the full degradation can be achieved. The latter task is commonly achieved with the use of either a linear degradation or an exponential degradation (see Fig. 5.45).

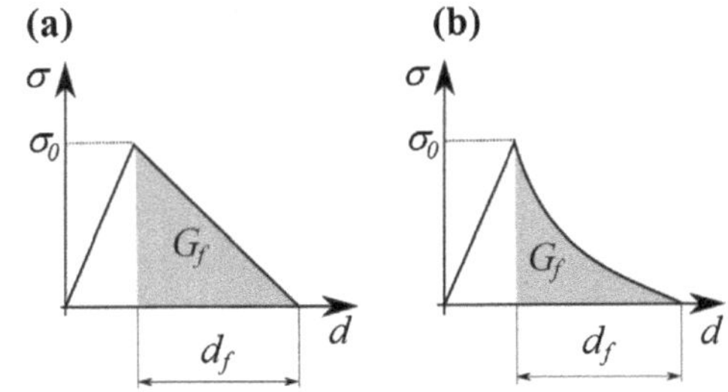

FIGURE 5.45 Stress-displacement softening (a) in linear form and (b) in exponential form.

The stress-displacement formulation of the softening material behavior previously provided can be implemented in such a way so that the stress value of the damage model that triggers the onset of damage is, at the same time, also its peak stress. It is further demonstrated that the evolution of the damage variable has to be postulated to guarantee that the stress follows a prescribed softening part.

In order to integrate the formulated softening rule given in the stress-displacement relationship within the continuum damage model based on the layout of Eq. (5.90) the displacement specified in the softening branch has to be converted into strain. For this purpose, a characteristic length L_c is defined. Since the energy release rate G_f is specified as a material parameter governing the softening part, we first need to compute the post-elastic displacement at failure. For example, by considering a linear softening this can be computed as:

$$u_f = \frac{2G_f}{\sigma_0} \tag{5.94}$$

The damage evolution is determined so that the stress follows the softening part of the material behavior. For the linear softening the relationship between the stress σ_s and the inelastic displacement d_s is given by (see Fig. 5.46):

$$\sigma_s = \sigma_0 \frac{d_f - d_s}{d_f} \tag{5.95}$$

Considering that, in the damage model, the stress is related to the total strain through Eq. (5.90), and recalling Eqs. (5.94) and (5.95), the evolution of the

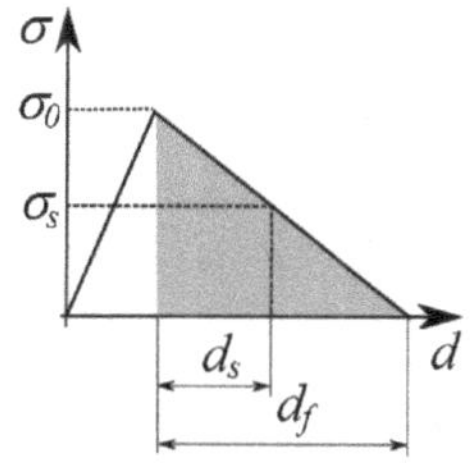

FIGURE 5.46 Stress degradation within linear softening.

damage variable ω is governed by the following relation:

$$\omega(\xi) = 1 - \frac{\sigma_s}{E\left(\frac{d_s}{L} + \varepsilon_{el}\right)} \tag{5.96}$$

where ε_{el} is the strain value corresponding to the stress at the damage initiation (i.e., $\varepsilon_{el} = \frac{\sigma_0}{E}$). The internal variable ξ here represents the maximum post-peak displacement exhibited by the previous load history:

$$\xi(t) = \max d_s(t) \tag{5.97}$$

An example of the implementation of this damage constitutive model is given in the MATLAB® listing below.

```
% Traction-separation damage evolution
clear
Gf=50; Sig0=100; % Setting material constitutive parameters
E=10000; L=1;
d_el=Sig0/E*L; % Displacement at the onset of damage
d_f=2*Gf/Sig0; % For linear degradation
steps=[0,0.02,0.01,0.04,0.01]; % Strain value for different steps
NoS=size(steps,2);
NInc=100; % Number of increments
omega=0; % Setting damage variable
nmb=1; epsilon=0;
ksi=Sig0/E; % Setting internal variable
ss=[0,0];
for j=1:NoS-1
    eps_inc=(steps(j+1)-steps(j))/NInc;
    for i=1:NInc
        epsilon=epsilon+eps_inc;
        if epsilon>ksi
            d_inel=(epsilon-Sig0/E)*L;
            omega=1-((d_f-d_inel)/d_f*Sig0)/((d_el+d_inel)/L)/E;
            if omega>1
                omega=1;
            end
            ksi=epsilon; %Updating internal variable
            sigma=E*(1-omega)*epsilon;
        else
```

```
            sigma=E*(1-omega)*epsilon;
        end
        nmb=nmb+1;
        ss(nmb,:)=[epsilon,sigma];
    end
end
```

Similar to the previous example, a multiple step analysis driven by the strain value at the end of each step is implemented and the corresponding solution determined with this damage model in Fig. 5.47. This response is now different from the one previously observed in Fig. 5.38 despite being calculated based on the same strain history. The peak stress exactly represents the value of the stress at the damage initiation that is one of the constitutive parameters specified in input, while the softening part is governed by the energy release rate G_f. The MATLAB® implementation has been used to determine the response by considering three different G_f values while keeping the same remaining material parameters. Fig. 5.48 shows the calculated results by considering G_f values of 2, 5, and 50 J/m^2, referred to in the figure as parameters 1, 2, and 3, respectively.

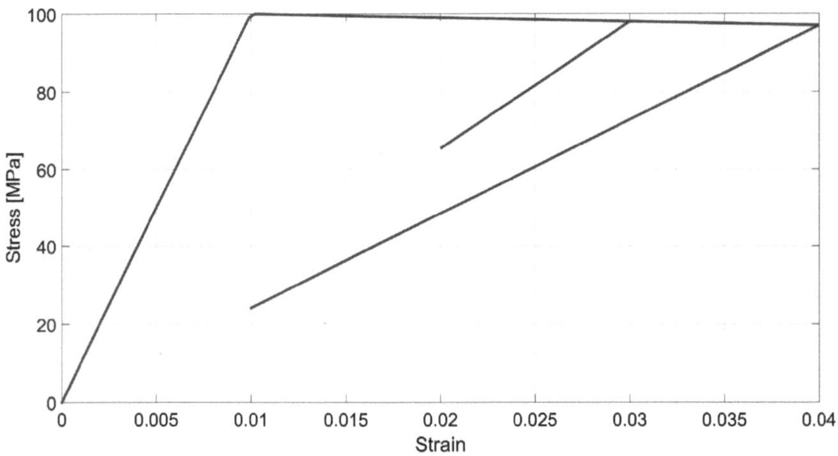

FIGURE 5.47 Stress–strain response of a damage model expressed with a stress-displacement formulation of the damage evolution.

5.3.1 Isotropic damage model for a general three-dimensional stress state

Generalization of the presented uniaxial damage model to the three-dimensional stress state is obtained by formulating the material stiffness tensor $\boldsymbol{D}$ as a function of internal variables that can include the scalar values ω and/or tensorial values $\boldsymbol{\omega}$. With such a formulation, we can express the stress as:

$$\boldsymbol{\sigma} = \boldsymbol{D}(\omega, \boldsymbol{\omega}) : \boldsymbol{\varepsilon} \tag{5.98}$$

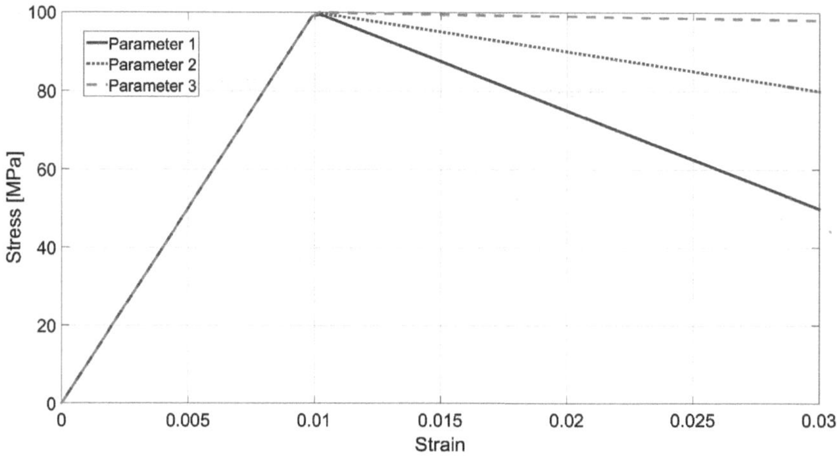

FIGURE 5.48 Stress–strain response by considering a damage model with three different values for the energy release rate.

The simplest damage model for the general three-dimensional stress state is represented by the *isotropic damage model.* The model is based on the simplification that the stiffness degradation is isotropic, i.e., stiffness moduli corresponding to different directions decrease proportionally, independently of the loading direction. Since an isotropic elastic constitutive model requires two independent constants, a general isotropic damage model should introduce two damage variables. A further simplification is usually made by assuming that the Poisson's ratio is not affected by damage. It is equivalent to the assumption that the shear and bulk moduli reduces in the same manner during damage evolution. This simplification leads to a damage model with a single damage variable. Within such a model the strain tensor is related to the stress tensor by the following relationship:

$$\boldsymbol{\sigma} = (1 - \omega)\boldsymbol{D}^e : \boldsymbol{\varepsilon} \tag{5.99}$$

where ω is the damage variable, and $\boldsymbol{D}^e$ is the elasticity tensor corresponding to the undamaged material.

The constitutive relation given by Eq. (5.99) has to be accompanied by a damage initiation criterion and damage evolution law. The damage evolution is usually postulated by defining a loading function $f(\boldsymbol{\varepsilon}, \xi)$ in terms of the strain tensor and of the internal variable ξ. For example, for a simple one-dimensional damage model, the loading function can be defined as $f(\varepsilon, \xi) = \varepsilon - \xi$, where ξ is the largest deformation ε_{max} reached during the previous loading history.

With this definition, the loading-unloading conditions are formalized within general inelastic constitutive models by using the Karush–Kuhn–Tucker conditions:

$$f \leq 0, \quad \dot{\xi} \geq 0, \quad f\dot{\xi} = 0 \tag{5.100}$$

The same formalization was already presented in plasticity in which case a yield function was used in place of the loading function, and a plastic load multiplier was considered instead of the internal variable ξ.

For the general three-dimensional stress state, it is convenient to postulate the loading function in the following form:

$$f(\boldsymbol{\varepsilon}, \xi) = \varepsilon^{eq}(\boldsymbol{\varepsilon}) - \xi \tag{5.101}$$

where ε^{eq} is a possible scalar measure of the strain level denoted as the *equivalent strain*, while ξ depicts physically a scalar measure of the largest deformation reached in the previous loading history. Different choices for the selection of the equivalent strain are available and these graphically correspond to different shapes of the elastic domain. The simplest one defines the equivalent strain as the Euclidean norm of the strain tensor:

$$\varepsilon^{eq} = \sqrt{\boldsymbol{\varepsilon} : \boldsymbol{\varepsilon}} \tag{5.102}$$

The selection of the equivalent strain measure ε^{eq} conditions the selection of the internal variable ξ, in what it represents the largest value of equivalent strain that has occurred within the previous deformation history:

$$\xi(t) = \max \varepsilon^{eq}(t) \tag{5.103}$$

Similarly, as for the yield function in plasticity, the expression for ε^{eq} defines the shape of the elastic domain. For example, for the equivalent strain given by Eq. (5.102) and for the two-dimensional deformation state in principal strain coordinates it represents a circle (see Fig. 5.49).

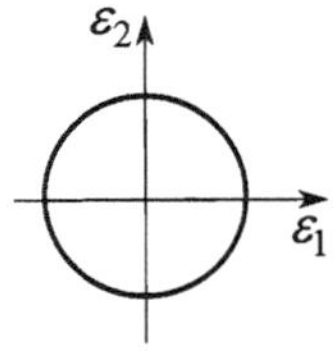

FIGURE 5.49 The elastic domain for an equivalent strain expressed as an Euclidean norm.

Damage models are usually adopted for modeling quasi-brittle materials. Common feature of these materials is that they provide less resistance when loaded in tension with respect to compressive loading. As a consequence the equivalent strain defined by Eq. (5.102) is not a good choice as it results in a symmetric behavior in tension and compression. A formulation of an equivalent strain that deals with this deficiency is proposed, for example, in [10] and is referred to as a *modified von Mises equivalent strain*. It modifies the standard von Mises definition by introducing the parameter k that represents the ratio between the uniaxial compressive strength and the uniaxial tensile strength. According to this formulation the equivalent strain is computed as:

$$\varepsilon^{eq} = \frac{(k-1)I_1^{\varepsilon}}{2k(1-2\nu)} + \frac{1}{2k}\sqrt{\frac{(k-1)^2}{(1-2\nu)^2}(I_1^{\varepsilon})^2 + \frac{12kJ_2^{\varepsilon}}{(1+\nu)^2}} \tag{5.104}$$

where I_1^{ε} and J_2^{ε} are the first invariant and the second deviatoric invariant of the strain tensor, respectively. For a two-dimensional deformation state the elastic domain corresponds to an ellipse that is shifted from the origin along the hydrostatic axis (see Fig. 5.50). By setting $k = 1$, all parts that depend on I_1^{ε} in Eq. (5.104) vanish and the equivalent strain corresponds to the standard von Mises one.

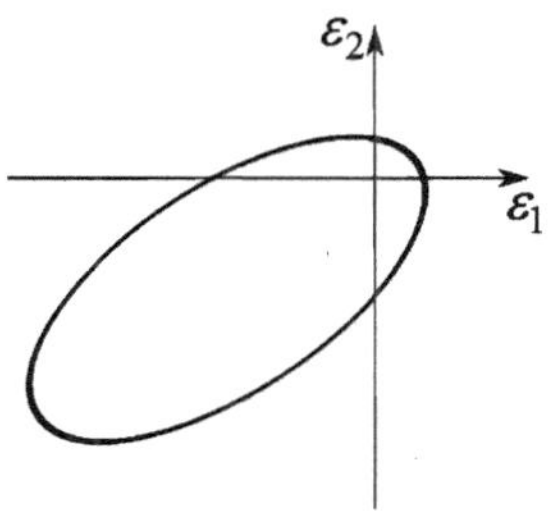

FIGURE 5.50 The elastic domain for the modified von Mises equivalent strain.

The isotropic damage models are commonly implemented so that the stress is evaluated explicitly (i.e., like the one-dimensional model previously presented). This implies that they do not perform an iterative strategy for solving the nonlinear equations. This simplifies the numerical implementation even if it usually requires smaller steps to yield satisfying results. The implementation algorithm can have the form described below.

1. Compute the strain increment $\Delta\varepsilon$
2. Update the total strain $\varepsilon_{i+1} = \varepsilon_i + \Delta\varepsilon$
3. Compute the equivalent strain ε^{eq} based on the adopted criterion
4. Evaluate the damage loading function f:
   if $f > 0$, update the internal variable: $\xi_{i+1} = \varepsilon^{eq}$
   else $\xi_{i+1} = \xi_i$
5. Update the damage variable $\omega_{i+1} = \omega(\xi_{i+1})$
6. Compute the new stress state: $\boldsymbol{\sigma}_{i+1} = (1 - \omega_{i+1})\boldsymbol{D}^e : \boldsymbol{\varepsilon}_{i+1}$

5.3.2 Anisotropic damage models for a general three-dimensional stress state

The assumption of isotropic damage models for which the stiffness degradation is isotropic and independent on the direction of loading can only be justified to a certain extent for materials that behave isotropically in the elastic domain. Such a simplification is usually accepted to deal with a significantly simpler constitutive model. In reality, if an isotropic material is loaded by an arbitrary three-dimensional stress state, the degradation of the stiffness (i.e., occurrence of damage) is usually faster in the direction along which the material exhibits larger deformations. As a consequence of this behavior, the material properties describing the damage conditions will start to be anisotropic.

Proper modeling of this behavior requires the introduction of damage through a fourth-order tensor. The formulation of this damage tensor for a general anisotropic case as well as the experimental calibration of the material parameters are not an easy task. For further details, reference should be made to [11], [12], [13]. Here, we intend to describe the nature of this tensor with the assistance of a simple example.

A possible choice of a fourth-order damage tensor stems from a straightforward generalization of the relationship between affective and nominal (i.e., macroscopic) stress, given for the uniaxial case by Eq. (5.91). Its generalization for a three-dimensional stress state can be expressed as:

$$\boldsymbol{\sigma} = (\boldsymbol{I} - \boldsymbol{\Omega}) : \boldsymbol{\sigma}_{ef} \tag{5.105}$$

where $\boldsymbol{\sigma}$ and $\boldsymbol{\sigma}_{ef}$ are the symmetric second-order tensors, and $\boldsymbol{I}$ and $\boldsymbol{\Omega}$ are the fourth-order unit tensor and fourth-order damage tensor, respectively. Instead of working with $\boldsymbol{\Omega}$, it is usual to work with tensor $\boldsymbol{M} = (\boldsymbol{I} - \boldsymbol{\Omega})^{-1}$, denoted as the *damage effect tensor*. Eq. (5.105) can therefore be rewritten as:

$$\boldsymbol{\sigma} = (\boldsymbol{M})^{-1} : \boldsymbol{\sigma}_{ef} \quad \text{or} \quad \boldsymbol{\sigma}_{ef} = \boldsymbol{M} : \boldsymbol{\sigma} \tag{5.106}$$

By adopting the Voigt reduced notation for symmetric tensors, the stress tensor is represented as a vector (i.e., column matrix) and the damage effect tensor can be expressed as a matrix. We now briefly discuss the particularization of the damage effect tensor for an orthotropic material.

If the material macroscopically exhibits an orthotropic behavior, it is usually due to a specific material microstructure. Typical examples include composite materials reinforced with long fibers in one direction. For such materials, there is no reasonable justification to adopt the assumption of an isotropic damage. In the direction parallel to the fibers the material response is primarily dictated by the mechanical properties of the fibers, while, in the perpendicular direction, their influence is relatively smaller as the material properties are more related to those of the matrix material.

In the following, we consider, for illustrative purposes, the case of plane-stress exhibited by an individual layer assumed to be part of a multiple laminate composite. The direct extension of the isotropic damage model to the orthotropic plane-stress case can be carried out by introducing separate diverse damage variables, that govern the degradation corresponding to different directions within the material. Specifically, three damage variables are introduced and these are associated to each stress component. The relationship between the nominal stress and the effective stress, as similarly specified for the one-dimensional case, is given by:

$$\boldsymbol{\sigma}_{ef} = \begin{bmatrix} \frac{1}{1-d_f} & 0 & \\ 0 & \frac{1}{1-d_m} & 0 \\ 0 & 0 & \frac{1}{1-d_s} \end{bmatrix} \boldsymbol{\sigma} \quad \rightarrow \quad \boldsymbol{\sigma}_{ef} = \boldsymbol{M}\boldsymbol{\sigma} \tag{5.107}$$

where d_f is the damage variable for the stiffness degradation in the longitudinal direction (governed by the fiber properties), d_m is damage variable in the transversal direction (governed by the matrix properties), and d_s is variable for shear damage.

The nonlinear stress-strain relationship for this model can be expressed as:

$$\boldsymbol{\sigma} = \boldsymbol{D}^s \cdot \boldsymbol{\varepsilon} \tag{5.108}$$

where the elasticity matrix $\boldsymbol{D}^s$ is a function of the above three damage variables and possesses the following form:

$$\boldsymbol{D}^s = \frac{1}{D}\begin{bmatrix} (1-d_f)E_1 & (1-d_f)(1-d_m)\nu_{21}E_1 & 0 \\ (1-d_f)(1-d_m)\nu_{12}E_2 & (1-d_m)E_2 & 0 \\ 0 & 0 & (1-d_s)GD \end{bmatrix} \tag{5.109}$$

with E_1 being the Young's modulus in the fiber direction, E_2 is the Young's modulus in the perpendicular direction (i.e., governed by the matrix properties), G is the shear modulus, while ν_{12} and ν_{21} are the Poisson's ratios. Variable D is computed based on:

$$D = 1 - (1-d_f)(1-d_m)\nu_{12}\nu_{21} \tag{5.110}$$

The definition of the failure criteria that has to be satisfied to initiate damage is not a trivial task for an orthotropic material. In the context of continuum damage mechanics, an adequate criterion should predict failure due to a specific (macroscopic) state of stress. Such criterion should not model the microstructural aspects of failure but it should be physically motivated to provide reasonable predictions.

5.4 Coupled damage plasticity models

When plastic deformation occurs in a material, its stress–strain response exhibits some degree of nonlinearity and permanent deformation. The stiffness of the material is not affected as it can be noted during the unloading by comparing the unloading linear curve with the initial linear-elastic response prior to plastic deformation (see Fig. 5.16). On the other hand, when during the loading gradual initiation and propagation of microcracks occur it leads macroscopically to a reduction of the material stiffness, described by a nonlinear stress–strain response. The unloading path does not follow the previous loading curve while, in general, no permanent deformation remains upon load removal as all microcracks are assumed to close. In the previous sections, we have seen that to model the former behavior (e.g., occurrence of permanent plastic deformation) an elasto-plastic model should be employed, while to predict the latter response (e.g., degradation of the elastic properties due to damage) it is required to implement a damage model.

Real materials, particularly if loaded to a large strain, often exhibit both permanent deformation and decrease of stiffness. The behavior is typically evidenced when dealing with cyclic loading, while it can also be observed with a single load-unload cycle (see Fig. 5.51). In such a case, neither the elasto-plastic model and nor the damage model can be used to correctly model the response. The preferable approach in this case is to combine the presented framework of damage and plasticity models.

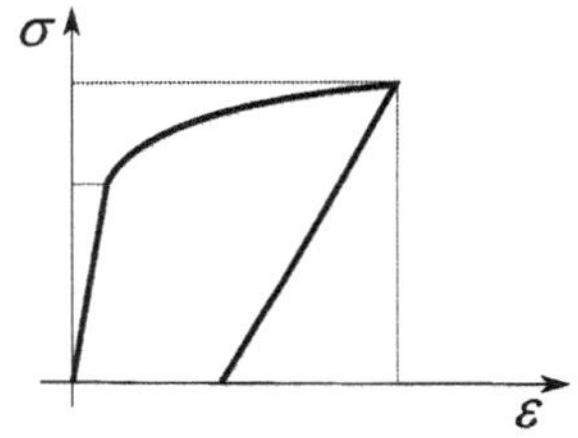

FIGURE 5.51 Uniaxial stress–strain curve for simultaneous damage and plasticity.

Different strategies can be used to achieve coupling of damage and plasticity. A possible, and arguably the most physically meaningful way, is to exploit the notion of effective stress, introduced within damage models. This quantity represents the stress transmitted by the bulk material existing between the formed microcracks within macroscopically observable stiffness degradation. Within classic elasto-damage models, the effective stress is computed by assuming a linear-elastic behavior of the intact material. Instead, within coupled damage-plasticity this stress should be computed on the basis of an elasto-plastic constitutive model.

The governing equation of such a coupled damage-plasticity model involves both a nominal stress (i.e., macroscopically observable) and an effective stress. In a typical damage-like form, the nominal stress is expressed as a nonlinear function of strain. In the context of the simplest isotropic damage model, this can be expressed as follows by adopting Voigt reduced notation:

$$\boldsymbol{\sigma} = (1-\omega)\boldsymbol{D}^e \cdot \boldsymbol{\varepsilon} \tag{5.111}$$

with ω being the damage variable and $\boldsymbol{D}^e$ the material stiffness tensor (expressed as a matrix) collecting the elastic constants corresponding to the intact material. The coupling is achieved by postulating that plasticity only applies to the intact material and not to the voids of the formed microcracks. Consequently, the elasto-plastic model equation should be formulated with reference to the effective stress. For an isotropic damage model with a single damage variable ω, the effective stress can be obtained as:

$$\boldsymbol{\sigma}_{ef} = \boldsymbol{D}^e \cdot \boldsymbol{\varepsilon}^{el} = \boldsymbol{D}^e \cdot (\boldsymbol{\varepsilon} - \boldsymbol{\varepsilon}^{pl}) \tag{5.112}$$

For a purely elasto-damage model, $\boldsymbol{\varepsilon}$ in Eq. (5.111) and $\boldsymbol{\varepsilon}_{el}$ in Eq. (5.112) represent the same quantity. For a coupled damage-plasticity model, the latter equation only contains the elastic part of the deformation tensor from the former equation and it relies on the additive decomposition of strains typical for elasto-plastic constitutive models. The yield function evaluation and flow-rule computation can then be performed on the basis of the effective stress:

$$f(\boldsymbol{\sigma}_{ef}, \xi) = \sigma_{eq} - \sigma_0(\xi), \quad \dot{\boldsymbol{\varepsilon}}^{pl} = \dot{\lambda}\frac{\partial f}{\partial \boldsymbol{\sigma}_{ef}} \tag{5.113}$$

where ξ is an internal variable that governs hardening, and σ_{eq} is the equivalent stress that depends on the adopted yield conditions (e.g., equivalent von Mises stress q for the von Mises yield condition).

A possible algorithm for the implementation of the coupled damage-plasticity constitutive model is given below.

At the beginning of the calculation set all updating variables equal to their initial values: current nominal stress $\boldsymbol{\sigma}^0$, current effective stress $\boldsymbol{\sigma}^0_{ef}$, current strain $\boldsymbol{\varepsilon}^0$, hardening internal variable ξ^0_p, yield limit $\sigma_y = \sigma_0$, damage variable ω^i, and damage evolution internal variable ξ^0_d.

1. Compute the strain increment $\Delta\boldsymbol{\varepsilon}$.
2. Update the strain $\boldsymbol{\varepsilon}^{i+1} = \boldsymbol{\varepsilon}^i + \Delta\boldsymbol{\varepsilon}$
3. Compute the trial nominal and effective stress:
   $\boldsymbol{\sigma}^{i+1} = (1-\omega^i)\boldsymbol{D}^e\boldsymbol{\varepsilon}^{i+1}$
   $\boldsymbol{\sigma}^{i+1}_{ef} = \boldsymbol{\sigma}^i_{ef} + \boldsymbol{D}^e\Delta\boldsymbol{\varepsilon}$
4. Evaluate the value of the yield function:
   $\bar{\sigma}_{eq} = \sigma_{eq}(\boldsymbol{\sigma}^{i+1}_{ef})$
   $f^p = \bar{\sigma}_{eq} - \sigma_y(\xi^i_p)$

If $f^p < 0$ accept the trial stress and move to point 5;
else calculate $\boldsymbol{\sigma}_{ef}^{i+1} = \boldsymbol{\sigma}_{ef}^{i+1} - \boldsymbol{D}^e \Delta \boldsymbol{\varepsilon}^{pl}$, with $\Delta \boldsymbol{\varepsilon}^{pl} = \Delta\lambda \frac{\partial f}{\partial \sigma_{ef}^i}$

5. Evaluate the damage loading function:
$\bar{\varepsilon}^{eq} = \varepsilon^{eq}(\varepsilon^{i+1})$
$f^d = \bar{\varepsilon}_{eq} - \xi_d^i$
If $f^d < 0$, $\xi_d^{i+1} = \xi_d^i$
else $\xi_d^{i+1} = \bar{\varepsilon}_{eq}$
6. Update the damage parameter: $\omega^{i+1} = \omega(\xi_d^{i+1})$
7. Compute the new stress: $\boldsymbol{\sigma}^{i+1} = (1 - \omega^{i+1})\boldsymbol{D}^e \boldsymbol{\varepsilon}^{i+1}$

The main characteristics of the outlined coupling strategy are that the damage model is implemented with reference to a nominal stress, while the plasticity model refers to the effective stress. The presented scheme assumes that the damage model (i.e., the onset and propagation of damage) is governed by the current state of total strain. The plasticity part of the model necessitates of the additive decomposition of strain into the elastic and plastic parts, typical of plasticity constitutive models. Based on the given strain increment, both effective and nominal trial stresses are computed (point 3). The effective trial stress is computed by assuming that all the strain increment is elastic. In the subsequent point (point 4), the plasticity model is employed to verify whether the yield condition is satisfied or a plastic strain has to be introduced to "relax" the stress to return it to the yield surface. This part requires an iterative solution not detailed here, but formally the same as the one presented for the elasto-plastic constitutive models in Section 5.2.

Upon reaching the convergence with the elasto-plastic part, the total strain is considered to evaluate the criterion for the damage evolution (point 5). This part is here presented with reference to the simplest, single parameter isotropic damage model. Despite this, the given coupling strategy is not limited to this case only, and more complex alternatives can be used instead (e.g., anisotropic damage models). The presented coupling strategy is used in a wide range of coupled elasto-damage models (see, e.g., [14], [15], [16], [17]).

An alternative coupling strategy consists in enriching the elasto-damage model by an additional irreversible strain. In general, the irreversible strain does not need to be a typical plastic strain produced by a plastic slip but it can represent a non-recoverable strain that remains upon unloading of the damaged material (e.g., due to the rough surface of the crack face). The stress–strain relation is therefore expressed in a manner typical for elasto-plastic models:

$$\boldsymbol{\sigma} = \boldsymbol{D} : (\boldsymbol{\varepsilon} - \boldsymbol{\varepsilon}^{pl}) \tag{5.114}$$

and, in this representation, $\boldsymbol{D}$ is the damage stiffness tensor that governs the evolution of damage through appropriate damage variables. The irreversible strain $\boldsymbol{\varepsilon}^{pl}$ is computed with the evolution equation of a plasticity flow rule or as required by the considered phenomena (e.g., frictional sliding along cracked surfaces). Further details of this strategy are not discussed here and reference for this purpose should be made to [18], [19], [20].

5.5 Modeling crack initiation and propagation

Structural analysis problems discussed within this book are expressed in terms of the identification of three fields: displacement, strain, and stress fields. The presented solution strategies are developed by relying on continuum hypothesis where all the three fields are assumed to be continuous over the domain (i.e., deformable body) for which the problem is solved. Finite element method, as numerical tool for generating an approximate solution, generally assures continuity for the displacement field, while strain, and consequently stress, in general can have "jumps" from one element to another.

When large, macroscopic cracks are formed within a deformable body the continuity is violated and, in the general case, the discontinuity in displacements appears over the crack mouth. The field equations based on continuum mechanics are not completely defined and, for the analysis of these problems, a different theoretical framework has to be used. Such framework is provided by *fracture mechanics*. From a numerical viewpoint, the classical displacement based FEM has to be modified to account for this displacement discontinuity. Typically used modification consists by enriching shape functions to allow for displacement discontinuity over a single element. An example of this strategy is provided with the *extended finite element method* (XFEM) [21], [22].

With constitutive modeling being the main topic of this book, hereafter we discuss only alternative strategies of modeling crack propagation which can be used in combination with standard finite element method. Commonly used counterpart of the discrete crack modeling is the so-called *smeared crack* concept. Smeared crack models can be seen as a particular group of constitutive models specifically developed for modeling crack propagation in quasi-brittle materials under dominantly tensile loading. The smear crack approach for the modeling of crack propagation first appeared in the late 1960s [23].

The basic idea behind the smeared crack concept consists in considering the cracked solid as a continuum with notions of stress and strain still defined. As in elasto-plastic constitutive models, smeared crack models decompose strain into two parts: the one that corresponds to the deformation of the uncracked material and the other part that represents the contribution of cracking:

$$\boldsymbol{\varepsilon} = \boldsymbol{\varepsilon}^{el} + \boldsymbol{\varepsilon}^{c} \tag{5.115}$$

Even though the first part can be modeled with any continuum constitutive model (including elasto-plastic models), it is commonly assumed to model them as linear-elastic, so the stresses can be related to strains by the fourth-order material stiffness tensor:

$$\boldsymbol{\sigma} = \boldsymbol{D}^{el} : \boldsymbol{\varepsilon}^{el} \tag{5.116}$$

as the smeared crack models are commonly used to model quasi-brittle materials where elastic deformations are generally small, and the inelastic behavior is dominantly due to material cracking.

The second part of the strain is the contribution of the crack opening. Although the crack represents a displacement discontinuity, within smeared crack models, it is assumed that the continuity is preserved, and the crack opening is smeared (i.e., averaged) over a certain distance. The crack is initiated when the stress state reaches a certain failure criterion, and it is further assumed that instead of a sudden loss of material integrity, a gradual degradation takes place. Locally, the material response is changed from an initial linear-elastic response to the one governed by the softening behavior introduced through the traction-separation law. The topology of the original finite element mesh remains the same, so the same model without any modification can be used. This approach makes the implementation computationally efficient. The crack propagation is modeled in the form of zone of elements in which the softening behavior is concentrated.

Quasi-brittle materials, as the main group of materials which are modeled by smeared crack models, are the most susceptible to fracture when loaded in tension. Therefore, the crack initiation criterion is commonly adopted as a Rankine criterion:

$$\sigma_1(\boldsymbol{\sigma}) = \sigma_m \tag{5.117}$$

where σ_1 is the maximum principal stress and σ_m is the uniaxial tensile strength.

The crack initiation criterion specifies also the orientation of the crack. According to the Rankine criterion, the crack plane is perpendicular to the direction of the maximum principal stress. Expressed in the context of inelastic deformation within the smeared crack model, the crack opening contributes to the normal strain perpendicular to the crack plane, while crack sliding contributes to the shear strain.

Let us, for simplicity, consider the 2D problem depicted in Fig. 5.52. The onset of the crack generates two strain components: ε_n^c and γ_{nl}^c, where n is the direction of the maximum principal stress and l is the direction of the crack. The two inelastic strain components are clearly expressed in a local coordinate system defined in the $n-l$ plane, and they have to be related to the strain components in the global coordinate system. This is considered by relying on the following transformation (see [3]):

$$\boldsymbol{E}_g^c = \boldsymbol{R}^T \cdot \boldsymbol{E}_l^c \cdot \boldsymbol{R} \tag{5.118}$$

where $\boldsymbol{R}$ is the transformation matrix defined by:

$$\boldsymbol{R} = \begin{bmatrix} \cos\varphi & \sin\varphi \\ -\sin\varphi & \cos\varphi \end{bmatrix} \tag{5.119}$$

and $\boldsymbol{E}_l^c$ and $\boldsymbol{E}_g^c$ are second-order strain tensors in local and global coordinate system, respectively. Considering that the only nonzero strain components due to cracking are ε_{nn}^c and ε_{nl}^c, the inelastic strain tensor in local coordinate system takes the following form:

$$\boldsymbol{E}_g^c = \begin{bmatrix} \varepsilon_{nn}^c & \varepsilon_{nl}^c \\ \varepsilon_{ln}^c & 0 \end{bmatrix} \tag{5.120}$$

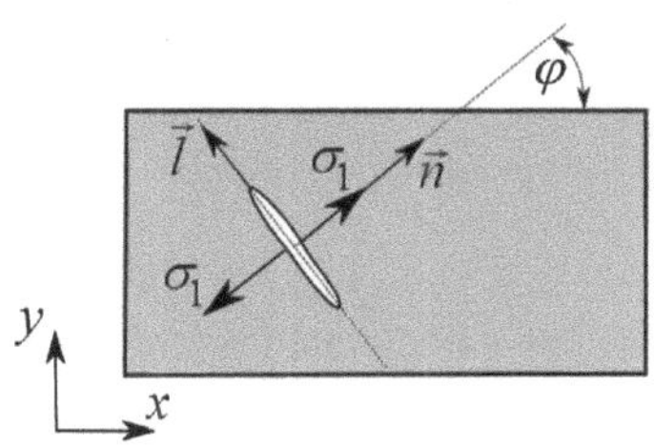

FIGURE 5.52 Local reference system $n - l$ related to the crack and its position with respect to the global $x - y$ system.

For the numerical implementation, it is more convenient to adopt the Voigt notation, and to represent strain as a vector (i.e., matrix-column). This changes the form of the transformation matrix as well. By writing Eq. (5.118) in expanded form (i.e., performing all the matrix multiplications) and then grouping all the coefficients which stand next to each strain component, the following transformation relation is obtained:

$$\boldsymbol{\varepsilon}_l^c = \boldsymbol{T}^T \cdot \boldsymbol{\varepsilon}_g^c, \quad \boldsymbol{\varepsilon}_g^c = \boldsymbol{T} \cdot \boldsymbol{\varepsilon}_l^c \tag{5.121}$$

where matrix $\boldsymbol{T}$ is:

$$\boldsymbol{T} = \begin{bmatrix} \cos^2\varphi & \sin^2\varphi & \sin\varphi\cos\varphi \\ \sin^2\varphi & \cos^2\varphi & -\sin\varphi\cos\varphi \\ -2\sin\varphi\cos\varphi & 2\sin\varphi\cos\varphi & \cos^2\varphi - \sin^2\varphi \end{bmatrix} \tag{5.122}$$

The components of $\boldsymbol{\varepsilon}_g^c$ are assumed to be related to the traction transmitted through the crack. In order to avoid mesh dependency of the solution (as observed for damage softening models), the traction is defined in the form of a stress transmitted by the cracking solid within the fracture process zone, i.e., as a function of crack opening displacement. The form is here denoted as *traction-separation law*. For purely Mode-I crack (i.e., crack initiates due to normal stress perpendicular to the crack path), it is specified that the stress normal to the crack path σ_{nn} can be expressed as a function of the crack opening:

$$\sigma_{nn} = f(w) \tag{5.123}$$

Similarly, as discussed for damage models dealing with softening behavior, the specified degradation can be of an arbitrary type (e.g., linear, bi-linear, exponential). The area beneath the traction-separation curve represents the specific fracture energy G_f.

Since smeared crack represents a continuum model, the traction-separation curve has to be transformed into a stress-strain law, and integrated with the remaining elastic response of the material. This is performed by dividing the crack opening value by the distance over which the crack is smeared, here denoted as h. Based on this, we can write the following expression for the normal strain perpendicular to the crack path:

$$\varepsilon_{nn}^{c} = \frac{w}{h} \tag{5.124}$$

The choice of the value for h is dictated by the size of the mesh of the finite elements and different approaches can be used for this purpose. For example, the finite element length projected on to the crack normal, or the square root for 2D problems, or the cube root for 3D problems, of the area bonded by the element's integration points. When selecting the value for h, care has to be taken as through its value some mesh-dependency of the solution remains.

By combining the elastic response of the traction-separation law given by Eq. (5.123) with the crack mouth opening strain, the component normal to the crack path is computed as follows:

$$\varepsilon_{nn} = \frac{\sigma_{nn}}{E} + \frac{f^{-1}(\sigma_{nn})}{h} \tag{5.125}$$

Transformation of the traction-separation law into its continuum representation in the form of stress–strain curve is plotted in Fig. 5.53.

The presented scheme assumes that the crack initiates within the continuum intact solid in the direction depicted by the maximum principal stress. Once a crack is initiated, its direction is defined. This formulation has the clear advantage of a simpler numerical implementation but presents a limitation that in certain situations is not justifiable. Alternatively, more refined approaches allow changes in the direction of the crack, for example, within the so-called rotating crack models [24].

Another formulation that is relatively simple to be implemented within conventional displacement-based FEM relies on the use of *cohesive crack models*. Most materials are not perfectly brittle in the sense that fracture can occur without any accumulation of damage. For most quasi-brittle materials, a certain

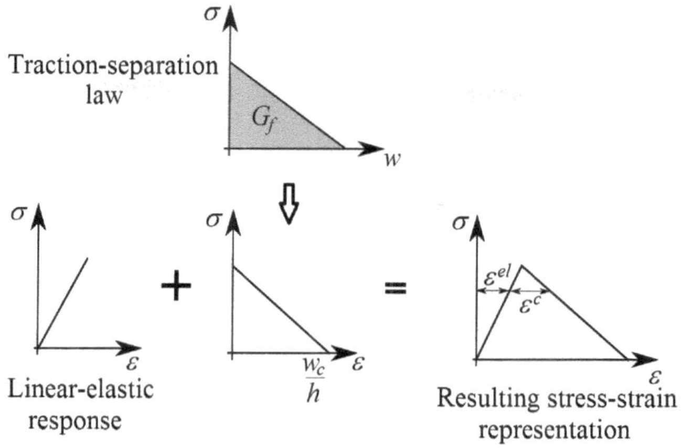

FIGURE 5.53 Transformation of the traction-separation curve into the corresponding stress–strain response.

nonlinear segment on the stress–strain curve precedes the rapture. This ductility is usually related to the zone in front of the crack tip in which yielding on a small scale, microcracking, void initiation or a combination of these (and similar phenomena) can take place. If this behavior is pronounced, there will be a zone within the fracture process in which cohesive forces, although not at the level of the intact material, contribute to the overall equilibrium and have to be taken into account in the predictions of the structural response. The most effective way of doing so is by employing cohesive crack models, first introduced by Dugdale [25] for elasto-plastic fracture and by Barenblatt for quasi-brittle fracture [26].

In the implementation of cohesive crack models over a crack propagation path, three zones can be identified (see Fig. 5.54): the zone where the crack is completely formed (i.e., no traction is transmitted), the zone in which degradation occurs but the material still possesses cohesive forces, and the zone with intact material. All dissipation processes which are taking place in front of the crack tip are lumped with these models into a single line for 2D problems or a plane for 3D problems. The degrading mechanism within a fracture zone is governed by the relation between the traction at the discontinuity and the relative displacement across the line or plane to which all dissipation processes are lumped. The relation is postulated in the form of a traction-separation law, in the same form as for a smeared crack models.

Although the cohesive crack models represent a discrete concept in which the crack path is modeled, its numerical implementation within standard FEM formulation is rather simple, and consist of the insertion of the so-called *cohesive elements* within the zone that is going to be intersected by the crack (see Fig. 5.55). In two-dimensional problems the nodes of these elements overlap before crack initiation. Upon satisfying a certain crack initiation criterion,

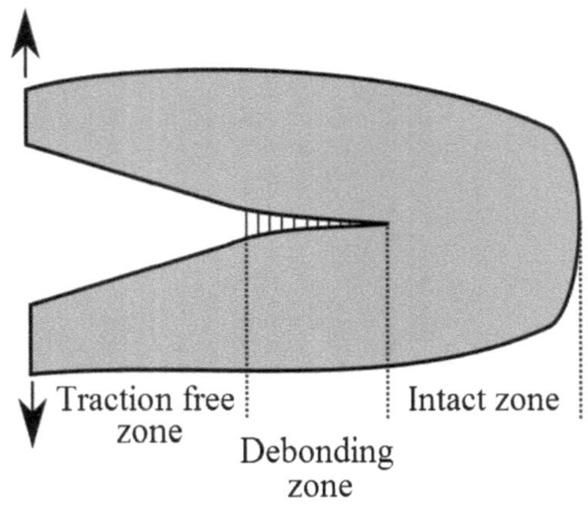

FIGURE 5.54 Different zones in crack propagation.

the elements start to open (i.e., have displacement components across the line). When this occurs, the constitutive behavior of the cohesive element is governed by a traction-separation law. Once the opening of the element reaches a critical value, prescribed in the model, forces are no more transmitted through the element, and the solid can be assumed to be fully cracked.

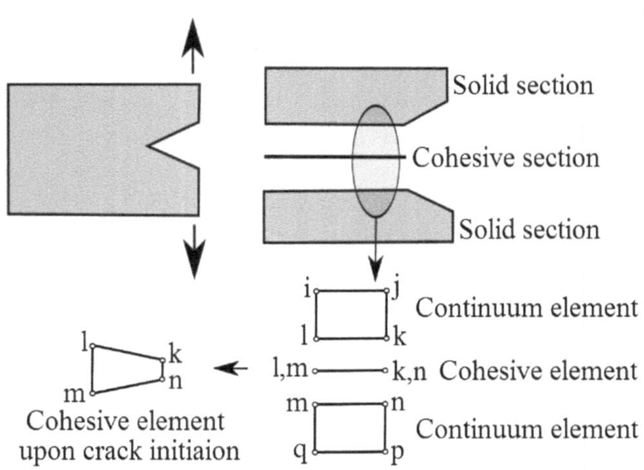

FIGURE 5.55 Modeling crack propagation with cohesive elements.

Within cohesive crack models the constitutive behavior is generally formulated by combining the elastic response with a traction-separation law. The elastic part of the response relates the tractions transmitted by the element with corresponding strain components in the element's local coordinate system. In the general 3D case, there are three stress components: traction in the normal direction and two shear components in the plane of the element. For 2D problems, these are reduced to two components, as there is only one shear component. For the elastic part of the response, these three components are related to the corresponding three strain components. Formulated in the local element reference system, this relation can be expressed as follows:

$$\begin{bmatrix} t_n \\ t_s \\ t_t \end{bmatrix} = \begin{bmatrix} K_{nn} & K_{ns} & K_{nt} \\ K_{sn} & K_{ss} & K_{st} \\ K_{tn} & K_{ts} & K_{tt} \end{bmatrix} \begin{bmatrix} \varepsilon_n \\ \varepsilon_s \\ \varepsilon_t \end{bmatrix} \quad \rightarrow \quad \boldsymbol{t} = \boldsymbol{K}\boldsymbol{\varepsilon} \tag{5.126}$$

where nominal strain components ε_n, ε_s, and ε_t are computed by dividing the crack opening and crack sliding displacements, by a specified parameter T_0. The parameter is commonly denoted as the *constitutive thickness* of the cohesive element, and influences the values of the stiffness coefficients entering into the matrix $\boldsymbol{K}$. In view of this parameter, the stiffness coefficients K_{ij} don't represent real elastic constants of the material (i.e., Young's modulus value) and need to be rescaled. The scaling can be easily explained by considering a truss element of length L and an initial cross-sectional area A_0. The elongation of such a truss is equal to:

$$\delta = \frac{FL}{EA} \tag{5.127}$$

Considering that the stress in a truss is constant and equal to the ratio of the force and the cross-sectional area, the relationship between the elongation (i.e., displacement) and stress is equal to:

$$\delta = \frac{\sigma}{K} \tag{5.128}$$

where $K = \frac{E}{L}$ is a stiffness coefficient that relates the nominal stress to the displacement. In a similar way, the stiffness coefficients of Eq. (5.126) have to be rescaled by the constitutive thickness T_0. For example, the stiffness coefficient that relates normal stress to normal strain is:

$$K_{nn} = \frac{E}{T_0} \tag{5.129}$$

where E depicts the Young's modulus of the material.

The stiffness matrix $\boldsymbol{K}$ in Eq. (5.126) provides a fully coupled behavior in which each traction component depends on all the components of the separation vector (i.e., on all strain components). When considering the uncoupled behavior, it is sufficient to set the off-diagonal terms to 0.

The response of the cohesive element is elastic until it reaches the crack initiation criterion. This can be implemented with different criteria. For example, with the maximum nominal stress criterion, the crack initiates once one of the three tractions (for 3D problems) reaches its critical value. This criterion can be represented as:

$$\max\left(\frac{t_n}{t_n^0}, \frac{t_l}{t_l^0}, \frac{t_s}{t_s^0}\right) = 1 \tag{5.130}$$

where different traction thresholds are specified for the normal direction (t_n^0) and the two shearing directions (t_l^0 and t_s^0).

Once the crack conditions are met, the response is governed by the softening branch of the traction-separation law that is generally defined on the basis of a specific fracture energy G_f. For a purely mode I crack (i.e., the one where the fracture process is driven only by crack opening without sliding), the traction separation law relates the normal traction t_n to the normal relative displacement taking place across the fracture process zone v_n. This relationship is arbitrary (e.g., expressed in a linear, bilinear or exponential form), while, in the general case, we can write:

$$G_f = \int_{v_n=0}^{v_n=v_n^f} t_n dv_n \tag{5.131}$$

where v_n^f depicts the crack opening corresponding to the complete failure of the cohesive element (i.e., no traction). In general, this parameter is computed on the basis of a specified softening law and the value of G_f.

If the material behavior is such that crack sliding (i.e., Mode II) plays an important role in the fracturing process, a more general version of Eq. (5.131) has to be used. A possible generalization of the concept is proposed by Tvergaard and Hutchinson [27] that can be described as follows:

$$G_f = \int_{v_n=0}^{v_n=v_n^f} t_n d\bar{v}_n \tag{5.132}$$

where $\bar{v}_n$ is called the *effective displacement* and is computed by:

$$\bar{v}_n = \sqrt{v_n^2 + \alpha\left(v_s^2 + v_l^2\right)} \tag{5.133}$$

with α being a mode-mixity parameter that establishes the ratio between mode I and mode II crack displacement components.

Alternatively, mode II fracture energy can be defined separately, that is used independently of the G_f given by Eq. (5.131) that in this case should be used strictly as the mode I fracture energy. For 3D problems, the mode II fracture energy can be defined as:

$$G_f^{II} = \int_{v_n=0}^{v_n=v_n^f} \tau d\bar{v} \tag{5.134}$$

with $\bar{v} = \sqrt{v_s^2 + v_t^2}$, and τ being the resultant shear stress in the plane of the crack. In addition to Eq. (5.134), Eq. (5.132) can be used when dealing with mode I fracture.

The numerical implementation of cohesive crack models through cohesive elements is relatively efficient and can be incorporated within classical displacement based FEM formulations. The major drawback is manifested in the limited flexibility in terms of the crack propagation paths. Clearly, the crack can only propagate along cohesive elements, thus their insertion within finite element mesh preconditions the cracking process by only allowing certain crack paths.

5.6 Summary

This chapter provided an introduction to constitutive models that account for inelastic deformation. Such material representations are useful in a wide range of engineering problems when a material exhausts its elastic capacity. In such situations, it was shown that, to obtain reliable results, it is important to model correctly the mechanisms that are at the origin of the inelastic deformation by employing appropriate constitutive models.

In general, the stress of a material that exhibits some degree of inelastic deformation depends on the current value of strain as well as on "something else." In this chapter, we introduced some of the possible representations that describe this "something else." It was also pointed out that it is convenient, in inelastic constitutive models, to keep track of the influence of the deformation history or of other variable changes by introducing the use of internal variables. The selection of these variables and the way they are included into the constitutive models distinguish one inelastic constitutive model from the another. For this purpose, we provided an overview of the main features exhibited by different typologies of inelastic constitutive models.

In the first part of the chapter, we covered visco-elastic models in which the response of the material is governed by time-dependent effects and by the loading/deformation history experienced by the material. We then moved to elasto-plastic materials and highlighted how the stress at a particular point depends not only on the deformation but also on its previous deformation history, once the material passes its elastic limit. A brief overview of the use and implementation of elasto-plastic constitutive models was provided for the particular case of one-dimensional models and for the general three-dimensional state of stress. The material modeling was complemented by the description of the flow rule and the yield functions that govern the plastic response. In the second part of the chapter, damage models were introduced to account for changes in the material elastic properties that can occur during different loading or deformation conditions. Also in this case the discussion considered the constitutive model implementation in both one- and three-dimensional structural domains. The coupling of damage and plasticity models was then discussed because useful for the modeling of a wide range of materials. In the final part of the chapter, formulations aimed at predicting crack initiation and propagation were presented with a discussion of some of the main features related to their numerical implementation, such as the mesh-dependency of local constitutive models.

References

[1] Erwin Kreyszig, Advanced Engineering Mathematics, second edition, John Willie and Sons, New York, 1967.

[2] Jacob Lubliner, Plasticity Theory, Dover Publications, INC, Mineola, New York, 2008.

[3] Rene de Borst, Mike A. Crisfield, Joris J.C. Remmers, Clements V. Verhoosel, Non-linear Finite Element Analysis of Solids and Structures, second edition, John Wiley and Sons, Ltd., Chichester, UK, 2012.

[4] Peter Wriggers, Nonlinear Finite Element Methods, Springer-Verlag, Berlin, 2008.

[5] Joachim Rosler, Harald Harders, Martin Backer, Mechanical Behavior of Engineering Materials, Springer Verlag, Heidelberg, 2007.

[6] M. Jirasek, Z. Bazant, Inelastic Analysis of Structures, Wiley, West Sussex, 2002.

[7] D. Radenkovic, Theoremes limite pour un materiau de Coulomb a dilatation non standardisee, Comptes Rendus Hebdomadaires des Seances de L'Academie des Sciences (1961) 4103–4104.

[8] E.A. de Souza Neto, D. Peric, D.R.J. Owen, Computational Methods for Plasticity - Theory and Applications, Wiley, Torquay, UK, 2008.

[9] G. Pijaudier-Cabot, Z. Bazant, Non-local damage theory, ASCE Journal of Engineering Mechanics 113 (1987) 1512–1533.

[10] J.H.P. de Vree, W.A.M. Brekelmans, M.A.J. van Gils, Comparison of nonlocal approaches in continuum damage mechanics, Computers and Structures 55 (1995) 581–588.

[11] F.A. Leckie, E.T. Onat, Tensorial nature of damage measuring internal variables, in: Physical Non-linearities in Structural Analysis, Springer, Berlin, 1981, pp. 140–155.

[12] J. Lemaitre, J.L. Chaboche, Mechanics of Solid Materials, Cambridge University Press, 1990.

[13] D. Krajcinovic, Continuous damage mechanics revisited: basic concepts and definitions, Journal of Applied Mechanics, ASME 48 (1981) 809–824.

[14] W. Ju, On energy-based coupled elastoplastic damage theories: constitutive modeling and computational aspects, International Journal of Solids and Structures 25 (1989) 803–833.

[15] J. Lemaitre, Coupled elasto-plasticity and damage constitutive equations, Computer Methods in Applied Mechanics and Engineering 51 (1985) 31–49.

[16] J.C. Simo, W. Ju, Stress and strain based continuum damage model: I. Formulation; II. Computational aspects, International Journal of Solids and Structures 23 (1987) 821–869.

[17] G.Z. Voyiadjis, T. Park, Anisotropic damage of fiber-reinforced MMC using overall damage analysis, Journal of Engineering Mechanics, ASCE 121 (1995) 1209–1217.

[18] M. Ortiz, A constitutive theory for the inelastic behavior of concrete, Mechanics of Materials 4 (1) (1985) 67–93.

[19] S. Yazdani, On a class of continuum damage mechanics theories, International Journal of Damage Mechanics 2 (1993) 162–176.

[20] B. Ayhan, P. Jehel, D. Brancherie, A. Ibrahimbegovic, Coupled damage-plasticity model for cyclic loading: theoretical formulation and numerical implementation, Engineering Structures 50 (2013) 30–42.

[21] S. Mohammadi, Extended Finite Element Method - for Fracture Analysis of Structures, Blackwell Publishing, 2008.

[22] N. Sukumar, J.J. Pervost, Modeling quasi-static crack growth with the extended finite element method Part I: computer implementation, International Journal for Solids and Structures 40 (2003) 7513–7537.

[23] Y.R. Rashid, Ultimate strength analysis of prestressed concrete pressure vessels, Nuclear Engineering and Design 7 (1968) 334–344.

[24] A.K. Gupta, H. Akbar, Cracking in reinforced concrete analysis, Journal of Structural Engineering, ASCE 110 (8) (1984) 1735–1746.
[25] D.S. Dugdale, Yielding of steel sheets containing slits, Journal of the Mechanics and Physics of Solids 8 (1960) 100–108.
[26] G.I. Barenblatt, The mathematical theory of equilibrium cracks in brittle materials, Advances in Applied Mechanics 7 (1962) 55–129.
[27] V. Tvergaard, J.W. Hutchinson, The relation between crack growth resistance and fracture process parameters in elastic-plastic solids, Journal of the Mechanics and Physics of Solids 41 (1993) 1119–1135.

Chapter 6

Numerical implementation of constitutive models with large deformation

In Chapter 1, we have seen how the analysis of a structural system can be performed by combining the compatibility, constitutive and equilibrium equations. For the determination of the solution, we considered a numerical approach based on the FEM. We also highlighted how the constitutive description plays a significant role, for example, in the definition of the stiffness matrix and, once the unknown displacements are calculated, in the evaluation of the stresses. The analysis procedure was formulated based on the assumption that deformations remain small with respect to the size of the domain over which the problem is solved. This assumption enables equilibrium conditions to be imposed in the undeformed configuration. The use of this assumption is satisfactory for engineering materials that undergo small deformations when subjected to loading.

There are other loading scenarios that induce significant deformation (elastic and/or inelastic) in the material when subjected to overloads and when approaching the material failure point. Examples of such engineering scenarios include metal forming simulations, collapse load calculations, and the analyses of slender members. In such cases, the analysis needs to account for large deformations and/or large displacements as a solution based on small deformation theory would lead to unacceptable errors.

The error induced by using a small deformation FE formulation is centered around the evaluation of the volume integrals required for the determination of the element stiffness matrices. Such error becomes important only in the presence of large deformation and when there is a significant difference between the deformed and undeformed configurations. If the integrals are to be solved over the deformed volume, the problem becomes nonlinear since this configuration depends on the displacements that are also the unknowns of the problem. Strictly speaking, this nonlinearity is connected only to the change of geometry of the domain and not to the material nonlinear behavior. Structural problems that fall in this category are referred to as *geometrically nonlinear problems* and these are the focus of this chapter. The introduction to geometric nonlinearity is given by presenting the main modifications to be applied to the formulation for small displacement and small strain to account for large strain in a FE formulation. This chapter will only cover the main concepts and features related

Constitutive Modeling of Engineering Materials. https://doi.org/10.1016/B978-0-12-814696-5.00011-3

to large deformation. For further details on this topic, reference should be made to dedicated textbooks available in the literature (some of these references are provided at the end of this chapter).

6.1 Geometrically nonlinear problems

Classical displacement-based FE formulations are centered on the minimization of the total potential energy, which involves the evaluation of the internally-stored energy within a deformed solid. The calculation of this energy involves the evaluation of deformation through certain adopted strain measures and subsequent calculations of volume integrals. An error is introduced if both the integral and strain measures are defined with respect to the undeformed configuration of the body and such error becomes non-negligible in the presence of significant differences between the deformed and undeformed configurations. Such error can be avoided if the deformed geometry is considered instead. This obviously leads to a nonlinear problem, as the deformation appears in the solution. The nonlinearity of the problem can be easily evidenced in a FE formulation, for example, in the computation of the stiffness matrix as shown in the evaluation of the integral (1.95) that now depends on the displacement vector $\boldsymbol{d}$.

When performing an analysis that involves large strain and/or large displacement, the geometry of the body must be updated throughout the analysis in order to determine the new position of the material points. The main difficulty in developing the solution strategy for this group of problems consists in making the correct choice of the formulation to describe this motion. This is because every formulation requires the use of a particular reference frame and, consequently, a suitable strain measure. This issue does not exist when dealing with small deformation and small displacement because, in this case, all strain measures are reduced to the infinitesimal (engineering) strains that represent the only available choice. The uniqueness of the strain definition does not occur when dealing with large deformation as, in the latter case, there is a theoretically infinite number of ways to define the strain measure.

There are two alternative descriptions of motion within the deformable body, e.g., [1], [2]. The first one is called the *Lagrangian* or *material* description, in which a material particle X is traced from the undeformed configuration C_0 to the deformed configuration C_1. The position of a material point is measured with, what is called, the *material coordinates* as $\boldsymbol{X} = [X, Y, Z]$ defining the position in the undeformed configuration and $\boldsymbol{x} = [x, y, z]$ being the position of the same material point in the deformed configuration. The motion of the body is referred with respect to a reference configuration that is usually taken as the undeformed configuration. In a Lagrangian description the current position of a material point can be expressed in terms of the reference (undeformed) position as:

$$\boldsymbol{x} = \boldsymbol{X} + \boldsymbol{u} \tag{6.1}$$

with $\boldsymbol{u}$ being a vector of displacements (see Fig. 6.1).

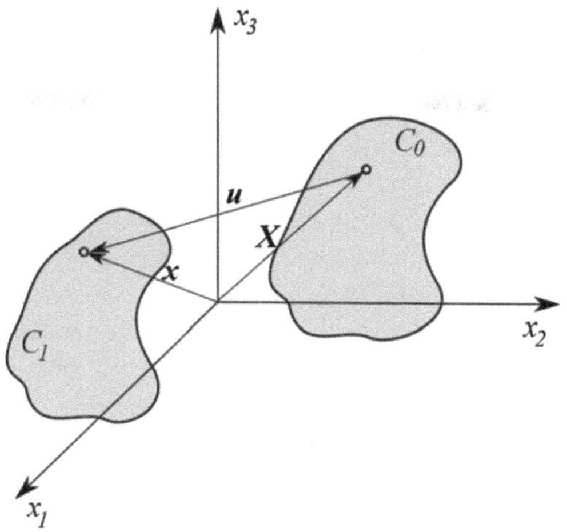

FIGURE 6.1 Material coordinates in the deformed and undeformed configurations.

In the *Eulerian* or *spatial* description, the motion is referred to the current configuration occupied by the body, and any particular quantity or interest ϕ (e.g., value of temperature, pressure, stress, etc.) is described with respect to the spatial coordinates $\boldsymbol{x} = [x, y, z]$ currently occupied by a certain material point. In this case the focus is not on a particular material point, but rather on the location of a point in space that, during the analysis, can be occupied by different material points. Within the Eulerian description the configuration can be seen as a "window" through which the motion of different material points is observed (see Fig. 6.2).

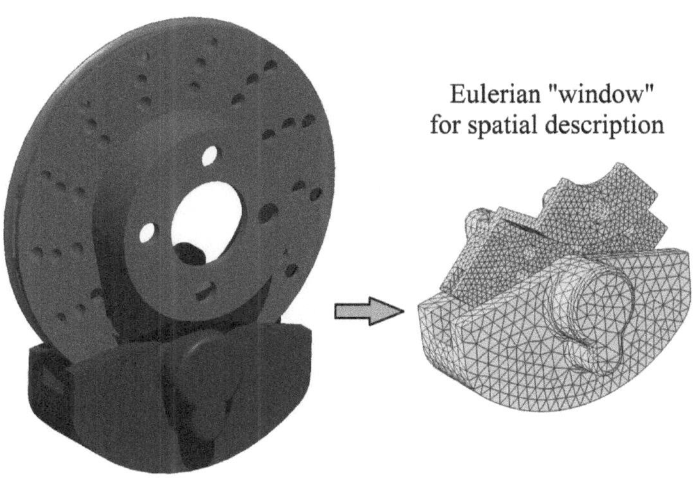

FIGURE 6.2 Spatial description of a rotating disk.

From the above description, it is clear that the Eulerian description is not the preferred one in the mechanics of deformable solids since the deformed configuration is unknown. On the contrary it is well suited to study the motion of fluids or to perform rolling contact analyses as the spatial configuration is known and it is not changing.

Throughout this book, we will consider only the Lagrangian description as our focus is to study the structural response of deformable solids.

In general, all nonlinear problems are solved incrementally. The simulation of a deformable body that undergoes large changes in geometry normally contains intermediate solutions when moving from the initial configuration to the final deformed one. The *total Lagrangian description* depicts the situation in which the initial configuration is used as the reference with respect to which all the quantities are measured. Alternatively, when using the *updated Lagrangian description*, the latest known configuration is used as the reference one (i.e. within the ith iteration it consists of the configuration $i - 1$). As it can be intuitively expected, the two descriptions differ in the selection of the strain measures, since they make reference to different configurations.

We will now consider different strain measures that are commonly used when employing the total Lagrangian approach. Detailed description of both approaches, i.e., the total Lagrangian description and the updated Lagrangian description, and their comparison in performance, are not covered here. For further information on these aspects, reference should be made, for example, to available textbooks dealing with nonlinear FEA (e.g., [1], [2]).

6.2 Strain and stress measures for the total Lagrangian description

When points of a deformable body undergo large displacements the measure of the engineering strain becomes inaccurate. The problem becomes apparent even when large displacements of material points are produced by a large rotation without material deformation. This can be observed by considering that when a single finite element is subjected to a large rotation (e.g., say of 45°), the two dilations, i.e., ε_x and ε_y, significantly differ from zero. In the FEA literature, geometrically nonlinear problems are decomposed in two groups: (1) problems of small deformation but large rotations; (2) problems of large deformation. The former one can occur for slender structural elements and, for a proper structural representation, requires a modification in the strain measure, while still using Hooke's law for defining the relationship between strains and stresses. It should be mentioned that, strictly speaking, different elastic constants should be used, since what is usually found in the literature refers to engineering strain and engineering stress. The difference however is not large. The latter occurs for hyper-elastic materials and for these problems the constitutive model should be formulated through the strain energy density.

In the following, we first consider the problem of large rotations, also referred to as the large displacement problem, even though such formulation is not sufficiently precise since only the rotation component of the displacement leads to inaccurate results when the engineering strain is used.

The deformation measure used in the total Lagrangian formulation is the Green–Lagrange strain tensor $\boldsymbol{E}$ (also denoted as the Green strain) that is expressed in terms of deformation gradient tensor as follows:

$$\boldsymbol{E} = \frac{1}{2}(\boldsymbol{F}^T \cdot \boldsymbol{F} - \boldsymbol{I}) \tag{6.2}$$

where $\boldsymbol{I}$ is the identity matrix, and $\boldsymbol{F}$ is the second-order deformation gradient tensor, which is a fundamental quantity used in different strain measures for large rotation and large deformation. $\boldsymbol{F}$ relates a material line $d\boldsymbol{X} = [dX, dY, dZ]$ before deformation to its corresponding material line after deformation $d\boldsymbol{x} = [dx, dy, dz]$ as follows:

$$d\boldsymbol{x} = \boldsymbol{F} \cdot d\boldsymbol{X} = d\boldsymbol{X} \cdot \boldsymbol{F}^T \rightarrow \boldsymbol{F} = \left(\frac{\partial \boldsymbol{x}}{\partial \boldsymbol{X}}\right)^T \tag{6.3}$$

where $\boldsymbol{F}$ can be written in matrix form as:

$$\boldsymbol{F} = \begin{bmatrix} \frac{\partial x}{\partial X} & \frac{\partial x}{\partial Y} & \frac{\partial x}{\partial Z} \\ \frac{\partial y}{\partial X} & \frac{\partial y}{\partial Y} & \frac{\partial y}{\partial Z} \\ \frac{\partial z}{\partial X} & \frac{\partial z}{\partial Y} & \frac{\partial z}{\partial Z} \end{bmatrix} \tag{6.4}$$

We now want to establish the relationship between strain components and displacements. For illustrative purposes and as previously considered in Chapter 1, we present the derivation for a 2D case. We start from the expanded form of the Green strain tensor that can be written as (see Eq. (6.2)):

$$\boldsymbol{E} = \frac{1}{2}\begin{bmatrix} \frac{\partial x}{\partial X} & \frac{\partial y}{\partial X} \\ \frac{\partial x}{\partial Y} & \frac{\partial y}{\partial Y} \end{bmatrix}\begin{bmatrix} \frac{\partial x}{\partial X} & \frac{\partial x}{\partial Y} \\ \frac{\partial y}{\partial X} & \frac{\partial y}{\partial Y} \end{bmatrix} - \frac{1}{2}\begin{bmatrix} 1 & 0 \\ 0 & 1 \end{bmatrix} \tag{6.5}$$

and, after performing the multiplication, $\boldsymbol{E}$ becomes:

$$\boldsymbol{E} = \frac{1}{2}\begin{bmatrix} (\frac{\partial x}{\partial X})^2 + (\frac{\partial y}{\partial X})^2 - 1 & \frac{\partial x}{\partial X}\frac{\partial x}{\partial Y} + \frac{\partial y}{\partial X}\frac{\partial y}{\partial Y} \\ \frac{\partial x}{\partial X}\frac{\partial x}{\partial Y} + \frac{\partial y}{\partial X}\frac{\partial y}{\partial Y} & (\frac{\partial x}{\partial Y})^2 + (\frac{\partial y}{\partial Y})^2 - 1 \end{bmatrix} \tag{6.6}$$

The Green strain tensor is symmetric and its components can be further represented, in Voigt notation, as a three-component vector based on the following notation:

$$\boldsymbol{E} = [E_{XX}, E_{YY}, E_{XY}]^T \tag{6.7}$$

The relationship relating the position of a material point before and after deformation previously introduced in Eq. (6.1) can be expressed in scalar form for 2D as follows:

$$\begin{aligned} x &= X + u \\ y &= Y + v \end{aligned} \tag{6.8}$$

By differentiation through the chain rule, the three components of the Green strain (see Eq. (6.7)) can be expressed for the 2D case as:

$$\begin{bmatrix} E_{XX} \\ E_{YY} \\ E_{XY} \end{bmatrix} = \frac{1}{2} \begin{bmatrix} 2\frac{\partial u}{\partial X} + (\frac{\partial u}{\partial X})^2 + (\frac{\partial v}{\partial X})^2 \\ 2\frac{\partial v}{\partial Y} + (\frac{\partial u}{\partial Y})^2 + (\frac{\partial v}{\partial Y})^2 \\ \frac{\partial u}{\partial Y} + \frac{\partial v}{\partial X} + \frac{\partial u}{\partial X}\frac{\partial u}{\partial Y} + \frac{\partial v}{\partial X}\frac{\partial v}{\partial Y} \end{bmatrix} \tag{6.9}$$

The capital letters for coordinates X, Y, Z are used here to highlight that the strain components are computed with respect to the initial, undeformed configuration. It is useful to perform the following further decomposition of Eq. (6.9) to separate the infinitesimal engineering strain from the nonlinear strain components:

$$\boldsymbol{E} = \begin{bmatrix} \frac{\partial u}{\partial X} \\ \frac{\partial v}{\partial Y} \\ \frac{1}{2}(\frac{\partial u}{\partial Y} + \frac{\partial v}{\partial X}) \end{bmatrix} + \frac{1}{2} \begin{bmatrix} (\frac{\partial u}{\partial X})^2 + (\frac{\partial v}{\partial X})^2 \\ (\frac{\partial u}{\partial Y})^2 + (\frac{\partial v}{\partial Y})^2 \\ (\frac{\partial u}{\partial X})(\frac{\partial u}{\partial Y}) + (\frac{\partial v}{\partial X})(\frac{\partial v}{\partial Y}) \end{bmatrix} \tag{6.10}$$

This decomposition is particularly useful for a FE implementation in which the first infinitesimal part, previously introduced in (1.88) and here denoted by $\boldsymbol{L}^L$, is combined with the second nonlinear part referred to as $\boldsymbol{L}^{NL}$.

We will now derive the expression of the strain matrix for the 2D case to illustrate the terms included in $\boldsymbol{L}^{NL}$. In the definition of $\boldsymbol{E}$ provided in (6.10) the out-of-diagonal strain term (i.e., the third term in the vector representation of the strain) needs to be multiplied by 2 to obtain the shear deformation. The modification is required to establish the equivalence between a double-dot product of two second-order tensors (i.e., stress and strain tensors) and a scalar product of two vectors. As previously explained in Chapter 1 (see Eq. (1.75)), by such representation of the out-of-diagonal terms the scalar product of the two vectors $\boldsymbol{\varepsilon} \cdot \boldsymbol{\sigma}$ represent the strain energy, i.e., the scaler quantity otherwise obtained as ${}^2\boldsymbol{\varepsilon} : {}^2\boldsymbol{\sigma}$, where stress and strain expressed as second-order tensors are here denoted by ${}^2\boldsymbol{\sigma}$ and ${}^2\boldsymbol{\varepsilon}$, respectively. Throughout this book, no extensive use of stress and strain given as second-order tensors will be used, therefore no specific notation is introduced for it. It will be generally clear from the context whether stress or strain tensor are given in Voigt or in full notation.

In order to derive $\boldsymbol{B}^{NL}$, we follow the same path as for the engineering strain, by expressing the strain-displacement relationship with the decomposition introduced in (6.10) for a point in the deformable body:

$$\begin{bmatrix} \varepsilon_{XX} \\ \varepsilon_{YY} \\ \gamma_{XY} \end{bmatrix} = \begin{bmatrix} \frac{\partial}{\partial X} & 0 \\ 0 & \frac{\partial}{\partial Y} \\ \frac{\partial}{\partial Y} & \frac{\partial}{\partial X} \end{bmatrix} \cdot \begin{bmatrix} u \\ v \end{bmatrix} + \frac{1}{2} \begin{bmatrix} \frac{\partial u}{\partial X}\frac{\partial}{\partial X} & \frac{\partial v}{\partial X}\frac{\partial}{\partial X} \\ \frac{\partial u}{\partial Y}\frac{\partial}{\partial Y} & \frac{\partial v}{\partial Y}\frac{\partial}{\partial Y} \\ \frac{\partial u}{\partial Y}\frac{\partial}{\partial X} + \frac{\partial u}{\partial X}\frac{\partial}{\partial Y} & \frac{\partial v}{\partial Y}\frac{\partial}{\partial X} + \frac{\partial v}{\partial X}\frac{\partial}{\partial Y} \end{bmatrix} \cdot \begin{bmatrix} u \\ v \end{bmatrix}$$

$$\rightarrow \boldsymbol{\varepsilon} = (\boldsymbol{L}^L + \frac{1}{2}\boldsymbol{L}^{NL}) \cdot \boldsymbol{u} \tag{6.11}$$

where the third row terms of Eq. (6.10) have been re-arranged as follows:

$$2\frac{\partial u}{\partial X}\frac{\partial u}{\partial Y} = \frac{\partial u}{\partial X}\frac{\partial u}{\partial Y} + \frac{\partial u}{\partial X}\frac{\partial u}{\partial Y} = (\frac{\partial u}{\partial Y}\frac{\partial}{\partial X} + \frac{\partial u}{\partial X}\frac{\partial}{\partial Y})u \tag{6.12}$$

This rearrangement enables to cast the nonlinear part of the matrix $\boldsymbol{L}$ into a form that premultiplies the vector of displacement of an arbitrary point $\boldsymbol{u}$, so that it can be added to the linear part $\boldsymbol{L}^L$, given by Eq. (1.86).

By recalling the four-node quadrilateral element previously introduced in Chapter 1 and its displacement in an arbitrary point interpolated through (1 84), we can write the Green strain in the form suitable for FE implementation as:

$$\boldsymbol{E} = (\boldsymbol{L}^L + \frac{1}{2}\boldsymbol{L}^{NL}) \cdot \boldsymbol{N} \cdot \boldsymbol{d}_e = (\boldsymbol{B}^L + \frac{1}{2}\boldsymbol{B}^{NL}) \cdot \boldsymbol{d}_e \tag{6.13}$$

tacitly assuming that in this formulation the third strain component represents the shear deformation (that can be calculated by multiplying by two the engineering shear strain). $\boldsymbol{B}^L$ is computed with Eq. (1.90), while $\boldsymbol{B}^{NL}$ is given by:

$$\boldsymbol{B}^{NL} = \boldsymbol{B}_u^{NL} + \boldsymbol{B}_v^{NL}$$

$$= \begin{bmatrix} \frac{\partial u}{\partial X}\frac{\partial N_1}{\partial X} & 0 & \frac{\partial u}{\partial X}\frac{\partial N_2}{\partial X} & 0 & \frac{\partial u}{\partial X}\frac{\partial N_3}{\partial X} & 0 & \frac{\partial u}{\partial X}\frac{\partial N_4}{\partial X} & 0 \\ \frac{\partial u}{\partial Y}\frac{\partial N_1}{\partial Y} & 0 & \frac{\partial u}{\partial Y}\frac{\partial N_2}{\partial Y} & 0 & \frac{\partial u}{\partial Y}\frac{\partial N_3}{\partial Y} & 0 & \frac{\partial u}{\partial Y}\frac{\partial N_4}{\partial Y} & 0 \\ \frac{\partial u}{\partial X}\frac{\partial N_1}{\partial Y} + \frac{\partial u}{\partial Y}\frac{\partial N_1}{\partial X} & 0 & \frac{\partial u}{\partial X}\frac{\partial N_2}{\partial Y} + \frac{\partial u}{\partial Y}\frac{\partial N_2}{\partial X} & 0 & \frac{\partial u}{\partial X}\frac{\partial N_3}{\partial Y} + \frac{\partial u}{\partial Y}\frac{\partial N_3}{\partial X} & 0 & \frac{\partial u}{\partial X}\frac{\partial N_4}{\partial Y} + \frac{\partial u}{\partial Y}\frac{\partial N_4}{\partial X} & 0 \end{bmatrix}$$

$$+ \begin{bmatrix} 0 & \frac{\partial v}{\partial X}\frac{\partial N_1}{\partial X} & 0 & \frac{\partial v}{\partial X}\frac{\partial N_2}{\partial X} & 0 & \frac{\partial v}{\partial X}\frac{\partial N_3}{\partial X} & 0 & \frac{\partial v}{\partial X}\frac{\partial N_4}{\partial X} \\ 0 & \frac{\partial v}{\partial Y}\frac{\partial N_1}{\partial Y} & 0 & \frac{\partial v}{\partial Y}\frac{\partial N_2}{\partial Y} & 0 & \frac{\partial v}{\partial Y}\frac{\partial N_3}{\partial Y} & 0 & \frac{\partial v}{\partial Y}\frac{\partial N_4}{\partial Y} \\ 0 & \frac{\partial v}{\partial X}\frac{\partial N_1}{\partial Y} + \frac{\partial v}{\partial Y}\frac{\partial N_1}{\partial X} & 0 & \frac{\partial v}{\partial X}\frac{\partial N_2}{\partial Y} + \frac{\partial v}{\partial Y}\frac{\partial N_2}{\partial X} & 0 & \frac{\partial v}{\partial X}\frac{\partial N_3}{\partial Y} + \frac{\partial v}{\partial Y}\frac{\partial N_3}{\partial X} & 0 & \frac{\partial v}{\partial X}\frac{\partial N_4}{\partial Y} + \frac{\partial v}{\partial Y}\frac{\partial N_4}{\partial X} \end{bmatrix} \tag{6.14}$$

The additional decomposition introduced for $\boldsymbol{B}^{NL} = \boldsymbol{B}_u^{NL} + \boldsymbol{B}_v^{NL}$ into the nonlinear part related to displacement component u and to displacement component v is here done for clarity and to simplify the writing. $\boldsymbol{B}^{NL}$ is nonlinear in displacements u and v as it contains partial derivatives of the displacement components u and v with respect to the material coordinates X and Y. For the quadrilateral element considered, these derivatives are computed using standard shape functions:

$$\begin{aligned}
\frac{\partial u}{\partial X} &= [\frac{\partial N_1}{\partial X}, \frac{\partial N_2}{\partial X}, \frac{\partial N_3}{\partial X}, \frac{\partial N_4}{\partial X}] \cdot [u_1, u_2, u_3, u_4]^T \\
\frac{\partial u}{\partial Y} &= [\frac{\partial N_1}{\partial Y}, \frac{\partial N_2}{\partial Y}, \frac{\partial N_3}{\partial Y}, \frac{\partial N_4}{\partial Y}] \cdot [u_1, u_2, u_3, u_4]^T \\
\frac{\partial v}{\partial X} &= [\frac{\partial N_1}{\partial X}, \frac{\partial N_2}{\partial X}, \frac{\partial N_3}{\partial X}, \frac{\partial N_4}{\partial X}] \cdot [v_1, v_2, v_3, v_4]^T \\
\frac{\partial v}{\partial Y} &= [\frac{\partial N_1}{\partial Y}, \frac{\partial N_2}{\partial Y}, \frac{\partial N_3}{\partial Y}, \frac{\partial N_4}{\partial Y}] \cdot [v_1, v_2, v_3, v_4]^T
\end{aligned} \tag{6.15}$$

A MATLAB® function performing the calculation of the strain in a single point of the quadrilateral element is presented in the following script. The function can compute the engineering strain and the Green strain by specifying the value of argument (NL) to be equal to 0 or 1, respectively.

```
function strain=greenstrain_in_point(COORD,de,pnt,NL)
% Function that computes strain components in a point
% Function arguments:
% COORD - matrix of nodal coordinates (physical)
% de - vector of nodal displacements
% pnt - coordinates of point where strain is to be computed (natural coordinates)
% NL - integer equal to 0 for engineering, or equal to 1 for Green strain
%------------------------------------------------------------------------
% Coordinates in natural space
ksi=pnt(1);
eta=pnt(2);
% Derivatives of shape functions with respect to natural coordinates
dN1ksi=-0.25*(1-eta);
dN1eta=-0.25*(1-ksi);
dN2ksi=0.25*(1-eta);
dN2eta=-0.25*(1+ksi);
dN3ksi=0.25*(1+eta);
dN3eta=0.25*(1+ksi);
dN4ksi=-0.25*(1+eta);
dN4eta=0.25*(1-ksi);
% Nodal coordinates
X1=COORD(1,1); Y1=COORD(1,2);
X2=COORD(2,1); Y2=COORD(2,2);
X3=COORD(3,1); Y3=COORD(3,2);
X4=COORD(4,1); Y4=COORD(4,2);
% Decomposing vector of nodal displacements to x-part and y-part
ue=[de(1);de(3);de(5);de(7)];
ve=[de(2);de(4);de(6);de(8)];
% Defining Jacobean and inverting it
JAK_col1=[dN1ksi,0,dN2ksi,0,dN3ksi,0,dN4ksi,0;...
          dN1eta,0,dN2eta,0,dN3eta,0,dN4eta,0]...
    *[X1;Y1;X2;Y2;X3;Y3;X4;Y4];
JAK_col2=[0,dN1ksi,0,dN2ksi,0,dN3ksi,0,dN4ksi;...
          0,dN1eta,0,dN2eta,0,dN3eta,0,dN4eta]...
    *[X1;Y1;X2;Y2;X3;Y3;X4;Y4];
```

```
JAK=[JAK_col1 , JAK_col2]
JAKin=inv(JAK);
% Derivatives  of shape function with respect to material coordinates:
dN1dX=JAKin(1,1)*dN1ksi+JAKin(1,2)*dN1eta;
dN2dX=JAKin(1,1)*dN2ksi+JAKin(1,2)*dN2eta;
dN3dX=JAKin(1,1)*dN3ksi+JAKin(1,2)*dN3eta;
dN4dX=JAKin(1,1)*dN4ksi+JAKin(1,2)*dN4eta;
dN1dY=JAKin(2,1)*dN1ksi+JAKin(2,2)*dN1eta;
dN2dY=JAKin(2,1)*dN2ksi+JAKin(2,2)*dN2eta;
dN3dY=JAKin(2,1)*dN3ksi+JAKin(2,2)*dN3eta;
dN4dY=JAKin(2,1)*dN4ksi+JAKin(2,2)*dN4eta;
% Matrix BL:
BBL=[dN1dX,0,dN2dX,0,dN3dX,0,dN4dX,0;
      0,dN1dY,0,dN2dY,0,dN3dY,0,dN4dY;
      dN1dY,dN1dX,dN2dY,dN2dX,dN3dY,dN3dX,dN4dY,dN4dX];
% Derivatives of displacement components:
dudX=[dN1dX,dN2dX,dN3dX,dN4dX]*ue;
dudY=[dN1dY,dN2dY,dN3dY,dN4dY]*ue;
dvdX=[dN1dX,dN2dX,dN3dX,dN4dX]*ve;
dvdY=[dN1dY,dN2dY,dN3dY,dN4dY]*ve;
BBNL=[dudX*dN1dX,dvdX*dN1dX,dudX*dN2dX,dvdX*dN2dX,...
    dudX*dN3dX,dvdX*dN3dX,dudX*dN4dX,dvdX*dN4dX;...
    dudY*dN1dY,dvdY*dN1dY,dudY*dN2dY,dvdY*dN2dY,...
    dudY*dN3dY,dvdY*dN3dY,dudY*dN4dY,dvdY*dN4dY;...
    dudX*dN1dY+dudY*dN1dX,dvdX*dN1dY+dvdY*dN1dX,dudX*dN2dY+dudY*dN2dX,...
    dvdX*dN2dY+dvdY*dN2dX,dudX*dN3dY+dudY*dN3dX,dvdX*dN3dY+dvdY*dN3dX,...
    dudX*dN4dY+dudY*dN4dX,dvdX*dN4dY+dvdY*dN4dX];
 % Total B matrix
BB=BBL+0.5*BBNL*NL;
strain_L=BBL*de'
strain_NL=0.5*BBNL*NL*de'
strain=BB*de';
```

We can now perform a simple, single element, calculation to demonstrate the error induced by the engineering strain when large rotations are present. For clarity, we consider a single rectangular element that undergoes a rotation of 45° (see Fig. 6.3) and that exhibits null deformation. The strain component for dilation along the X- and Y-axis of the engineering strain significantly differ from zero, therefore creating a fictitious strain and associated stress, while the Green strain recognizes that all strain components are equal to zero. The results presented in Fig. 6.3 have been obtained by calling the MATLAB® function `greenstrain_in_point`. The listing of the script used to calculate the results of Fig. 6.3 is given below.

```
rotation=pi/4
COORD=[ 0 0 ; 1 0 ; 1 1 ; 0 1 ]
pnt=[ 0 0 ]
NL=1
coord(1,1)=cos(rotation)*COORD(1,1)-sin(rotation)*COORD(1,2)
coord(2,1)=cos(rotation)*COORD(2,1)-sin(rotation)*COORD(2,2)
coord(3,1)=cos(rotation)*COORD(3,1)-sin(rotation)*COORD(3,2)
```

```
coord(4,1)=cos(rotation)*COORD(4,1)-sin(rotation)*COORD(4,2)
coord(1,2)=sin(rotation)*COORD(1,1)+cos(rotation)*COORD(1,2)
coord(2,2)=sin(rotation)*COORD(2,1)+cos(rotation)*COORD(2,2)
coord(3,2)=sin(rotation)*COORD(3,1)+cos(rotation)*COORD(3,2)
coord(4,2)=sin(rotation)*COORD(4,1)+cos(rotation)*COORD(4,2)
de(1)=coord(1,1)-COORD(1,1)
de(2)=coord(1,2)-COORD(1,2)
de(3)=coord(2,1)-COORD(2,1)
de(4)=coord(2,2)-COORD(2,2)
de(5)=coord(3,1)-COORD(3,1)
de(6)=coord(3,2)-COORD(3,2)
de(7)=coord(4,1)-COORD(4,1)
de(8)=coord(4,2)-COORD(4,2)
strain=greenstrain_in_point_rev(COORD,de,pnt,NL)
```

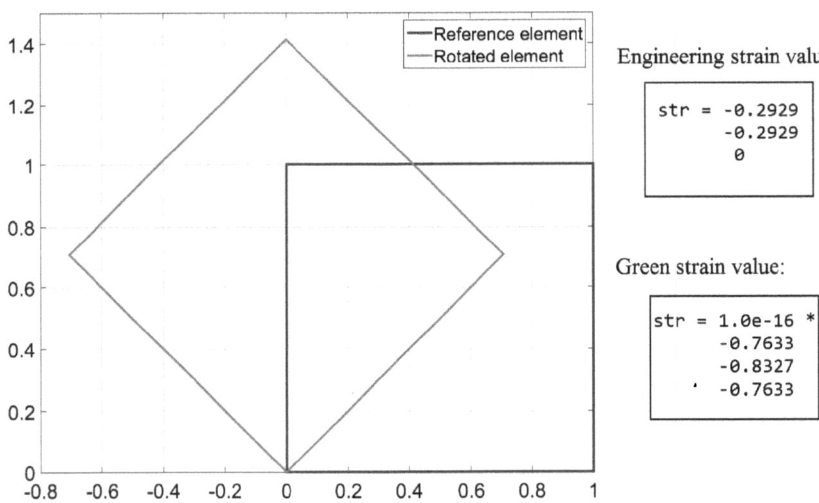

FIGURE 6.3 Pure rotation and corresponding engineering and Green strains.

Even though the main focus of this chapter is on the total Lagrangian approach, we will now demonstrate that only a marginal modification is required within the presented implementation to formulate the function that computes the Almansi strain that relies for reference on the deformed configuration and that is used with the updated Lagrangian formulation. The Almansi strain, also denoted as the Euler strain, is defined as follows:

$$\boldsymbol{e} = \frac{1}{2}\left[\boldsymbol{I} - (\boldsymbol{F}^{-1})^T \cdot \boldsymbol{F}^{-1}\right] \tag{6.16}$$

By recalling that:

$$d\boldsymbol{X} = \boldsymbol{F}^{-1} \cdot d\boldsymbol{x} \rightarrow (\boldsymbol{F}^{-1})^T = \left(\frac{\partial \boldsymbol{X}}{\partial \boldsymbol{x}}\right) \tag{6.17}$$

and by manipulating the expressions considering the coordinate relationship previously introduced in Eq. (6.8), the components of the Almansi strain can be written as:

$$\begin{bmatrix} e_{xx} \\ e_{yy} \\ e_{xy} \end{bmatrix} = \frac{1}{2} \begin{bmatrix} 2\frac{\partial u}{\partial x} - (\frac{\partial u}{\partial x})^2 - (\frac{\partial v}{\partial x})^2 \\ 2\frac{\partial v}{\partial y} - (\frac{\partial u}{\partial y})^2 - (\frac{\partial v}{\partial y})^2 \\ \frac{\partial u}{\partial y} + \frac{\partial v}{\partial x} - \frac{\partial u}{\partial x}\frac{\partial u}{\partial y} - \frac{\partial v}{\partial x}\frac{\partial v}{\partial y} \end{bmatrix} \tag{6.18}$$

This expression can be further rearranged to separate the infinitesimal strain from the nonlinear part as follows:

$$\boldsymbol{e} = \begin{bmatrix} \frac{\partial u}{\partial x} \\ \frac{\partial v}{\partial y} \\ \frac{1}{2}(\frac{\partial u}{\partial y} + \frac{\partial v}{\partial x}) \end{bmatrix} - \frac{1}{2} \begin{bmatrix} (\frac{\partial u}{\partial x})^2 + (\frac{\partial v}{\partial x})^2 \\ (\frac{\partial u}{\partial y})^2 + (\frac{\partial v}{\partial y})^2 \\ (\frac{\partial u}{\partial x})(\frac{\partial u}{\partial y}) + (\frac{\partial v}{\partial x})(\frac{\partial v}{\partial y}) \end{bmatrix} \tag{6.19}$$

All partial derivatives included in Eq. (6.19) are expressed with respect to the current, deformed, configuration as depicted by the use of the small letters x and y.

By comparing Eq. (6.19) with Eq. (6.10), it may be seen that they differ by one sign: the nonlinear part of the strain is added to the infinitesimal strain to obtain the Green strain, while it is subtracted from it, to get the Almansi strain. It is also important to emphasize that the components of the Green strain are obtained by performing the derivatives with respect to the undeformed configuration, while in the case of the Almansi strain the derivatives are determined with respect to the deformed (current) configuration. Based on these considerations, the Green and the Almansi strains are expected to differ in an arbitrary case.

The fact that the difference is only in one sign makes the numerical implementation easily achieved by modification of the previously presented MATLAB® function developed to compute the Green strain at a point (i.e., `greenstrain_in_point`). Keeping the rest of the function the same, for the implementation of the Almansi strain calculation it is required to change the line related to the evaluation of the matrix $\boldsymbol{B}^{NL}$, as shown below, while considering all variables determined with respect to the current, deformed, configuration as a function of x and y. For this reason, the coordinates of the element passed as the argument of the function in this case have to be the deformed ones.

```
...
% Total B matrix
BB=BBL-0.5*BBNL*NL;
...
```

The same, previously discussed example of a single element rotated by 45° can be used to show that also the Almansi strain can detect that no deformation are induced in an element subjected to a large rotation.

In an arbitrary case, the values of the Green and Almansi strain components are different and this is shown with the following example. For this purpose, we consider a single rectangular element with sides equal to 1, and we apply a node displacement of $u = 0.2$ and $v = 0.35$ only to the top right node as depicted in Fig. 6.4. For this deformed element, the three different strain measures, i.e., the infinitesimal, the Green, and the Almansi strains, are computed in the central point (i.e., with natural coordinates $\xi = 0, \eta = 0$). The three values are reported in the figure, and the difference between them is notable.[1] Later in the chapter, we highlight how such difference has an important consequence in the definition of the elastic constants of the constitutive model.

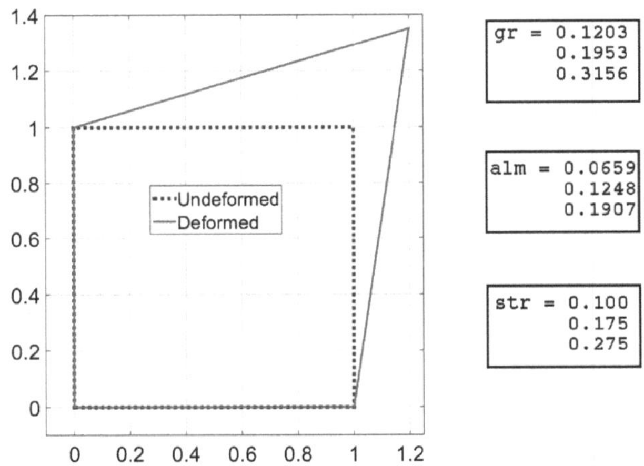

FIGURE 6.4 Single rectangular element deformed and deformation calculated with the infinitesimal, the Green, and the Almansi strains in the central point.

The presented implementation computes the strain an a single point. When calculating the stiffness matrix of an element, the solution requires the determination of the strain values at different points to evaluate the integral of Eq. (1.95). This is usually performed based on a Gauss numerical integration (that requires strains to be computed at Gauss points). For the bilinear quadrilateral element considered here, the "full" integration requires the use of four Gauss points. When dealing with small deformation and small rotations, the engineering strain can be used so only the part here denoted by $\boldsymbol{B}^L$ is considered both for the evaluation of the integral (1.95) (i.e., to compute the stiffness matrix) and for the computation of the strain at a point. For geometrically nonlinear problems, the situation is somewhat different since, due to the nonlinearity, the solution is sought iteratively, with the use of the *tangent stiffness matrix*. Unlike in linear problems, two different matrices $\boldsymbol{B}$ are used to compute strains at a

[1] The calculation of the strains can be performed with the MATLAB® functions previously shown while making sure to consider the undeformed configuration in the evaluation of the infinitesimal and Green strains, and the deformed configuration for the Almansi strain.

point and to evaluate the integral for the tangent stiffness matrix. We now illustrate this concept with a simple, 1-dimensional example whose solution strategy makes use of the tangent stiffness matrix.

6.3 Tangent stiffness matrix for geometrically nonlinear problems: a simple one-dimensional example

Let us consider the inclined bar element loaded with a vertical force, applied to its end B that is allowed to move vertically (see Fig. 6.5), while the other end A is pinned, allowing only rotation. The material of the bar is linear-elastic with known values of the Young's modulus E and the cross-section area of the bar, denoted as A. Upon application of the load, the bar is compressed and, to satisfy equilibrium, the vertical projection of this compressive force has to equal the external force F. As the applied force increases, the bar shortens and a vertical movement takes place at the end B. In this process, the vertical projection of the compressive force varies due to the rotation of the bar. These aspects define the nonlinear nature of the problem, even if the bar behaves in a linear-elastic fashion. A diagram of the system's response is illustrated in Fig. 6.6, while the main variables considered are depicted in Fig. 6.5.

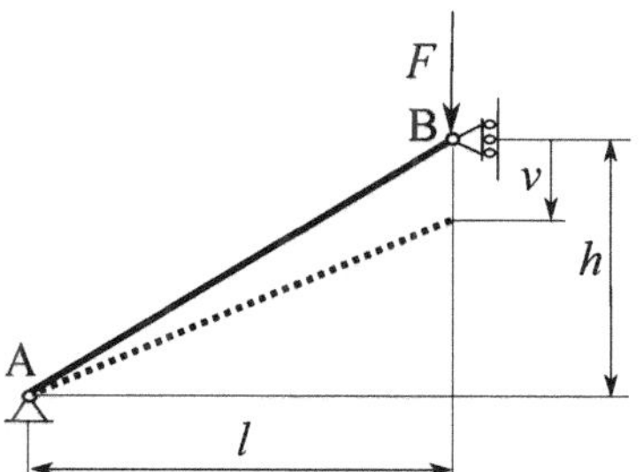

FIGURE 6.5 A 1-dimensional geometrically nonlinear problem for an inclined bar loaded with a vertical force.

With reference to the notation of Fig. 6.5 and based on geometrical considerations, we can calculate the strain in the bar, in an arbitrary deformed position as follows:

$$\varepsilon = \frac{\sqrt{(h-v)^2+l^2}-\sqrt{h^2+l^2}}{\sqrt{h^2+l^2}} = \frac{\sqrt{(\frac{h-v}{l})^2+1}}{\sqrt{(\frac{h}{l})^2+1}} - 1 = \left[1+\left(\frac{h-v}{l}\right)^2\right]^{\frac{1}{2}}\left[1+\left(\frac{h}{l}\right)^2\right]^{-\frac{1}{2}} - 1 \tag{6.20}$$

that lends itself for approximation in the following form[2]:

$$\varepsilon \approx \left[1+\frac{1}{2}\left(\frac{h-v}{l}\right)^2\right]\left[1-\frac{1}{2}\left(\frac{h}{l}\right)^2\right]-1= \\ =\frac{1}{2}\left(\frac{h-v}{l}\right)^2-\frac{1}{2}\left(\frac{h}{l}\right)^2-\frac{1}{4}\left(\frac{h}{l}\right)^2\left(\frac{h-v}{l}\right)^2 \tag{6.21}$$

This expression can be further simplified as follows by neglecting the last term because it is of a higher order than the order of the first two terms:

$$\varepsilon \approx \frac{1}{2}\left(\frac{v}{l}\right)^2-\left(\frac{h}{l}\right)\left(\frac{v}{l}\right) \tag{6.22}$$

based on which the force in the compressed bar can be approximated as:

$$N = EA\varepsilon \approx \frac{EA}{l^2}\left(\frac{1}{2}v^2-hv\right) \tag{6.23}$$

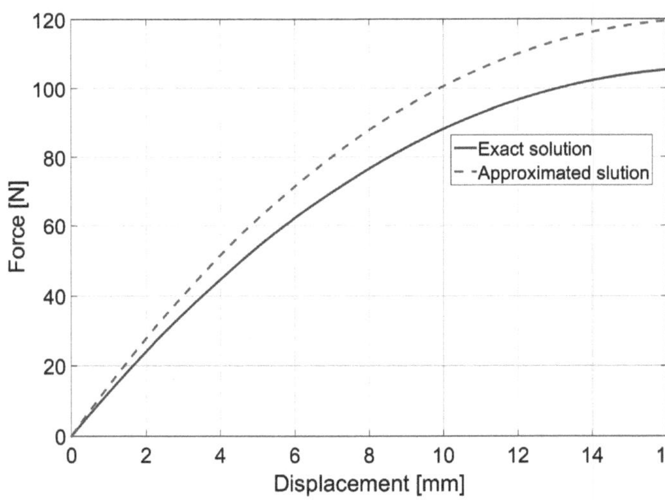

FIGURE 6.6 Force displacement response: exact and approximated.

The force N in Eq. (6.23) is compressive and, therefore, negative, when the bar undergoes a positive v. In this case the strain computed with (6.20) depicts a shortening and is therefore negative. When writing the force-displacement relation for the system, it is convenient to associate a positive force F with a

---

[2] The approximation is obtained by applying the expansion through the binomial series of the function $(1+x)^a = \sum_{k=1}^{\infty}\binom{a}{k}x^k = 1+ax+\frac{a(a-1)}{2!}x^2+\ldots$ for the cases of $a=\frac{1}{2}$ and $a=-\frac{1}{2}$, and by keeping the first two terms.

positive (i.e., downwards) displacement v that can be expressed as a function of the bar force N as follows:

$$F = -N \sin\theta \tag{6.24}$$

where θ is the inclination angle which is changing with the deformation and is equal to:

$$\sin\theta = \frac{h - v}{\sqrt{h^2 + l^2}} \tag{6.25}$$

Strictly speaking, the denominator in Eq. (6.25) should contain the deformed length of the bar and, therefore, should be equal to $\sqrt{(h-v)^2 + l^2}$. We accept this simplification in the expression for θ in Eq. (6.25) for the purpose of this illustrative example that aims at demonstrating the iterative solution strategy. The force-displacement curve obtained with the adopted approximations is compared against the curve determined with exact closed-form solution of the problem in Fig. 6.5.

By inserting Eqs. (6.23) and (6.25) into Eq. (6.24), we can derive an expression for the approximated force-displacement relationship in the following form:

$$F = C_0(\frac{1}{2}v^3 - \frac{3}{2}hv^2 + h^2 v) \tag{6.26}$$

which is clearly nonlinear as it contains both square and cubic terms of v. Constant C_0 collects all geometrical and material properties which are not changing during the analysis and is equal to:

$$C_0 = \frac{EA}{l^2\sqrt{h^2 + l^2}} \tag{6.27}$$

When dealing with nonlinear problems, it is usually desirable to trace the solution in terms of the force-displacement response, rather than to determine the displacement for a specific value of applied force. This is achieved by applying the load incrementally and the solution for each increment is typically obtained based on an iterative Newton–Raphson procedure. This incremental-iterative procedure is centered on the minimization of a residual that, in our example, represents the out-of-balance force defined as follows:

$$r = C_0(\frac{1}{2}v^3 - \frac{3}{2}hv^2 + h^2 v) - i \cdot \Delta F \tag{6.28}$$

where, for ease of notation, ΔF represents the increment of load applied at each increment step, and i depicts the load increment number.

Eq. (6.28) is used to find the displacement v for which the equilibrium condition is satisfied (i.e., the residual is equal to zero). The iterative procedure to be implemented for this minimization is obtained on the basis of a truncated

Taylor series based on:

$$r_n = r_0 + \frac{dr}{dv}\delta v + \frac{1}{2}\left[\frac{d^2 r}{dv^2}\right](\delta v)^2 + ... \tag{6.29}$$

In order to track the solution for increasing levels of load, the load increments are generally small. Without introducing a significant error, the Taylor expansion can be truncated after the first derivatives to get the following approximation:

$$r_n \approx r_0 + \frac{dr}{dv}\delta v \tag{6.30}$$

By setting the residual to zero, the following expression can be used to solve for the unknown displacement increment δv:

$$\delta v = -r_0 \cdot \left(\frac{dr}{dv}\right)_0^{-1} \tag{6.31}$$

where the (subscript) index 0 indicates that the residual and the derivative on the right-hand side of equation are computed for v_0. The equation represents the basis for the iterative solution where a refined solution, i.e., with a smaller residual, is obtained for $v = v_0 + \delta v$. The so-calculated value of v would then become the new v_0 for the successive iteration.

The iterative sequence adopted for the iterative solution of the proposed problem is detailed below.

```
F, NI                          ▷ Set maximal force and desired number of increments
TOLres = 10^-5                 ▷ Set tolerance for residual minimization
ΔF = F/NI                      ▷ Compute value of increment
v = 0
for i = 1, NI do               ▷ Cycle for increments
    res = 1
    while |res| > TOLres do    ▷ Iterative, Newton-Raphson cycle
        res = C0(1/2 v^3 - 3/2 h v^2 + h^2 v) - i · ΔF
        dr/dv = C0(3/2 v^2 - 3hv + h^2)
        v = v - res · (dr/dv)^-1
    end while
    FD[i, 1] = v               ▷ Updating force-displacement curve
    FD[i, 2] = i · ΔF          ▷ in two-column matrix
end for
```

In the example presented here the term $(\frac{dr}{dv})$ turned out to be a scalar. In general, we deal with multidimensional problems and this term would be represented by a matrix denoted as the *tangent stiffness matrix*. The term *tangent stiffness* is used as it represents the stiffness exhibited by the structure at that particular equilibrium point of the loading path, accounting for the tangent response of the material.

The following MATLAB® script can be used to generate the results of Fig. 6.6.

```
E=80000
b=20
d=b
EA=b*d*E
l=1000
h=50
NI=10
v=0
F=130 % kN
deltaF=F/NIchapter6_bar
TOLres=10^(-5)
I_Pcr=b*d^3/12
Pcr=pi*E*I_Pcr/l^2
for iOption=1:3
    for iteration=1:NI
        '>>>>>> iteration = ',iteration
        res=1
        icount=0
        while abs(res)>TOLres
            icount=icount+1
            C0a=EA/(l^2*sqrt(h^2+l^2))
            C0b=EA/(l^2*sqrt((h-v)^2+l^2))
            eps_CFS=sqrt( ((h-v)/l)^2 + 1) / sqrt( (h/l)^2 + 1) - 1
            eps_approx=1/2*(v/l)^2 - (h/l)*(v/l)
            sin_theta_a=(h-v)/sqrt(h^2+l^2)
            sin_theta_b=(h-v)/sqrt((h-v)^2+l^2)
            res_CFS_a=-EA*eps_CFS*sin_theta_a-iteration*deltaF
            res_CFS_b=-EA*eps_CFS*sin_theta_b-iteration*deltaF
            res_approx_a=-EA*eps_approx*sin_theta_a-iteration*deltaF
            res_approx_b=-EA*eps_approx*sin_theta_b-iteration*deltaF
            if(iOption==1)
                % CFS
                dres_CFS_a=(EA*(((h - v)^2/l^2 + 1)^(1/2)/(h^2/l^2 + 1)^(1/2) - 1)
  )/(h^2 + l^2)^(1/2) + (EA*(h - v)*(2*h - 2*v))/(2*l^2*(h^2/l^2 + 1)^(1/2)*((h
   - v)^2/l^2 + 1)^(1/2)*(h^2 + l^2)^(1/2))
                deltav_CFS_a=-res_CFS_a/dres_CFS_a
                v=v+deltav_CFS_a
                res=res_CFS_a
            elseif(iOption==2)
                % CFS
                dres_CFS_b=(EA*(((h - v)^2/l^2 + 1)^(1/2)/(h^2/l^2 + 1)^(1/2) - 1)
  )/(l^2 + (h - v)^2)^(1/2) - (EA*(((h - v)^2/l^2 + 1)^(1/2)/(h^2/l^2 + 1)
  ^(1/2) - 1)*(h - v)*(2*h - 2*v))/(2*(l^2 + (h - v)^2)^(3/2)) + (EA*(h - v)
  *(2*h - 2*v))/(2*l^2*(l^2 + (h - v)^2)^(1/2)*(h^2/l^2 + 1)^(1/2)*((h - v)^2/l
  ^2 + 1)^(1/2))
                deltav_CFS_b=-res_CFS_b/dres_CFS_b
```

```
                v=v+deltav_CFS_b
                res=res_CFS_b
            elseif(iOption==3)
                res=C0a*(1/2*v^3-3/2*h*v^2+h^2*v)-iteration*deltaF
                dr_over_dv=C0a*(3/2*v^2-3*h*v+h^2)
                deltav=-res*1/dr_over_dv
                v=v+deltav
            end
        end
        FD(iteration,1)=v;
        FD(iteration,2)=iteration*deltaF;
    end
    FDplot=[0 0 ; FD]
    plot(FDplot(:,1),FDplot(:,2))
    hold on
end
```

6.4 Tangent stiffness matrix for a two-dimensional displacement-based finite element

In the previous section, we determined the tangent stiffness response for the example of the inclined bar (see Fig. 6.5) by differentiating the out-of-balance force, i.e., the residual within the iterative minimization procedure. The same approach can be implemented with the finite element formulation. The out-of-balance forces are evaluated in a FE approach by computing the difference between the applied external forces and the equivalent nodal forces obtained by integrating the internal stresses resisted within a finite element (see Fig. 1.16).

A simple and effective procedure to determine the equivalent nodal forces that are statically equivalent to the stress distribution induced in the finite element is to make use of the principle of virtual work. In doing so, after applying an arbitrary, virtual, nodal displacement to the element, the external and internal work produced by such displacement are equated. Let us denote a virtual displacement attributed to the element's nodes as $\delta \boldsymbol{d}^{v}$. During this displacement the external nodal forces acting on the element perform the following work:

$$\delta W = \delta \boldsymbol{d}^{vT} \cdot \boldsymbol{f}^{eq} \tag{6.32}$$

The virtual strain can be determined from the virtual displacement $\delta \boldsymbol{d}^{v}$. For an infinitesimal, engineering strain, the strain energy accumulated over a single element is determined by calculating the following integral:

$$\delta U = \int_{V^e} \delta \boldsymbol{\varepsilon}^T \boldsymbol{\sigma} dV = \delta \boldsymbol{d}^{vT} \int_{V^e} \boldsymbol{B}^T \boldsymbol{\sigma} dV \tag{6.33}$$

Eqs. (6.32) and (6.33) have to be identical to ensure that the nodal forces $\boldsymbol{f}^{eq}$ are statically equivalent to the stress distribution as follows:

$$f^{eq} = \int_{V^e} \boldsymbol{B}^T \boldsymbol{\sigma} dV \tag{6.34}$$

where $\boldsymbol{\sigma}$ is an engineering stress (i.e., force over the undeformed initial area) as the strain measure considered here is the infinitesimal, engineering strain.

For a geometrically nonlinear problem, as demonstrated, we need to change the strain measure for the one that is suitable for large displacement (and /or deformation). In reality, the choice of a strain measure is connected to the selection of the stress to be used in the calculation of the strain energy, previously formulated with the integral (6.33). By changing only the strain in performing this calculation, the result would change which is not acceptable as the strain energy absorbed in a deformable body cannot depend on the selection of stress and strain measures. Any pair of stress and strain measures whose product results in a strain energy are said to be *work-conjugates*. Detailed discussion of various stress-strain pairs and proofs of being work-conjugates can be found elsewhere (e.g., see [1], [2], [4]). In order to outline the main features of the numerical implementation for the solution of geometrically nonlinear problems, we consider a strategy adopted for the total Lagrangian approach that requires to switch to the Green strain measure, whose work conjugate is the second Piola–Kirchhoff stress. In this case, the internal strain energy is computed as follows:

$$\delta U = \int_{V^e} \delta \boldsymbol{E}^T \boldsymbol{S} dV \tag{6.35}$$

where $\boldsymbol{E}$ and $\boldsymbol{S}$ are the Green strain and the second Piola–Kirchhoff stress, respectively, written in Voigt notation, therefore in vector form.

The change in the Green strain is different from the change undergone by an engineering strain, as the former contains nonlinear terms. Let us first decompose the Green strain again into a first, linear part, and a second, nonlinear part (slightly rearranged from the expression previously presented in Eq. (6.14)):

$$\boldsymbol{E} = \begin{bmatrix} \frac{\partial u}{\partial X} \\ \frac{\partial v}{\partial Y} \\ \frac{\partial u}{\partial Y} + \frac{\partial v}{\partial X} \end{bmatrix} + \frac{1}{2} \begin{bmatrix} \frac{\partial u}{\partial X} & 0 & \frac{\partial v}{\partial X} & 0 \\ 0 & \frac{\partial u}{\partial Y} & 0 & \frac{\partial v}{\partial Y} \\ \frac{\partial u}{\partial Y} & \frac{\partial u}{\partial X} & \frac{\partial v}{\partial Y} & \frac{\partial v}{\partial X} \end{bmatrix} \begin{bmatrix} \frac{\partial u}{\partial X} \\ \frac{\partial u}{\partial Y} \\ \frac{\partial v}{\partial X} \\ \frac{\partial v}{\partial Y} \end{bmatrix} = \boldsymbol{\varepsilon} + \frac{1}{2} \boldsymbol{A} \boldsymbol{B} \tag{6.36}$$

With the help of the differentiation, we can then write that the change in the Green strain is:

$$\delta \boldsymbol{E} = \delta \boldsymbol{\varepsilon} + \frac{1}{2} (\delta \boldsymbol{A}) \boldsymbol{B} + \frac{1}{2} \boldsymbol{A} (\delta \boldsymbol{B}) + (\delta \boldsymbol{H}) \tag{6.37}$$

where $\delta \boldsymbol{H}$ symbolically represents all higher order terms, which, for small virtual displacements, can be neglected. $\delta \boldsymbol{\varepsilon}$, $\delta \boldsymbol{A}$, and $\delta \boldsymbol{B}$ are the same terms used in Eq. (6.36) except that the terms $\frac{\partial u}{\partial X}$ are replaced by $\frac{\partial \delta u}{\partial X}$. The same applies to the other partial derivatives of the virtual displacements.

From the expressions for $\boldsymbol{A}$ and $\boldsymbol{B}$, it is clear that $(\delta \boldsymbol{A})\boldsymbol{B}$ and $\boldsymbol{A}(\delta \boldsymbol{B})$ are the same and by neglecting higher order terms, we can write:

$$\delta \boldsymbol{E} \approx \delta \boldsymbol{\varepsilon} + \boldsymbol{A}(\delta \boldsymbol{B}) \tag{6.38}$$

By considering the interpolation of displacements adopted for the quadrilateral element, the last equation can be re-arranged as:

$$\delta \boldsymbol{E} = (\boldsymbol{B}^L + \boldsymbol{B}^{NL})\delta \boldsymbol{d} \tag{6.39}$$

where the approximation sign is replaced by equality, since for small virtual displacements, consistent with the incremental solution strategy the expression is sufficiently accurate even without higher order terms.

The equivalent nodal forces are then computed for the Green strain and the Second Piola–Kirchhoff stress as:

$$\boldsymbol{f}^{eq} = \int_{V^e} (\boldsymbol{B}^L + \boldsymbol{B}^{NL})^T \boldsymbol{S} dV \tag{6.40}$$

with $\boldsymbol{B}^L$ and $\boldsymbol{B}^{NL}$ being defined by Eq. (1.90) and Eq. (6.14), respectively. It should be noted that, when considering a small strain formulation, the same matrix $\boldsymbol{B}^L$ is used to determine the strains from displacements and the equivalent nodal forces. This is not the case when dealing with geometric nonlinearity in which case we use $\boldsymbol{B}^L + \boldsymbol{B}^{NL}$ in the integral to compute the equivalent nodal forces, while the strains are evaluated through Eq. (6.13) by employing the $\boldsymbol{B}^L + \frac{1}{2}\boldsymbol{B}^{NL}$ matrix.

Similarly, for the computation of the stiffness matrix when dealing with a small strain formulation, the same matrix $\boldsymbol{B}^L$ is used to perform the integral (1.95). For the geometrically nonlinear case, additional complexity is involved in the evaluation of the tangent stiffness matrix that is obtained by differentiating the out-of-balance force that can be represented by a *residual* vector defined as:

$$\boldsymbol{r} = \boldsymbol{f}^{eq} - \boldsymbol{F} \tag{6.41}$$

where $\boldsymbol{F}$ depicts a vector of externally applied nodal forces and $\boldsymbol{f}^{eq}$ represents the equivalent nodal forces through Eq. (6.40).

When differentiating the residual vector, the vector of external forces $\boldsymbol{F}$ is considered as a constant and the change in the internal forces can be written as:

$$\delta \boldsymbol{f}^{eq} = \int_{V^e} \left[(\boldsymbol{B}^L + \boldsymbol{B}^{NL})^T \delta \boldsymbol{S} + \delta (\boldsymbol{B}^L + \boldsymbol{B}^{NL})^T \boldsymbol{S} \right] dV \tag{6.42}$$

By performing the differentiation of the residual vector with respect to the displacement vector $\boldsymbol{d}$ (as previously discussed), the first part of Eq. (6.42), denoted as the *standard tangent stiffness matrix*, can be expressed as:

$$\boldsymbol{K}_{t1} = \int_{V^e} \left[(\boldsymbol{B}^L + \boldsymbol{B}^{NL})^T \delta \boldsymbol{S} \right] dV = \int_{V^e} \boldsymbol{B}^T \cdot \boldsymbol{D} \cdot \boldsymbol{B} dV \tag{6.43}$$

where $B = B^L + B^{NL}$. The form of the standard stiffness matrix given in Eq. (6.43) can be obtained by considering that stresses are linearly related through the material elasticity matrix D to strains. In this manner, we can write $\delta S = D \cdot \delta E$, while the change in strain is given by Eq. (6.39).

The second part of Eq. (6.42) is referred to as the *geometric* or *initial stress* contribution to the tangent stiffness matrix.[3] The following expression is obtained after the differentiation of the geometric tangent stiffness:

$$K_{t\sigma} = \int_{V^e} \left[\delta(B^L + B^{NL})^T S\right] dV = \int_{V^e} (B_G^{NL})^T \cdot {}_2S \cdot B_G^{NL} dV \qquad (5.44)$$

where ${}_2S$ collects the components of the second Piola–Kirchhoff stress arranged in matrix form as follows:

$${}_2S = \begin{bmatrix} S_{11} & S_{12} & 0 & 0 \\ S_{12} & S_{22} & 0 & 0 \\ 0 & 0 & S_{11} & S_{12} \\ 0 & 0 & S_{12} & S_{22} \end{bmatrix} \qquad (5.45)$$

while B_G^{NL} is computed as:

$$B_G^{NL} = \begin{bmatrix} \frac{\partial N_1}{\partial X} & 0 & \frac{\partial N_2}{\partial X} & 0 & \frac{\partial N_3}{\partial X} & 0 & \frac{\partial N_4}{\partial X} & 0 \\ \frac{\partial N_1}{\partial Y} & 0 & \frac{\partial N_2}{\partial Y} & 0 & \frac{\partial N_3}{\partial Y} & 0 & \frac{\partial N_4}{\partial Y} & 0 \\ 0 & \frac{\partial N_1}{\partial X} & 0 & \frac{\partial N_2}{\partial X} & 0 & \frac{\partial N_3}{\partial X} & 0 & \frac{\partial N_4}{\partial X} \\ 0 & \frac{\partial N_1}{\partial Y} & 0 & \frac{\partial N_2}{\partial Y} & 0 & \frac{\partial N_3}{\partial Y} & 0 & \frac{\partial N_4}{\partial Y} \end{bmatrix} \qquad (5.46)$$

The detailed derivation of the final form of the geometric tangent stiffness matrix are not provided here and reference should be made to dedicated textbooks for further details, e.g., [1] or [3].

6.5 Numerical example: total Lagrangian iterative procedure with one finite element

We consider a rectangular domain, which is going to be modeled with a single quadrilateral finite element and with the boundary conditions depicted in Fig. 6.7. It is assumed that stresses depend linearly on strains, and that the problem is solved based on plane-stress. Based on this, the stress–strain relationship can be written as:

$$\begin{bmatrix} \sigma_{XX} \\ \sigma_{YY} \\ \sigma_{XY} \end{bmatrix} = \frac{E}{1-\nu^2} \begin{bmatrix} 1 & \nu & 0 \\ \nu & 1 & 0 \\ 0 & 0 & \frac{1-\nu}{2} \end{bmatrix} \begin{bmatrix} \varepsilon_{XX} \\ \varepsilon_{YY} \\ \gamma_{XY} \end{bmatrix} \rightarrow \boldsymbol{\sigma} = \mathbf{D} \cdot \boldsymbol{\epsilon} \qquad (5.47)$$

[3] As it comes from $\delta B \cdot S$, it depends on the current value of the stress.

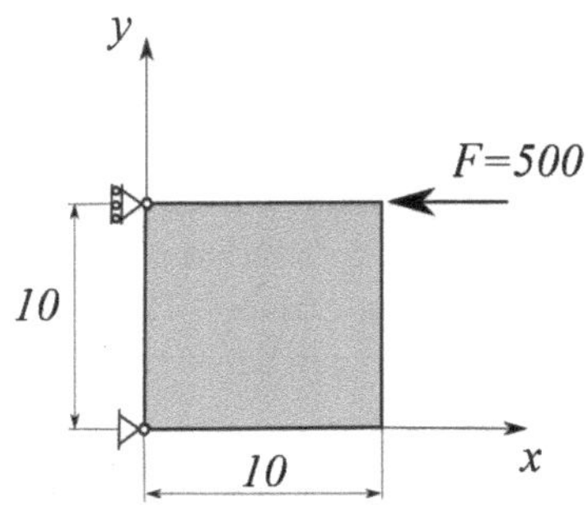

FIGURE 6.7 Rectangular domain loaded with a concentrated force (note: dimensions are in mm and forces are in N).

The elastic constants are assumed to have the following values:

$$E = 1000\ \text{MPa} \quad \nu = 0.25 \tag{6.48}$$

With a linear relationship adopted for the material stress–strain relationship the problem is solved within a geometrically nonlinear regime and, consequently, the solution accounts for the change in shape through a suitable strain measure, here taken as the Green strain. The result is obviously different from the linear solution as shown below. The following MATLAB® script provides the code for the implementation of the example. The main iterative procedure is developed following the procedure previously introduced for the 1-dimensional example of Section 6.3.

```
clear
% Setting the problem ~~~~~~~~~~~~~~~~~~~~~~~~~~~~~~~~~~~~~~~~~~~~~~~~
% Nodal coordinates and element connections
elm=[1,2,3,4];
coo=[0,0;...
    10,0;...
    10,10;...
    0,10];
% Material properties
prop=[1000,0.25];
% Constraining (for Lagrange multipliers method)
CON=[1,0,0,0,0,0,0,0;...
    0,1,0,0,0,0,0,0;...
    0,0,0,0,0,0,1,0];
Q=[0;0;0];
% Total Force vector
dF(8,1)=0;
dF(5)=-500;
FTOT=dF;
% Iterative Total Lagrangian solution ~~~~~~~~~~~~~~~~~~~~~~~~~~~~~~~
d(8,1)=0; % For first iteration, zero nodal displacements
RESIDUAL=1;
iter=0;
```

```
24 while RESIDUAL>1e-10
25     iter=iter+1;
26     KNL=stiffnlgreen(coo,d,prop(1),prop(2)); % Stiffness matrix
27     KGST=geostiffgr(coo,d,prop(1),prop(2)); % Tangent stiffness
28     KTOT=KNL+KGST;
29     dk=lagmult(KTOT,CON,dF,Q);
30     d=d+dk;
31     FND=nodforcTL(coo,d,prop(1),prop(2));
32     RES=FTOT-FND;
33     dF=RES;
34     dF(1)=0; % Reactions are excluded from comparison
35     dF(2)=0;
36     dF(7)=0;
37     RESIDUAL=dF'*dF;
38 end
```

Within the MATLAB® script 6.5, the iterative Newton–Raphson procedure is provided between lines 24 and 38, and convergence is achieved once the residual becomes smaller than 1^{-10}. The standard tangent stiffness matrix of Eq. (6.43) is computed within the iterative loop by calling the MATLAB® function `stiffnlgreen` provided in the MATLAB® script below.

```
function KNL=stiffnlgreen(X,d,E,pois)
% Function that computes nonlinear stiffness matrix for Green strain
% measurement for four node element.
% Syntax:
%   [knl]=stiffnlgreen(X,d,E,pois)
%   where:
%   X - coordinates (non deformed)
%   d - nodal displacements
%   E - Young's modulus
%   pois - Poisson's ratio

% Decomposing vector of displacements to x-part and y-part
dx=[d(1);d(3);d(5);d(7)];
dy=[d(2);d(4);d(6);d(8)];
% Defining material stiffness matrix
D=[1,pois,0;pois,1,0;0,0,(1-pois)/2];
D=E/(1-pois^2)*D;
% Gauss coordinates
gausscoor=[-sqrt(3)/3,-sqrt(3)/3;
            sqrt(3)/3,-sqrt(3)/3;
            sqrt(3)/3,sqrt(3)/3;
           -sqrt(3)/3,sqrt(3)/3];
% Components of KNL matrix
nmb=0;
for i=1:4 % Cycle for the Gauss points
    ksi=gausscoor(i,1); %setting the coordinates of the point
    eta=gausscoor(i,2);
    % Derivatives of shape functions with respect to natural coordinates
    dN1ksi=-0.25*(1-eta);
    dN1eta=-0.25*(1-ksi);
```

```
    dN2ksi=0.25*(1-eta);
    dN2eta=-0.25*(1+ksi);
    dN3ksi=0.25*(1+eta);
    dN3eta=0.25*(1+ksi);
    dN4ksi=-0.25*(1+eta);
    dN4eta=0.25*(1-ksi);
    % Defining Jacobean and inverting it
    JAK=[dN1ksi,dN2ksi,dN3ksi,dN4ksi;dN1eta,dN2eta,dN3eta,dN4eta]*X;
    JAKin=inv(JAK);
    % Derivatives of shape function with respect to material coordinates:
    dN1dx=dN1ksi*JAKin(1,1)+dN1eta*JAKin(1,2);
    dN2dx=dN2ksi*JAKin(1,1)+dN2eta*JAKin(1,2);
    dN3dx=dN3ksi*JAKin(1,1)+dN3eta*JAKin(1,2);
    dN4dx=dN4ksi*JAKin(1,1)+dN4eta*JAKin(1,2);
    dN1dy=dN1ksi*JAKin(2,1)+dN1eta*JAKin(2,2);
    dN2dy=dN2ksi*JAKin(2,1)+dN2eta*JAKin(2,2);
    dN3dy=dN3ksi*JAKin(2,1)+dN3eta*JAKin(2,2);
    dN4dy=dN4ksi*JAKin(2,1)+dN4eta*JAKin(2,2);
    % Matrix BL:
    BBL=[dN1dx,0,dN2dx,0,dN3dx,0,dN4dx,0;
      0,dN1dy,0,dN2dy,0,dN3dy,0,dN4dy;
      dN1dy,dN1dx,dN2dy,dN2dx,dN3dy,dN3dx,dN4dy,dN4dx];
    % Derivatives of displacement components and matrix BNL:
    dudx=[dN1dx,dN2dx,dN3dx,dN4dx]*dx;
    dudy=[dN1dy,dN2dy,dN3dy,dN4dy]*dx;
    dvdx=[dN1dx,dN2dx,dN3dx,dN4dx]*dy;
    dvdy=[dN1dy,dN2dy,dN3dy,dN4dy]*dy;
    BBNL=[dudx*dN1dx,dvdx*dN1dx,dudx*dN2dx,dvdx*dN2dx,dudx*dN3dx,dvdx*dN3dx,dudx*dN4dx,dvdx*dN4dx;
        dudy*dN1dy,dvdy*dN1dy,dudy*dN2dy,dvdy*dN2dy,dudy*dN3dy,dvdy*dN3dy,dudy*dN4dy,dvdy*dN4dy;
        dudx*dN1dy+dudy*dN1dx,dvdx*dN1dy+dvdy*dN1dx,dudx*dN2dy+dudy*dN2dx,dvdx*dN2dy+dvdy*dN2dx,dudx*dN3dy+dudy*dN3dx,dvdx*dN3dy+dvdy*dN3dx,dudx*dN4dy+dudy*dN4dx,dvdx*dN4dy+dvdy*dN4dx];
    BB=BBL+BBNL; % Total B matrix
    % Building KKNL components for numerical integration
    nmb=nmb+1;
    KKNL(:,:,nmb)=BB'*D*BB*det(JAK)*1*1; % Weight coefficients are 1
end
KNL=KKNL(:,:,1);
for i=2:nmb
    KNL=KNL+KKNL(:,:,i);
end
```

The geometric (i.e., initial stress) tangent stiffness matrix given by Eq. (6.44) is computed within the iterative loop by calling the MATLAB® function `geostiffgr` that is described in the listing of MATLAB® script given below.

```
function KGST=geostiffgr(X,d,E,pois)
% Function that computes geometrical stiffness for Total Lagrangian
% approximation and green strain measurements for four node element.
% Syntax:
%   [KG]=geostiffgr(coo,u,prop,plane)
```

```
% where,
%   coo - nodal coordinates coordinates
%   d - nodal displacements
%   E - Young's modulus
%   pois - Poisson's ratio

% Decomposing vector of displacements to x-part and y-part
dx=[d(1);d(3);d(5);d(7)];
dy=[d(2);d(4);d(6);d(8)];
% Defining material elasticity matrix
D=[1,pois,0;pois,1,0;0,0,(1-pois)/2];
D=E/(1-pois^2)*D;
% Gauss coordinates
gausscoor=[-sqrt(3)/3,-sqrt(3)/3;
            sqrt(3)/3,-sqrt(3)/3;
            sqrt(3)/3,sqrt(3)/3;
            -sqrt(3)/3,sqrt(3)/3];
% Components of KGST matrix
nmb=0;
for i=1:4 % Cycle for the Gauss points
    ksi=gausscoor(i,1); %setting the coordinates of the point
    eta=gausscoor(i,2);
    % Derivatives of shape functions with respect to natural coordinates
    dN1ksi=-0.25*(1-eta);
    dN1eta=-0.25*(1-ksi);
    dN2ksi=0.25*(1-eta);
    dN2eta=-0.25*(1+ksi);
    dN3ksi=0.25*(1+eta);
    dN3eta=0.25*(1+ksi);
    dN4ksi=-0.25*(1+eta);
    dN4eta=0.25*(1-ksi);
    % Defining Jacobean and inverting it
    JAK=[dN1ksi,dN2ksi,dN3ksi,dN4ksi;dN1eta,dN2eta,dN3eta,dN4eta]*X;
    JAKin=inv(JAK);
    % Derivatives of shape function with respect to material coordinates:
    dN1dx=dN1ksi*JAKin(1,1)+dN1eta*JAKin(1,2);
    dN2dx=dN2ksi*JAKin(1,1)+dN2eta*JAKin(1,2);
    dN3dx=dN3ksi*JAKin(1,1)+dN3eta*JAKin(1,2);
    dN4dx=dN4ksi*JAKin(1,1)+dN4eta*JAKin(1,2);
    dN1dy=dN1ksi*JAKin(2,1)+dN1eta*JAKin(2,2);
    dN2dy=dN2ksi*JAKin(2,1)+dN2eta*JAKin(2,2);
    dN3dy=dN3ksi*JAKin(2,1)+dN3eta*JAKin(2,2);
    dN4dy=dN4ksi*JAKin(2,1)+dN4eta*JAKin(2,2);
    % Geometric stiffness B matrix
    BNLG=[dN1dx,0,dN2dx,0,dN3dx,0,dN4dx,0;
          dN1dy,0,dN2dy,0,dN3dy,0,dN4dy,0;
          0,dN1dx,0,dN2dx,0,dN3dx,0,dN4dx;
          0,dN1dy,0,dN2dy,0,dN3dy,0,dN4dy];
    % Components for the Gauss points used to compute Green strain
    dudx=[dN1dx,dN2dx,dN3dx,dN4dx]*dx;
    dudy=[dN1dy,dN2dy,dN3dy,dN4dy]*dx;
    dvdx=[dN1dx,dN2dx,dN3dx,dN4dx]*dy;
```

```
    dvdy=[dN1dy,dN2dy,dN3dy,dN4dy]*dy;
    % Green strain
    EGR(1,1)=dudx+0.5*(dudx^2+dvdx^2);
    EGR(2,1)=dvdy+0.5*(dudy^2+dvdy^2);
    EGR(3,1)=(dudy+dvdx+dudx*dudy+dvdx*dvdy);
    SII=D*EGR; % Second Piola-Kirchoff stress
    STR_STIFF=[SII(1),SII(3),0,0;... % S2 reshaped
               SII(3),SII(2),0,0;...
               0,0,SII(1),SII(3);...
               0,0,SII(3),SII(2)];
    nmb=nmb+1;
    KKNL(:,:,nmb)=BNLG'*STR_STIFF*BNLG*det(JAK)*1*1; % Weight coefficients are 1
end
KGST=KKNL(:,:,1);
for i=2:nmb
    KGST=KGST+KKNL(:,:,i);
end
```

Incremental displacements are solved by using the Lagrange multipliers technique through the MATLAB® function `lagmult` depicted in the MATLAB® script below. This is a classic method frequently adopted to solve a system of linear algebraic equations whose description can be found, for example, in textbooks dealing with applied mathematics or finite element method, e.g., [5]. Alternatively, a reduced stiffness matrix can be used for this purpose, by excluding the rows and columns from the full stiffness matrix that correspond to constrained degrees of freedom.

```
function dk=lagmult(STF,C,F,Q)
% Function that solves the constrained problem by Lagrange multipliers
% Syntax:
%   langmult(STF,C,F,Q)
%   -STF - stiffness matrix
%   -C - constrain matrix
%   -F - force vector
%   -Q - right-hand side of constrain system
DOF1=size(STF);
DOF=DOF1(1);
const1=size(C);
const=const1(1);
null(1:const,1:const)=0;
STFm=[STF,C';C,null];
dkm=inv(STFm)*[F;Q];
dk=dkm(1:DOF,1);
```

The equivalent nodal forces, given by Eq. (6.40), are computed by employing the MATLAB® function `nodforcTL`.

```
function FND=nodforcTL(X,d,E,pois)
% Function that computes nodal forces for the deformed geometry for one
% element. Arguments of the function are the following:
%   coo - matrix of nodal coordinate
%   u - vector of nodal displacements
```

```
%   E - Young's modulus
%   pois - Poisson's ratio

% Decomposing vector of displacements to x-part and y-part
dx=[d(1);d(3);d(5);d(7)];
dy=[d(2);d(4);d(6);d(8)];
% Defining material elasticity matrix
D=[1,pois,0;pois,1,0;0,0,(1-pois)/2];
D=E/(1-pois^2)*D;
% Gauss coordinates
gausscoor=[-sqrt(3)/3,-sqrt(3)/3;
            sqrt(3)/3,-sqrt(3)/3;
            sqrt(3)/3,sqrt(3)/3;
           -sqrt(3)/3,sqrt(3)/3];
% Integrating BTOT * SII
nmb=0;
for i=1:4 % Cycle for the Gauss points
    ksi=gausscoor(i,1); %setting the coordinates of the point
    eta=gausscoor(i,2);
    % Derivatives of shape functions with respect to natural coordinates
    dN1ksi=-0.25*(1-eta);
    dN1eta=-0.25*(1-ksi);
    dN2ksi=0.25*(1-eta);
    dN2eta=-0.25*(1+ksi);
    dN3ksi=0.25*(1+eta);
    dN3eta=0.25*(1+ksi);
    dN4ksi=-0.25*(1+eta);
    dN4eta=0.25*(1-ksi);
    % Defining Jacobean and inverting it
    JAK=[dN1ksi,dN2ksi,dN3ksi,dN4ksi;dN1eta,dN2eta,dN3eta,dN4eta]*X;
    JAKin=inv(JAK);
    % Derivatives of shape function with respect to material coordinates:
    dN1dx=dN1ksi*JAKin(1,1)+dN1eta*JAKin(1,2);
    dN2dx=dN2ksi*JAKin(1,1)+dN2eta*JAKin(1,2);
    dN3dx=dN3ksi*JAKin(1,1)+dN3eta*JAKin(1,2);
    dN4dx=dN4ksi*JAKin(1,1)+dN4eta*JAKin(1,2);
    dN1dy=dN1ksi*JAKin(2,1)+dN1eta*JAKin(2,2);
    dN2dy=dN2ksi*JAKin(2,1)+dN2eta*JAKin(2,2);
    dN3dy=dN3ksi*JAKin(2,1)+dN3eta*JAKin(2,2);
    dN4dy=dN4ksi*JAKin(2,1)+dN4eta*JAKin(2,2);
    % Matrix BL:
    BBL=[dN1dx,0,dN2dx,0,dN3dx,0,dN4dx,0;
      0,dN1dy,0,dN2dy,0,dN3dy,0,dN4dy;
      dN1dy,dN1dx,dN2dy,dN2dx,dN3dy,dN3dx,dN4dy,dN4dx];
    % Derivatives of displacement components:
    dudx=[dN1dx,dN2dx,dN3dx,dN4dx]*dx;
    dudy=[dN1dy,dN2dy,dN3dy,dN4dy]*dx;
    dvdx=[dN1dx,dN2dx,dN3dx,dN4dx]*dy;
    dvdy=[dN1dy,dN2dy,dN3dy,dN4dy]*dy;
    BBNL=[dudx*dN1dx,dvdx*dN1dx,dudx*dN2dx,dvdx*dN2dx,dudx*dN3dx,dvdx*dN3dx,dudx*
     dN4dx,dvdx*dN4dx;
```

```
        dudy*dN1dy,dvdy*dN1dy,dudy*dN2dy,dvdy*dN2dy,dudy*dN3dy,dvdy*dN3dy,dudy*
     dN4dy,dvdy*dN4dy;
        dudx*dN1dy+dudy*dN1dx,dvdx*dN1dy+dvdy*dN1dx,dudx*dN2dy+dudy*dN2dx,dvdx*
     dN2dy+dvdy*dN2dx,dudx*dN3dy+dudy*dN3dx,dvdx*dN3dy+dvdy*dN3dx,dudx*dN4dy+dudy*
     dN4dx,dvdx*dN4dy+dvdy*dN4dx];
    BB=BBL+BBNL; % Total B matrix
    % Green strain
    EGR(1,1)=dudx+0.5*(dudx^2+dvdx^2);
    EGR(2,1)=dvdy+0.5*(dudy^2+dvdy^2);
    EGR(3,1)=(dudy+dvdx+dudx*dudy+dvdx*dvdy);
    SII=D*EGR; % Second Piola-Kirchoff stress
    nmb=nmb+1;
    FF(:,:,nmb)=BB'*SII*det(JAK)*1*1;
end
FND=FF(:,:,1);
for i=2:nmb
    FND=FND+FF(:,:,i);
end
```

The solution of the example presented in Fig. 6.7 can be carried out using the proposed solution and the numerical solution is obtained after 5 iterations. Table 6.1 summarizes the values calculated for the different iterations until convergence is achieved. In particular, the table depicts the displacement component of the point where the load is applied, the internal force at the same point (i.e., at this point equilibrium requires that the internal force equals the external force of 500 N), and the residual.

TABLE 6.1 Results of iterative solution for considered problem.

Iteration	Displacement (mm)	Internal force (N)	Residual (N)
1	-1.5227	383.0238	2.54E4
2	-2.1139	484.1313	560.3370
3	-2.2452	499.3218	1.5037
4	-2.2512	499.9986	6.723E-6
5	-2.2512	500.0000	1.775E-16

The solution reached after the first iteration corresponds to the linear solution since it is achieved by attributing to all initial displacement components a value of zero, thus the contributions of the tangent stiffness and of the nonlinear part of the Green strain are nil. Fig. 6.8 depicts the linear solution as well as the nonlinear solution obtained after 5 iterations, highlighting the significant differences between the two sets of results.

6.5.1 The influence of the elastic constants on the solution

The previous example has shown the significant differences between the results obtained by applying the linear and nonlinear approaches to an example that undergoes large deformation (or displacement). In this case the linear solution

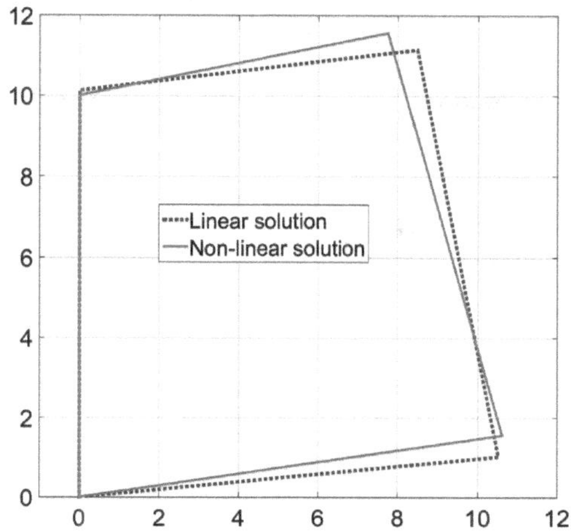

FIGURE 6.8 Comparison of the linear and nonlinear solutions obtained for the example of Fig. 6.7 (note: dimensions are in mm).

does not capture the response well, and a nonlinear solution has to be sought for an adequate representation of the structural response. The "correct" solution can be reached by implementing either the total or the updated Lagrangian procedure, and one would expect that the solutions obtained with these approaches are identical. The premises for producing the same results relies on a suitable selection of the elastic constants.

We previously outlined that, under certain conditions, there is a need to change the adopted strain measure and the corresponding stress (i.e., the work-conjugate). Both the stress and the strain represent normalized quantities of load and deformation. Such consideration does not imply that, by changing the strain measure, the corresponding stress could be obtained through the same elastic constants. This is outlined with the following example.

We consider again a rectangular domain, and enforce boundary conditions to reproduce an uniaxial state of stress (see Fig. 6.9). Using the same elastic constants in input, we analyze the structure by considering the small deformation regime and the total Lagrangian formulation to account for large displacements.

The linear solution leads to the following displacements (that can be verified by simple hand calculation):

$$\boldsymbol{d}^{lin} = \begin{bmatrix} 0 & 0 & 0.100 & 0 & 0.100 & -0.025 & 0 & -0.025 \end{bmatrix}^T \text{ mm} \tag{6.49}$$

based on which we can write the horizontal infinitesimal (i.e., engineering) strain component as:

$$\varepsilon_{xx} = \frac{\Delta L}{L_0} = \frac{0.1}{1} = 0.1 \tag{6.50}$$

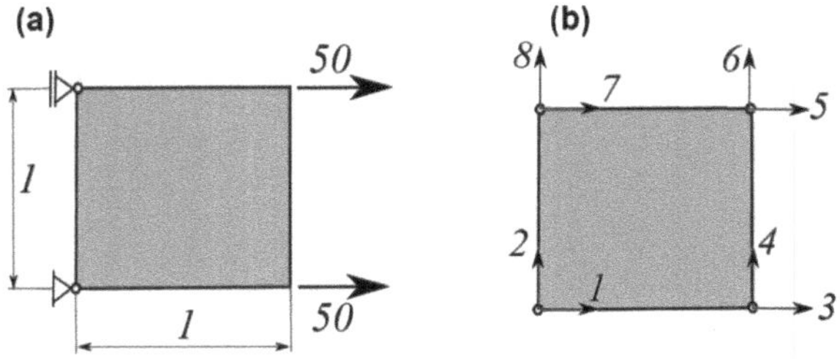

FIGURE 6.9 Single finite element: (a) element subjected to an uniaxial stress condition (note: dimensions are in mm and forces are in N) and (b) degrees of freedom at the nodes.

By considering the assumed value of the Young's modulus, we can also write that the normal stress in the horizontal direction is equal to:

$$\sigma_{xx} = E \cdot \varepsilon_{xx} = 1000 \cdot 0.1 = 100 \text{ MPa} \tag{6.51}$$

For the linear solution, this represents the engineering stress (i.e., force over the undeformed area). We can then compute the internal force by multiplying the stress with the initial area of the element (obtained by the product of the height of the element and the unitary depth):

$$F^{int} = \sigma_{xx} \cdot A_0 = 100 \cdot 1 \cdot 1 = 100 \text{ N} \tag{6.52}$$

which equals the total applied load, therefore confirming that equilibrium is satisfied.

The nonlinear solution obtained with the total Lagrangian approach is achieved after four iterations. Table 6.2 summarizes key results calculated for the displacement component $d(3)$ and the residual value through the iterations.

TABLE 6.2 Results of the iterative solution for the considered problem.

Iteration	Displacement $d(3)$ (mm)	Residual (N)
1	0.1000	129.3147
2	0.0882	0.0287
3	0.0880	1.641E-9
4	0.0880	5.366E-24

Upon convergence, the final displacement vector becomes:

$$\mathbf{d}^{nl} = \begin{bmatrix} 0 & 0 & 0.0880 & 0 & 0.0880 & -0.0232 & 0 & -0.0232 \end{bmatrix}^T \text{ mm} \tag{6.53}$$

As both stress and strain are uniform over the element, we may use the previous MATLAB® function to compute the Green strain at an arbitrary point of the element based on the computed displacements. The resulting Green strain components are as follows:

$$E = \begin{bmatrix} 0.0919 \\ 0.0230 \\ 0.0000 \end{bmatrix} \tag{6.54}$$

The problem is solved considering the case of plane-stress, and the second Piola–Kirchhoff stress components can be calculated by considering the calculated Green strain components with Eq. (6.47). By using the values of the elastic constants specified in Eq. (6.48), the values of the second Piola–Kirchhoff stress components are (in MPa):

$$S = D \cdot E = \begin{bmatrix} 91.9089 \\ -2.7756E-15 \\ -2.7756E-15 \end{bmatrix} \tag{6.55}$$

For the given boundary conditions the state of stress is uniaxial and, therefore, the only nonzero component is S_{xx}, while the other two stress components are numerical zeros.

At this point, we now calculate the true, Cauchy stress. This can be done by employing the deformation gradient on the basis of the computed second Piola–Kirchhoff stress (in MPa) as (for details, see [1]):

$$\sigma = \frac{1}{J} F \cdot S \cdot F^T = \begin{bmatrix} 102.3801 & -3.918E-15 \\ -3.918E-15 & 2.328E-13 \end{bmatrix} \tag{6.56}$$

where F is the deformation gradient given by Eq. (6.4), and J stands for the determinant of F.[4]

It is important to note that the relationship (6.56) stems from tensor calculus, therefore the stresses should be strictly represented as a second-order tensor, not as a vector, which is a common way of representing a symmetric second-order tensors in structural calculations, as also adopted in this chapter. To emphasize this, the resulting Cauchy stress in Eq. (6.56) is given by a matrix (i.e., second-order tensor).

The solution obtained through the total Lagrangian procedure accounts for the change in geometry and, therefore, the equilibrium should be computed on the basis of the true, Cauchy stress. In this case, the state of stress is uniaxial, therefore it is easy to compute the stress component σ_{xx} as the total force

4 The numerical implementation for the calculation of the stress is not presented here as it can be obtained by combining pieces of MATLAB® functions previously introduced to obtain the deformation gradient.

divided by the actual area, which is obtained by considering the deformed element height, multiplied by the unitary depth:

$$\sigma_{xx} = \frac{100}{(1 - 0.0232) \cdot 1} = 102.3801 \text{ MPa} \tag{6.57}$$

The obtained value is the same as the one given in Eq. (6.56), which concludes the verification.

On the other hand, the nodal displacements obtained from the nonlinear solution can also be used to calculate the Almansi strain. By using the previously introduced MATLAB® function while making sure that deformed coordinates are used as the reference one the following result is obtained:

$$\boldsymbol{e} = \begin{bmatrix} 0.0776 \\ -0.0241 \\ -6.054E - 18 \end{bmatrix} \tag{6.58}$$

that is the work conjugate to the Cauchy stress and, therefore, it should be linearly related to it. However, if the same elastic constants as those that were employed to obtain the above result with the total Lagrangian procedure are used also here, the following results are obtained:

$$\begin{bmatrix} \sigma_{xx} \\ \sigma_{yy} \\ \sigma_{xy} \end{bmatrix} = \frac{E}{1-\nu^2} \begin{bmatrix} 1 & \nu & 0 \\ \nu & 1 & 0 \\ 0 & 0 & \frac{1-\nu}{2} \end{bmatrix} \begin{bmatrix} e_{xx} \\ e_{yy} \\ e_{xy} \end{bmatrix} = \begin{bmatrix} 76.3912 \\ -4.9862 \\ 2.421E - 15 \end{bmatrix} \tag{6.59}$$

and these differ from the Cauchy stress components computed with Eq. (6.56) and, in addition, these values do not satisfy equilibrium. This result clearly indicates that we cannot use the same elastic constants for the total and the updated Lagrangian procedures, or, alternatively, that we cannot yield the same results if we adopt the same elastic constants with the two procedures.

It is common in practice to use the same elastic coefficients when using the total and the updated Lagrangian formulations. This is usually a consequence of the lack of data, since the elastic constants readily available in engineering handbooks usually refer to the infinitesimal strain and the engineering stress. This situation is acceptable by recognizing that most engineering materials can develop only small elastic deformation and, therefore, the error between various strain measures can be neglected. This difference may become important when dealing with large elastic deformation. In these situations, it is more convenient to use the formulation of the constitutive model expressed by means of the elastic potential. If a problem is characterized by small deformation and large displacements, either the total or the updated Lagrange approach implemented with the same elastic constants can lead to satisfying results, while the difference between the two solutions appears in the results of the displacement field (not those associated to the stress field) that can often be of secondary importance when dealing with engineering applications.

6.6 Hyperelastic models

The most general geometrically nonlinear problem represents the one in which both large deformation and large displacements occur. This is encountered, for example, when modeling rubber-like materials that are capable of developing large elastic deformation. A typical uniaxial stress–strain curve for rubbers has the so-called S-shape and looks as depicted in Fig. 4.9, with almost all the deformation being elastic. Generated stresses are not linearly related to the elastic strain and, therefore, the linearized Hooke's law cannot be used here as the constitutive model. Instead, the direct formulation of the constitutive model is usually employed by defining a scalar elastic potential. As demonstrated in Chapter 4, stresses are obtained as partial derivatives of such a potential that represents the major difference in the numerical implementation of the hyperelastic constitutive models, when compared to the elastic models presented until now.

A review of the theoretical modeling of hyperelastic materials is beyond the scope of this book and reference for detailed theoretical discussions can be found in the early works of Mooney [6], Rivlin [7], Treloar [8], while early finite element implementation can be found in [9], [10], followed by further developments introducing incompressibility (see, e.g., [11], [12]). In the following, we will cover only the main concepts required for the numerical implementation of constitutive models for which stresses are derivable from a scalar elastic potential.

In Section 4.4, we demonstrated that for isotropic, linear elastic materials the strain energy density can be written as a function of two strain invariants. Such result is reasonable considering that the isotropic material behavior is independent of the direction in which the material is loaded, thus also the elastic potential should be independent of the reference frame. We further highlighted that the elastic potential for such materials consists in a quadratic form of the strain components and, therefore, for its full definition it requires the quantification of two independent constants. We will now revisit the elastic potential for a linear elastic, isotropic material that has the following form:

$$U = 2\mu \bar{I}_2 + \frac{\lambda}{2} I_1^2 \tag{6.60}$$

where μ and λ are the two independent material constants (denoted as Lame constants) that relate to the more commonly used Young's modulus and Poisson's ratio as follows:

$$\mu = \frac{E}{2(1+\nu)}, \quad \lambda = \frac{E\nu}{(1-2\nu)(1+\nu)} \tag{6.61}$$

In Eq. (6.60) I_1 is the first invariant of the strain tensor:

$$I_1 = \varepsilon_{11} + \varepsilon_{22} + \varepsilon_{33} \tag{6.62}$$

while $\bar{I}_2$ is equal to:

$$\bar{I}_2 = \frac{1}{2}\varepsilon_{11}^2 + \frac{1}{2}\varepsilon_{22}^2 + \frac{1}{2}\varepsilon_{33}^2 + \varepsilon_{12}\varepsilon_{21} + \varepsilon_{13}\varepsilon_{31} + \varepsilon_{23}\varepsilon_{32} \tag{6.63}$$

and it relates to the second invariant based on the following relationship:

$$I_2 = \frac{1}{2}I_1^2 - \bar{I}_2 \tag{6.64}$$

The corresponding stress components can be obtained by differentiating the potential with respect to the strain components:

$$\begin{aligned}\sigma_{xx} = \frac{\partial U}{\partial \varepsilon_{xx}} &= 2\mu\varepsilon_{xx} + \lambda\varepsilon_{xx} + \lambda\varepsilon_{yy} + \lambda\varepsilon_{zz} = \\ \left(\frac{E}{1+\nu} + \frac{E\nu}{(1-2\nu)(1+\nu)}\right)\varepsilon_{xx} &+ \frac{E\nu}{(1-2\nu)(1+\nu)}\varepsilon_{yy} + \frac{E\nu}{(1-2\nu)(1+\nu)}\varepsilon_{zz} = \\ &\frac{E}{(1-2\nu)(1+\nu)}\left[(1-\nu)\varepsilon_{xx} + \nu\varepsilon_{yy} + \nu\varepsilon_{zz}\right]\end{aligned} \tag{6.65}$$

that corresponds to the first algebraic equation of the system described in Eq. (4.28). By performing the differentiation with respect to the remaining 5 components of strain, the complete system of 6 linear equations, representing the extended Hooke's law for the isotropic linear elastic material is obtained. With this process, we presented another way of formulating Hooke's law through the elastic potential given in Eq. (6.60).

In the case of hyperelastic isotropic materials, the same invariants can be used to construct higher-order (i.e., not only linear) material models. In this process, care should be placed for the selection of a suitable strain measure. It was already pointed out earlier that the engineering strain in the presence of large rotations leads to fictitious strains. In such a case, the invariants, as well as the strain energy function would differ from zero, therefore incorrectly leading to an accumulated strain energy produced by a large rotation. Such a situation can occur for materials that undergo large deformation and a suitable strain measure needs to be adopted in formulating the strain energy density function. We still consider an isotropic material, but the strain energy is defined with respect to different strain measures.

For large deformation, the selection of a strain measure is not easy because, unlike in the case of small deformation where only one strain measure can be defined, we can, theoretically, define an infinite number of strain measures. It is therefore convenient to define the elastic potential as a function of a certain measure of deformation that is invariant with respect to both coordinate system and strain measure. This is true for the principal stretch ratios, denoted as λ_1, λ_2, and λ_3, as it will be illustrated in the following.

According to the polar decomposition theorem, the deformation gradient can be decomposed as follows (for details see, e.g., [13]):

$$\boldsymbol{F} = \boldsymbol{R}\boldsymbol{U} = \boldsymbol{V}\boldsymbol{R} \tag{6.66}$$

where $\boldsymbol{R}$ is an orthogonal rotation tensor, while $\boldsymbol{U}$ and $\boldsymbol{V}$ are symmetrical right and left stretch tensors, respectively. This can lead to the interpretation that any deformation can be described by a deformation defined through a stretch tensor $\boldsymbol{U}$ that is then followed by a rigid body rotation defined by the tensor $\boldsymbol{R}$ The same deformation can also be described by first performing the rigid body rotation defined by $\boldsymbol{R}$, followed by the deformation of the rotated volume defined by $\boldsymbol{V}$. The two stretch tensors are obviously related through a pure rotation as:

$$\boldsymbol{V} = \boldsymbol{F}\boldsymbol{R}^T = \boldsymbol{R}\boldsymbol{U}\boldsymbol{R}^T \tag{6.67}$$

Considering that any rigid body rotation is not generating stresses, it is therefore convenient to define the elastic potential only as a function of the stretch tensors.

In elementary one-dimensional considerations a stretch is defined as the ratio between current length and the original one. Therefore, for the undeformed state, it has the value of 1. In an arbitrary three-dimensional case, the three eigenvalues of $\boldsymbol{U}$, λ_1, λ_2, and λ_3, are called the principal stretches, and they clearly represent a measure of deformation that is invariant with respect to the coordinate frame.

The polar decomposition is a computationally expensive operation, and it is usually avoided [14]. It is possible to compute a measure of the deformation, that is independent of the rotation, without actually computing the rotation matrix. By relying on the polar decomposition theorem, we can write:

$$\boldsymbol{F}^T\boldsymbol{F} = (\boldsymbol{R}\boldsymbol{U})^T\boldsymbol{R}\boldsymbol{U} = \boldsymbol{U}^T\boldsymbol{R}^T\boldsymbol{R}\boldsymbol{U} \tag{6.68}$$

and by considering that the rotation matrix $\boldsymbol{R}$ is orthogonal, its transpose is equal to its inverse:

$$\boldsymbol{R}^T\boldsymbol{R} = \boldsymbol{R}^{-1}\boldsymbol{R} = \boldsymbol{I} \tag{6.69}$$

By combining the last two equations, we can write:

$$\boldsymbol{F}^T\boldsymbol{F} = \boldsymbol{U}^T\boldsymbol{U} = \boldsymbol{U}^2 = \boldsymbol{C} \tag{6.70}$$

in which tensor $\boldsymbol{C}$ is denoted as the *right Cauchy–Green deformation tensor*. By definition, tensor $\boldsymbol{C}$ has the same principal directions as $\boldsymbol{U}$, while its eigenvalues are the following squared terms: λ_1^2, λ_2^2, and λ_3^2.

In a similar way, we can write the following for the left stretch tensor:

$$\boldsymbol{F}\boldsymbol{F}^T = \boldsymbol{V}\boldsymbol{R}(\boldsymbol{V}\boldsymbol{R})^T = \boldsymbol{V}\boldsymbol{R}\boldsymbol{R}^T\boldsymbol{V}^T = \boldsymbol{V}^T\boldsymbol{V} = \boldsymbol{V}^2 = \boldsymbol{B} \tag{6.71}$$

Tensor $\boldsymbol{B}$ is called the *left Cauchy-Green deformation tensor*. The eigenvalues on the left and right Cauchy–Green deformation tensors are identical, and they are easier to compute than the stretch tensor $\boldsymbol{U}$. They do represent a measure of pure deformation since they do not depend on rotations. By considering that tensor $\boldsymbol{C}$ is used to construct the Green strain (see Section 6.2), while tensor $\boldsymbol{B}$ is used to define the Almansi strain, the principal stretches can be treated as being independent on the selection of the strain measure. Such statement can be strengthened by formulating other strain measures as a function of principal stretches (see [3] and [14]). Therefore, λ_1, λ_2, and λ_3 represent a good choice for defining the elastic potential. Indeed, many different models formulate the elastic potential as a function of principal stretches only.

For an isotropic material, the strain energy density function has to satisfy the following criteria:

1. the energy function U is always greater than zero;
2. the energy function U is zero for $\lambda_1 = \lambda_2 = \lambda_3 = 1$, that depicts a non-deformed state; and
3. the energy function is symmetric in $\lambda_1, \lambda_2, \lambda_3$ since, within the isotropic material, there is no preferable direction and the material has the same behavior in any direction.

Considering that it is easier to compute the right Cauchy–Green deformation tensor $\boldsymbol{C}$ than the stretch tensor $\boldsymbol{U}$, it is useful to establish a relationship between their eigenvalues. Since the eigenvalues of the former are squares of the eigenvalues of the latter, we can write:

$$\begin{aligned} I_{1C} &= \lambda_1^2 + \lambda_2^2 + \lambda_3^2 = C_{11} + C_{22} + C_{33} \\ I_{2C} &= \lambda_1^2\lambda_2^2 + \lambda_2^2\lambda_3^2 + \lambda_1^2\lambda_3^2 = \\ &C_{11}C_{22} + C_{22}C_{33} + C_{33}C_{11} - C_{12}C_{21} - C_{23}C_{32} - C_{13}C_{31} \\ I_{3C} &= \lambda_1^2\lambda_2^2\lambda_3^2 = det(\boldsymbol{C}) = \\ &C_{11}C_{22}C_{33} + 2C_{12}C_{23}C_{13} - C_{11}C_{23}C_{32} - C_{22}C_{13}C_{31} - C_{33}C_{12}C_{21} \end{aligned} \tag{6.72}$$

Based on this, we can write the strain energy density as a function of the invariants of the $\boldsymbol{C}$ tensor, that can be calculated without the actual computation of principal stretches λ_1, λ_2, and λ_3. For an undeformed body, stretches are all equal to 1, so the invariants I_{1C}, I_{2C}, and I_{3C} are equal to 3, 3, and 1, respectively. The form of the strain energy density function which satisfies all the above criteria, from which the family of different models are derived is given by:

$$U = \sum_{p,q,r=1}^{\infty} = C_{pqr}(I_{1C} - 3)^p (I_{2C} - 3)^q (I_{3C} - 1)^r \tag{6.73}$$

where C_{pqr}, p, q, and r are material parameters that should be identified experimentally. Higher order polynomials can better capture the nonlinear behavior over the wider range of possible deformation, even if the calibration of their larger number of material constants may become more difficult.

For the modeling of the incompressible materials, the form given by Eq. (6.73) can be further simplified. Considering that the incompressibility implies that $\det(\boldsymbol{F}) = \lambda_1\lambda_2\lambda_3 = 1$ and, consequently, $I_{3C} = 1$, the elastic potential for this case becomes:

$$U = \sum_{p,q=1}^{\infty} = C_{pq}(I_{1C} - 3)^p (I_{2C} - 3)^q \tag{6.74}$$

One popular model, obtained as the truncated version of Eq. (6.74), is the Mooney–Rivlin strain energy density function given by:

$$U = C_{10}(I_{1C} - 3) + C_{01}(I_{2C} - 3) \tag{6.75}$$

that is governed by two material parameters only, i.e., C_{10} and C_{01}. Considering that it is defined by a relatively low order, the model cannot capture well the typical S-shape and it is generally good up to strain of about 100%.

Even simpler form with only one material constant can be obtained as the so called "neo-Hookean" function given by:

$$U = C_1(I_{1C} - 3) \tag{6.76}$$

that are good for strains up to about 30%.

It is important to mention that, even though certain rubbers behave almost like incompressible materials, pure incompressibility may lead to severe numerical difficulties, as discussed in the following. For this reason, even if a very small compressibility is present within the material, it may be useful to include it in the material model for numerical stability purposes.

6.6.1 Calculating stresses and the tangent stiffness matrix for hyperelastic models

When dealing with geometrically nonlinear problems, care should be placed to adopt stress and strain measures that are work-conjugates. The same care should be taken for large deformation problems when the constitutive model is formulated through the elastic potential. Recalling that the Green strain is a work-conjugate of the second Piola–Kirchhoff stress, for a given model function of the strain energy density, we can then compute components of the second Piola–Kirchhoff stress as:

$$\boldsymbol{S} = \frac{\partial U}{\partial \boldsymbol{E}} \tag{6.77}$$

In the previous section, we showed that, even though formally the elastic potential is defined as a function of principal stretches that are the eigenvalues of the stretch matrix $\boldsymbol{U}$, we do not have to compute them when relying on the components of the left Cauchy–Green tensor $\boldsymbol{C}$, which is computationally more efficient. The same applies when calculating the components of the second Piola-Kirchhoff stress. Considering that, by definition, the Green strain is defined as (see Section 6.2):

$$\boldsymbol{E} = \frac{1}{2}(\boldsymbol{C} - \boldsymbol{I}) \tag{6.78}$$

and the stress tensor can be expressed as:

$$\boldsymbol{S} = 2\frac{\partial U}{\partial \boldsymbol{C}} \tag{6.79}$$

For illustrative purposes, we can briefly consider the incompressible Mooney–Rivlin strain density function, given by Eq. (6.75). The invariants of tensor $\boldsymbol{C}$ are expressed in Eq. (6.72). The components of the second Piola–Kirchhoff stress, given as a function of $\boldsymbol{C}$ tensor components, can then be calculated as:

$$\begin{aligned}
S_{11} &= 2\frac{\partial U}{\partial C_{11}} = 2C_{10} + 2C_{01}(C_{22} + C_{33}) \\
S_{22} &= 2\frac{\partial U}{\partial C_{22}} = 2C_{10} + 2C_{01}(C_{11} + C_{33}) \\
S_{33} &= 2\frac{\partial U}{\partial C_{33}} = 2C_{10} + 2C_{01}(C_{11} + C_{22}) \\
S_{12} &= 2\frac{\partial U}{\partial C_{12}} = -2C_{01}C_{21} \\
S_{13} &= 2\frac{\partial U}{\partial C_{13}} = -2C_{01}C_{31} \\
S_{23} &= 2\frac{\partial U}{\partial C_{23}} = -2C_{01}C_{32}
\end{aligned} \tag{6.80}$$

where C_{10} and C_{01} are material constants.

The tangent stiffness matrix required for the iterative solution of the nonlinear problem is computed by further derivation of the stress tensor with respect to the Green strain:

$$\boldsymbol{D}_t = \frac{\partial \boldsymbol{S}}{\partial \boldsymbol{E}} = 2\frac{\partial \boldsymbol{S}}{\partial \boldsymbol{C}} = 4\frac{\partial^2 U}{\partial \boldsymbol{C}^2} \tag{6.81}$$

For the numerical implementation of the calculation of both stresses and tangent stiffness, it is required to compute the deformation gradient on the basis of the vector of displacements. The derivatives are then computed with respect to the right Cauchy–Green deformation tensor $\boldsymbol{C}$.

The numerical implementation for the calculation of stresses in a point within an 8-node, hexahedral element is given in the following MATLAB® script.

```
function [S]=hyper_elastic(coo,de,pnt,C01,C10)
% Function that computes Second Piola-Kirchhoff stress components
% in a point for Mooney-Rivlin strain energy density function
ksi=pnt(1); eta=pnt(2); zeta=pnt(3);
% Decomposing vector of nodal displacements to x- y- and z-part
dx=[de(1);de(4);de(7);de(10);de(13);de(16);de(19);de(22)];
dy=[de(2);de(5);de(8);de(11);de(14);de(17);de(20);de(23)];
dz=[de(3);de(6);de(9);de(12);de(15);de(18);de(21);de(24)];
% Nodal coordinates
x1=coo(1,1); y1=coo(1,2); z1=coo(1,3);
x2=coo(2,1); y2=coo(2,2); z2=coo(2,3);
x3=coo(3,1); y3=coo(3,2); z3=coo(3,3);
x4=coo(4,1); y4=coo(4,2); z4=coo(4,3);
x5=coo(5,1); y5=coo(5,2); z5=coo(5,3);
x6=coo(6,1); y6=coo(6,2); z6=coo(6,3);
x7=coo(7,1); y7=coo(7,2); z7=coo(7,3);
x8=coo(8,1); y8=coo(8,2); z8=coo(8,3);
% Derivatives of shape function
dN1ksi=-0.125*(1-eta)*(1-zeta);
dN1eta=-0.125*(1-ksi)*(1-zeta);
dN1zeta=-0.125*(1-ksi)*(1-eta);
dN2ksi=0.125*(1-eta)*(1-zeta);
dN2eta=-0.125*(1+ksi)*(1-zeta);
dN2zeta=-0.125*(1+ksi)*(1-eta);
dN3ksi=0.125*(1+eta)*(1-zeta);
dN3eta=0.125*(1+ksi)*(1-zeta);
dN3zeta=-0.125*(1+ksi)*(1+eta);
dN4ksi=-0.125*(1+eta)*(1-zeta);
dN4eta=0.125*(1-ksi)*(1-zeta);
dN4zeta=-0.125*(1-ksi)*(1+eta);
dN5ksi=-0.125*(1-eta)*(1+zeta);
dN5eta=-0.125*(1-ksi)*(1+zeta);
dN5zeta=0.125*(1-ksi)*(1-eta);
dN6ksi=0.125*(1-eta)*(1+zeta);
dN6eta=-0.125*(1+ksi)*(1+zeta);
dN6zeta=0.125*(1+ksi)*(1-eta);
dN7ksi=0.125*(1+eta)*(1+zeta);
dN7eta=0.125*(1+ksi)*(1+zeta);
dN7zeta=0.125*(1+ksi)*(1+eta);
dN8ksi=-0.125*(1+eta)*(1+zeta);
dN8eta=0.125*(1-ksi)*(1+zeta);
dN8zeta=0.125*(1-ksi)*(1+eta);
% Defining Jacobean and inverting it
JAK=[dN1ksi,dN2ksi,dN3ksi,dN4ksi,dN5ksi,dN6ksi,dN7ksi,dN8ksi;
    dN1eta,dN2eta,dN3eta,dN4eta,dN5eta,dN6eta,dN7eta,dN8eta;
    dN1zeta,dN2zeta,dN3zeta,dN4zeta,dN5zeta,dN6zeta,dN7zeta,dN8zeta]...
    *[x1,y1,z1;x2,y2,z2;x3,y3,z3;x4,y4,z4;
    x5,y5,z5;x6,y6,z6;x7,y7,z7;x8,y8,z8];
JAKin=inv(JAK);
% Derivatives with respect to material coordinates
dN1dx=dN1ksi*JAKin(1,1)+dN1eta*JAKin(1,2)+dN1zeta*JAKin(1,3);
dN2dx=dN2ksi*JAKin(1,1)+dN2eta*JAKin(1,2)+dN2zeta*JAKin(1,3);
```

```
dN3dx=dN3ksi*JAKin(1,1)+dN3eta*JAKin(1,2)+dN3zeta*JAKin(1,3);
dN4dx=dN4ksi*JAKin(1,1)+dN4eta*JAKin(1,2)+dN4zeta*JAKin(1,3);
dN5dx=dN5ksi*JAKin(1,1)+dN5eta*JAKin(1,2)+dN5zeta*JAKin(1,3);
dN6dx=dN6ksi*JAKin(1,1)+dN6eta*JAKin(1,2)+dN6zeta*JAKin(1,3);
dN7dx=dN7ksi*JAKin(1,1)+dN7eta*JAKin(1,2)+dN7zeta*JAKin(1,3);
dN8dx=dN8ksi*JAKin(1,1)+dN8eta*JAKin(1,2)+dN8zeta*JAKin(1,3);
dN1dy=dN1ksi*JAKin(2,1)+dN1eta*JAKin(2,2)+dN1zeta*JAKin(2,3);
dN2dy=dN2ksi*JAKin(2,1)+dN2eta*JAKin(2,2)+dN2zeta*JAKin(2,3);
dN3dy=dN3ksi*JAKin(2,1)+dN3eta*JAKin(2,2)+dN3zeta*JAKin(2,3);
dN4dy=dN4ksi*JAKin(2,1)+dN4eta*JAKin(2,2)+dN4zeta*JAKin(2,3);
dN5dy=dN5ksi*JAKin(2,1)+dN5eta*JAKin(2,2)+dN5zeta*JAKin(2,3);
dN6dy=dN6ksi*JAKin(2,1)+dN6eta*JAKin(2,2)+dN6zeta*JAKin(2,3);
dN7dy=dN7ksi*JAKin(2,1)+dN7eta*JAKin(2,2)+dN7zeta*JAKin(2,3);
dN8dy=dN8ksi*JAKin(2,1)+dN8eta*JAKin(2,2)+dN8zeta*JAKin(2,3);
dN1dz=dN1ksi*JAKin(3,1)+dN1eta*JAKin(3,2)+dN1zeta*JAKin(3,3);
dN2dz=dN2ksi*JAKin(3,1)+dN2eta*JAKin(3,2)+dN2zeta*JAKin(3,3);
dN3dz=dN3ksi*JAKin(3,1)+dN3eta*JAKin(3,2)+dN3zeta*JAKin(3,3);
dN4dz=dN4ksi*JAKin(3,1)+dN4eta*JAKin(3,2)+dN4zeta*JAKin(3,3);
dN5dz=dN5ksi*JAKin(3,1)+dN5eta*JAKin(3,2)+dN5zeta*JAKin(3,3);
dN6dz=dN6ksi*JAKin(3,1)+dN6eta*JAKin(3,2)+dN6zeta*JAKin(3,3);
dN7dz=dN7ksi*JAKin(3,1)+dN7eta*JAKin(3,2)+dN7zeta*JAKin(3,3);
dN8dz=dN8ksi*JAKin(3,1)+dN8eta*JAKin(3,2)+dN8zeta*JAKin(3,3);
dudx=[dN1dx,dN2dx,dN3dx,dN4dx,dN5dx,dN6dx,dN7dx,dN8dx]*dx;
dudy=[dN1dy,dN2dy,dN3dy,dN4dy,dN5dy,dN6dy,dN7dy,dN8dy]*dx;
dudz=[dN1dz,dN2dz,dN3dz,dN4dz,dN5dz,dN6dz,dN7dz,dN8dz]*dx;
dvdx=[dN1dx,dN2dx,dN3dx,dN4dx,dN5dx,dN6dx,dN7dx,dN8dx]*dy;
dvdy=[dN1dy,dN2dy,dN3dy,dN4dy,dN5dy,dN6dy,dN7dy,dN8dy]*dy;
dvdz=[dN1dz,dN2dz,dN3dz,dN4dz,dN5dz,dN6dz,dN7dz,dN8dz]*dy;
dwdx=[dN1dx,dN2dx,dN3dx,dN4dx,dN5dx,dN6dx,dN7dx,dN8dx]*dz;
dwdy=[dN1dy,dN2dy,dN3dy,dN4dy,dN5dy,dN6dy,dN7dy,dN8dy]*dz;
dwdz=[dN1dz,dN2dz,dN3dz,dN4dz,dN5dz,dN6dz,dN7dz,dN8dz]*dz;
% Deformation gradient
FGR=[1+dudx, dudy, dudz;
          dvdx, 1+dvdy, dvdz;
          dwdx, dwdy, 1+dwdz];
C=FGR'*FGR;
% Stress components:
S(1)=2*C01+2*C10*(C(2,2)+C(3,3));
S(2)=2*C01+2*C10*(C(1,1)+C(3,3));
S(3)=2*C01+2*C10*(C(1,1)+C(2,2));
S(4)=-2*C10*C(2,1);
S(5)=-2*C10*C(3,1);
S(6)=-2*C10*C(3,2);
```

The tangent stiffness matrix calculated for the iterative Newton–Raphson procedure involves the computation of the second derivatives with respect to the components of tensor $\boldsymbol{C}$. The calculated stresses are used to compute the equivalent nodal forces that are further employed for the computation of the residual.

Even though the assumption of pure incompressibility may be unrealistic for certain industrial rubbers, the outlined constitutive model finds application in a wide range of scenarios owing to its simplicity as it depends only on two material parameters. As mentioned before, pure incompressibility leads to severe numerical difficulties, except for plane stress problems. When dealing with 3D problems and plane strain, an additional numerical treatment is required if completely incompressible, or close to incompressible models, are required. The main difficulties related to this issue, and possible ways to treat them, are briefly discussed in the following.

6.6.2 Incompressibility and mesh locking

The deformation that changes only the shape of a body, while preserving the material volume possessed by the material before the application of loads, is said to be *incompressible*. There are many physical phenomena, which include motion that preserves the volume. For illustrative purposes, we will consider the incompressible deformation that can occur for linear-elastic isotropic materials. It was previously shown that for this class of materials the elastic potential is given by Eq. (6.60). Considering that the Lame parameter λ is computed through Eq. (6.61), it can be observed that the value of λ tends to infinity if the Poisson's ratio approaches the value of 0.5 that represents its limit value.

In linear theory of elasticity, it is usually demonstrated that, for a pure hydrostatic load (i.e., defined as the case when all principal stresses are equal: $\sigma_1 = \sigma_2 = \sigma_3 = p$) applied to an isotropic material, the volume dilation e (i.e., the change in volume per unit volume) is linearly related to the corresponding stress (i.e., pressure) by:

$$e = \frac{p}{B} \tag{6.82}$$

where B is denoted as the bulk modulus of elasticity and is defined as:

$$B = \frac{E}{3(1-2\nu)} \tag{6.83}$$

From the definition of the bulk modulus given in Eq. (6.83), it can be noted that the limit value of the Poisson's ratio of $\nu = 0.5$ corresponds to the case of incompressible deformation because, for such a limit value, B tends to infinity meaning that an infinitely large pressure needs to be applied to change the volume of the material.

The nature of the numerical difficulty that arise when modeling incompressible materials can be illustrated by considering a two-dimensional problem. Let us consider a portion of a material modeled by means of triangular finite elements as depicted in Fig. 6.10. The bottom edge and the left edge of the meshed region are blocked, while the load is applied along the upper edge and the right edge is left unrestrained. The incompressibility in this two-dimensional problem is translated into a criterion of preserving the area for each triangular element during the deformation.

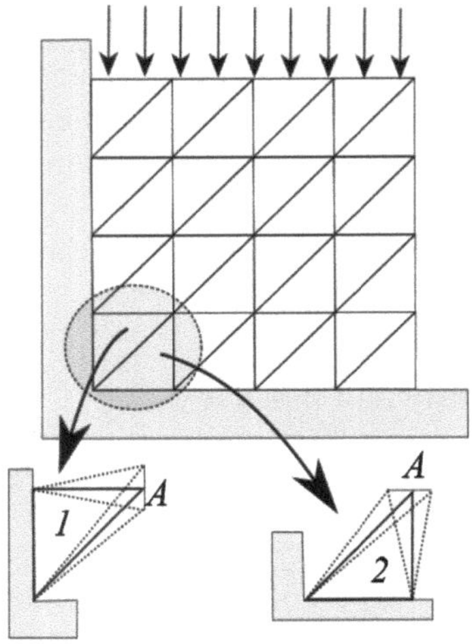

FIGURE 6.10 Rectangular incompressible component.

We may now have a closer look at two corner elements. Based on the prescribed boundary conditions, the only movable node is node "A." In order to preserve their areas, the triangular element "1" allows only a vertical movement of this node, while the triangular element "2" requires only a horizontal displacement. To satisfy both requirements, the only feasible position for the node is to remain in its position occupied in the undeformed configuration. Therefore, regardless of the amount of applied load, no displacement of any of the nodes will occur for a completely incompressible material. This aspect is usually denoted in the finite element literature as "mesh-locking," since the nodes are "locked" in their current position. The same difficulty is encountered even for nearly incompressible materials (i.e., materials with Poisson's ratio approaching the value of 0.5).

Within the standard displacement formulation of finite element method (referred hereafter as the irreducible formulation), the problem is formulated by considering only a displacement field as the unknown. Once the displacements are determined, the kinematic equations are used to obtain the strains, that are then used to determine stresses based on the constitutive equations. Considering that, for the discussed problem of incompressibility, this scheme faces the above mentioned locking problem, we now consider what is required to be modified within the solution procedure to remove this difficulty.

We previously saw that, when a locking problem is encountered, the displacements, and consequently strains, remain zero, regardless of the increment in loads. An alternative formulation of the same problem is that only a small amount of strains can result in infinitely large stresses. Therefore we may seek for the solution of the problem in the calculation of stresses.

A linear-elastic isotropic constitutive model is depicted in Eq. (6.60) and its stresses can be computed as partial derivatives of the elastic potential with respect to the strain. We can then write for each stress component:

$$\sigma_{ij} = \frac{\partial U}{\partial \varepsilon_{ij}} = 2\mu \frac{\partial}{\partial \varepsilon_{ij}} \left(\bar{I}_2\right) + \frac{\lambda}{2} \frac{\partial}{\partial \varepsilon_{ij}} \left(I_1^2\right) \tag{6.84}$$

The incompressibility is characterized by the Lame constant λ approaching infinity. It is therefore clear that the second term in Eq. (6.84) cannot be used for the stress calculation as it would tend to infinity as the material approaches the incompressible condition. The first term may still remain and be used as commonly done in the general irreducible formulation. Hence, for linear, small strain theory, we can write:

$$2\mu \frac{\partial}{\partial \varepsilon_{ij}} \left(\bar{I}_2\right) = 2\mu\varepsilon_{ij} = 2\mu d_{i,j} \tag{6.85}$$

where $u_{i,j}$ represents the derivative of displacement component i with respect to the spatial coordinate j (e.g., $d_{x,y} = \frac{\partial u}{\partial y}$).

The second term in Eq. (6.84) requires a different treatment. Considering Eq. (6.62) and in a view of small strain theory, we can write:

$$\frac{\lambda}{2} \frac{\partial}{\partial \varepsilon_{ij}} \left(I_1^2\right) = \lambda(\varepsilon_{xx} + \varepsilon_{yy} + \varepsilon_{zz}) = \lambda \cdot e \tag{6.86}$$

where e is the volume dilation that depicts the part of stress related to pressure and that cannot be calculated for the incompressible case.

A possible remedy to this inconvenience is to treat the pressure as an independent unknown together with the other unknown displacements. In a finite element approach, this can be achieved by using the so-called mixed formulation. Unlike the irreducible formulation, where only a displacement field is considered as the independent field, mixed formulations rely on independent approximation of either the stress or the strain fields, or even both when dealing with three-field approximation techniques. For details regarding two- and three-field approximations, reference should be made to [15].

For incompressible, or nearly incompressible materials, commonly two-field approximations are used where the displacements and pressure are treated as independent variables. The weak formulation of the mixed problem is similar to the one for the irreducible case, except that an independently approximated field of pressure is now introduced and the pressure is interpolated in the same fashion as the displacements:

$$p(\boldsymbol{x}) = \sum_{i=1}^{m} N_i^p(\boldsymbol{x}) \cdot p_i \tag{6.87}$$

where N_i^p are the pressure shape functions associated to the ith node with pressure p_i.

Shape functions used for the interpolation of the pressure do not have to be (and in general they are not) those specified for the displacement interpolations.

In general, a wider range of interpolation functions is available to depict the variations of pressure than those available for displacements, as the former do not have to be C^0 continuous (i.e., continuous and differentiable over the elements) because no derivatives of these functions are present in the formulation. Also the position of the pressure nodes and of the displacement nodes does not need to be the same.

Despite the relaxation of the shape function continuity requirements for the pressure interpolation functions, certain choices of combination for the interpolation of the two fields may not give meaningful or stable results. The selection of an adequate interpolation function is generally limited by the stability conditions of the mixed field approximations (for details, see [15] and [16]). Some of the combinations that yield good results are depicted in Fig. 6.11.

An alternative way to reach the same result within a single field approximation is to modify the finite element. This can be achieved by means of two approaches which exhibit an equivalent performance: (i) the use of the so-called *incompatible* or *nonconforming* elements that violate the C^0 continuity, and (ii) the modification of the Gauss quadrature by employing reduced or selective reduced integration. The latter technique is particularly easy to implement and requires only marginal modification of the standard, irreducible formulation. This technique is demonstrated in the following subsection.

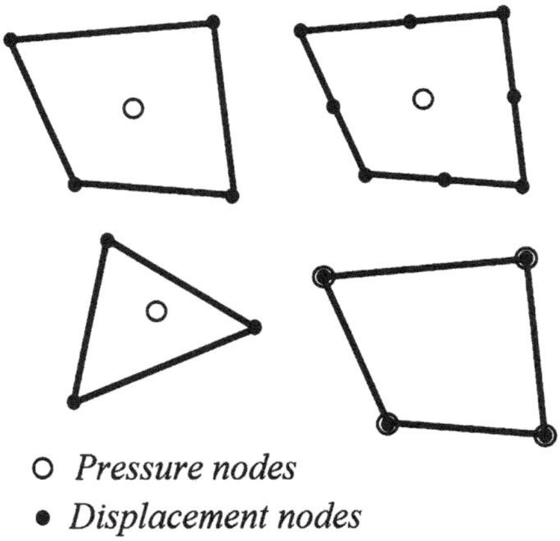

FIGURE 6.11 Combination of the pressure and displacement interpolations for a two-field approximation for incompressible materials.

6.6.3 Reduced and selective reduced integration

It was shown in Chapter 1 that the stiffness matrix within an irreducible formulation is computed as follows:

$$K = \int_V B^T D B dV \tag{6.88}$$

where matrix $\boldsymbol{D}$ collects the material properties. Here, we want to discuss the problem of locking, related to the difficulty to compute stresses when ν tends to 0.5, within the case of plane-strain (as it will not occur for plane stress) The reason for this becomes evident by observing the form that matrix $\boldsymbol{D}$ takes in the two cases, i.e., the plane-strain and the plane-stress scenarios.

Let us consider the expression for $\boldsymbol{D}$ related to the plane-strain scenario:

$$\boldsymbol{D} = \frac{E(1-\nu)}{(1+\nu)(1-2\nu)} \begin{bmatrix} 1 & \frac{\nu}{1-\nu} & 0 \\ \frac{\nu}{1-\nu} & 1 & 0 \\ 0 & 0 & \frac{1-2\nu}{2(1-\nu)} \end{bmatrix} \tag{6.89}$$

By comparing Eq. (6.89) with Eq. (1.97), it is clear that here (i.e., plane-strain case) the scalar in front of the matrix tends to infinity when considering an incompressible material. From Eq. (1.97), it can be seen that this is not the case when dealing with plane stress. By recalling Eq. (6.61), matrix $\boldsymbol{D}$ can be written as a function of the Lame constants as follows:

$$\boldsymbol{D} = \mu \begin{bmatrix} 2 & 0 & 0 \\ 0 & 2 & 0 \\ 0 & 0 & 1 \end{bmatrix} + \lambda \begin{bmatrix} 1 & 1 & 0 \\ 1 & 1 & 0 \\ 0 & 0 & 0 \end{bmatrix} = \bar{\boldsymbol{D}} + \bar{\bar{\boldsymbol{D}}} \tag{6.90}$$

At the limit of incompressibility, as λ tends to infinity, the second term, denoted by $\bar{\bar{\boldsymbol{D}}}$, requires a special treatment as it would tend to infinity as well.

If we perform the same separation also in the description of the stiffness matrix we can write:

$$\begin{aligned} \boldsymbol{K} &= \bar{\boldsymbol{K}} + \bar{\bar{\boldsymbol{K}}} \\ \bar{\boldsymbol{K}} &= \int_V \boldsymbol{B}^T \bar{\boldsymbol{D}} \boldsymbol{B} dV \\ \bar{\bar{\boldsymbol{K}}} &= \int_V \boldsymbol{B}^T \bar{\bar{\boldsymbol{D}}} \boldsymbol{B} dV \end{aligned} \tag{6.91}$$

With this formulation, $\bar{\boldsymbol{K}}$ is proportional to μ while $\bar{\bar{\boldsymbol{K}}}$ is proportional to λ. At the incompressibility limit the numerical values of the second part of the stiffness matrix tend to be very large when compared to those of the first part and, therefore, attempts to maintain the volumetric stiff behavior, so the irreducible formulation tends to lock.

A simple and practical way to overcome this difficulty is to reduce the order of the numerical integration employed to evaluate $\bar{\bar{\boldsymbol{K}}}$. The method that reduces the order of integration only on the second, volumetric part is called a *selective reduced integration*. If the reduction is performed on both parts the procedure is referred to as an *uniformly reduced integration*. The latter has the advantage that a shorter computing time is required when forming the stiffness matrix, but generally reduces the rank of the stiffness matrix and may result in singularity of the global stiffness matrix. Selective reduced integration retains the correct rank and overcomes the difficulty encountered by the irreducible formulation in the vicinity of incompressibility.

Malkus and Hughes [17] demonstrated that there is an equivalence between many mixed and reduced and selective integration elements. The equivalence theorem states that the element stiffness matrix of the mixed method is identical to the element stiffness of the reduced or selective integration approach, and, consequently, the displacements are also identical. Some elements, which yield good results for nearly incompressible materials and for which the equivalence holds, are depicted in Fig. 6.12.

	Mixed formulation		*Equivalent selective reduced integration*	
	Displacement interpolation	*Pressure interpolation*	*Normal Gauss integraion for* $\bar{\mathbf{K}}$	*Reduced Gauss integration for* $\bar{\bar{\mathbf{K}}}$
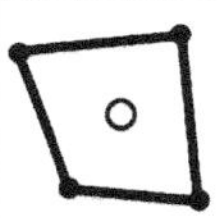	*Bilinear*	*Constant*	2 x 2	*One point*
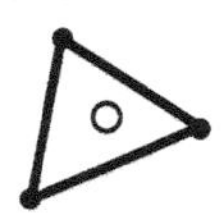	*Linear*	*Constant*	*One point*	*One point*

FIGURE 6.12 Compatibility between mixed formulation and selective reduced integration.

The modification is therefore simple to implement: the function which is used to compute the stiffness matrix for the element has to be modified to perform a separate integration, i.e., a full integration for $\bar{K}$ and a reduced integration for $\bar{\bar{K}}$.

The MATLAB® function that computes the stiffness matrix on the basis of this procedure for the quadrilateral element derived in plane strain is given below.

```
function stf=shpql4nsel(n1,n2,n3,n4,E,poas)
% Computing element stiffness matrix with selective reduced integration
%~~~~~~~~~~~~~~~~~~~~~~~~~~~~~~~~~~~~~~~~~~~~~~~~~~~~
x1=n1(1);y1=n1(2);
x2=n2(1);y2=n2(2);
x3=n3(1);y3=n3(2);
x4=n4(1);y4=n4(2);
Lambda=poas*E/((1+poas)*(1-2*poas));
mu=E/(2*(1+poas));
db=mu*[2,0,0;0,2,0;0,0,1]; % d_bar
dbb=Lambda*[1,1,0;1,1,0;0,0,0]; % d_barbar
```

```
cnt=0;
% Numerical integration related to D_bar part
gausscoor=[-sqrt(3)/3,-sqrt(3)/3; % Gauss coordinates
            sqrt(3)/3,-sqrt(3)/3;
            sqrt(3)/3,sqrt(3)/3;
            -sqrt(3)/3,sqrt(3)/3];
for i=1:2
  for j=1:2
      ksi=gausscoor(j);
      eta=gausscoor(i);
      dN1ksi=-0.25*(1-eta);
      dN1eta=-0.25*(1-ksi);
      dN2ksi=0.25*(1-eta);
      dN2eta=-0.25*(1+ksi);
      dN3ksi=0.25*(1+eta);
      dN3eta=0.25*(1+ksi);
      dN4ksi=-0.25*(1+eta);
      dN4eta=0.25*(1-ksi);
      JAK=[dN1ksi,dN2ksi,dN3ksi,dN4ksi;dN1eta,dN2eta,dN3eta,dN4eta]*[x1,y1;x2,y2;
     x3,y3;x4,y4];
      JAKin=inv(JAK);
      cnt=cnt+1;
     %Derivatives of shape function with respect to material coordinates:
      dN1dx=dN1ksi*JAKin(1,1)+dN1eta*JAKin(1,2);
      dN2dx=dN2ksi*JAKin(1,1)+dN2eta*JAKin(1,2);
      dN3dx=dN3ksi*JAKin(1,1)+dN3eta*JAKin(1,2);
      dN4dx=dN4ksi*JAKin(1,1)+dN4eta*JAKin(1,2);
      dN1dy=dN1ksi*JAKin(2,1)+dN1eta*JAKin(2,2);
      dN2dy=dN2ksi*JAKin(2,1)+dN2eta*JAKin(2,2);
      dN3dy=dN3ksi*JAKin(2,1)+dN3eta*JAKin(2,2);
      dN4dy=dN4ksi*JAKin(2,1)+dN4eta*JAKin(2,2);
      BB=[dN1dx,0,dN2dx,0,dN3dx,0,dN4dx,0; % Matrix B:
          0,dN1dy,0,dN2dy,0,dN3dy,0,dN4dy;
          dN1dy,dN1dx,dN2dy,dN2dx,dN3dy,dN3dx,dN4dy,dN4dx];
      kkb(:,:,cnt)=BB'*db*BB*det(JAK)*1*1; % Weight coefficients
  end
end
keb=kkb(:,:,1);
for i=2:cnt
    keb=keb+kkb(:,:,i);
end
% % Numerical integration related to D_bar_bar part
gausscoor=[0,0]; % Gauss coordinates
for i=1:1
  for j=1:1
      ksi=gausscoor(j);
      eta=gausscoor(i);
      dN1ksi=-0.25*(1-eta);
      dN1eta=-0.25*(1-ksi);
      dN2ksi=0.25*(1-eta);
      dN2eta=-0.25*(1+ksi);
      dN3ksi=0.25*(1+eta);
```

```
        dN3eta=0.25*(1+ksi);
        dN4ksi=-0.25*(1+eta);
        dN4eta=0.25*(1-ksi);
        JAK=[dN1ksi,dN2ksi,dN3ksi,dN4ksi;dN1eta,dN2eta,dN3eta,dN4eta]*[x1,y1;x2,y2;
       x3,y3;x4,y4];
        JAKin=inv(JAK);
       %Derivativies of shape function with respect to material coordinates:
        dN1dx=dN1ksi*JAKin(1,1)+dN1eta*JAKin(1,2);
        dN2dx=dN2ksi*JAKin(1,1)+dN2eta*JAKin(1,2);
        dN3dx=dN3ksi*JAKin(1,1)+dN3eta*JAKin(1,2);
        dN4dx=dN4ksi*JAKin(1,1)+dN4eta*JAKin(1,2);
        dN1dy=dN1ksi*JAKin(2,1)+dN1eta*JAKin(2,2);
        dN2dy=dN2ksi*JAKin(2,1)+dN2eta*JAKin(2,2);
        dN3dy=dN3ksi*JAKin(2,1)+dN3eta*JAKin(2,2);
        dN4dy=dN4ksi*JAKin(2,1)+dN4eta*JAKin(2,2);
        BB=[dN1dx,0,dN2dx,0,dN3dx,0,dN4dx,0;% Matrix B:
            0,dN1dy,0,dN2dy,0,dN3dy,0,dN4dy;
            dN1dy,dN1dx,dN2dy,dN2dx,dN3dy,dN3dx,dN4dy,dN4dx];
        cnt=cnt+1;
        kkbb(:,:,cnt)=BB'*dbb*BB*det(JAK)*2*2; % Weight coefficients are 2
   end
end
kebb=kkbb(:,:,1);
for i=2:cnt
    kebb=kebb+kkbb(:,:,i);
end
stf=keb+kebb;
```

To illustrate how the selective integration works, we can solve the problem depicted in Fig. 6.10 that is generally used as a benchmark problem to test particular procedure in handling incompressible or nearly incompressible materials. The MATLAB® script given below can be used as the main code to solve this problem. In the code, the data related to the mesh and to the boundary conditions is loaded. In particular, matrices with elements and coordinates are loaded first, followed by the three matrices required to solve the constrained problem through the Lagrange multipliers' technique. The code then invokes the function `asmbstifsel` that assemblies the global stiffness matrix by performing the selective reduced integration.

```
% Loading mesh
load elm.txt;
load coo.txt;
load cc.txt;
load ff.txt;
load qq.txt;
% Asigning the properties of the material
prop=[500,0.49999];
% Assembling of stiffness matrix
STFF=asmbstifsel(elm,coo,prop(1),prop(2));
dof=size(KRUT,1);
% Solving for displacements
```

```
dk=lagmult(STFF,cc,100*ff,qq);
dy=dk(2:2:dof,1); % Displacements in y direction
% Calculating the deforemd coordinates
def=deformedsh(dk,coo);
figure(1)
trisurf(elm,def(:,1),def(:,2),def(:,1)-def(:,1),dy,'FaceColor','interp')
title('Deformed shape')
colorbar
```

```
function STF=asmbstifsel(ELM,COO,E,poas)
% Assembling global stiffness matrix.
% Inputs:
% ELM matrix; COO matrix; E - Young's modulus; poas - Poisson's ratio
Nel=size(ELM,1);
Nnd=size(COO,1);
ZERO=zeros(2*Nnd,2*Nnd); % Setting zero stiffness matrix
STF=ZERO;
for i=1:Nel % Look for assembling stiffness matrix
    n1=COO(ELM(i,1),:);
    n2=COO(ELM(i,2),:);
    n3=COO(ELM(i,3),:);
    n4=COO(ELM(i,4),:);
    krel=shpql4nsel(n1,n2,n3,n4,E,poas);
    K11=krel(1:2,1:2); % Forming blocks of stiffness matrix
    K12=krel(1:2,3:4);
    K13=krel(1:2,5:6);
    K14=krel(1:2,7:8);
    K21=krel(3:4,1:2);
    K22=krel(3:4,3:4);
    K23=krel(3:4,5:6);
    K24=krel(3:4,7:8);
    K31=krel(5:6,1:2);
    K32=krel(5:6,3:4);
    K33=krel(5:6,5:6);
    K34=krel(5:6,7:8);
    K41=krel(7:8,1:2);
    K42=krel(7:8,3:4);
    K43=krel(7:8,5:6);
    K44=krel(7:8,7:8);
    NOD1=ELM(i,1);
    NOD2=ELM(i,2);
    NOD3=ELM(i,3);
    NOD4=ELM(i,4);
    STF1=ZERO; % Stiffness matrix for the current element
    STF1(2*NOD1-1:2*NOD1,2*NOD1-1:2*NOD1)=K11;
    STF1(2*NOD2-1:2*NOD2,2*NOD2-1:2*NOD2)=K22;
    STF1(2*NOD3-1:2*NOD3,2*NOD3-1:2*NOD3)=K33;
    STF1(2*NOD4-1:2*NOD4,2*NOD4-1:2*NOD4)=K44;
    STF1(2*NOD1-1:2*NOD1,2*NOD2-1:2*NOD2)=K12;
    STF1(2*NOD1-1:2*NOD1,2*NOD3-1:2*NOD3)=K13;
    STF1(2*NOD1-1:2*NOD1,2*NOD4-1:2*NOD4)=K14;
    STF1(2*NOD2-1:2*NOD2,2*NOD1-1:2*NOD1)=K21;
```

```
    STF1(2*NOD2-1:2*NOD2,2*NOD3-1:2*NOD3)=K23;
    STF1(2*NOD2-1:2*NOD2,2*NOD4-1:2*NOD4)=K24;
    STF1(2*NOD3-1:2*NOD3,2*NOD1-1:2*NOD1)=K31;
    STF1(2*NOD3-1:2*NOD3,2*NOD2-1:2*NOD2)=K32;
    STF1(2*NOD3-1:2*NOD3,2*NOD4-1:2*NOD4)=K34;
    STF1(2*NOD4-1:2*NOD4,2*NOD1-1:2*NOD1)=K41;
    STF1(2*NOD4-1:2*NOD4,2*NOD2-1:2*NOD2)=K42;
    STF1(2*NOD4-1:2*NOD4,2*NOD3-1:2*NOD3)=K43;
    STF=STF+STF1;
end
```

Upon finishing the assembling of global stiffness matrix, the boundary value problem is solved by means of the Lagrange multipliers' technique. A simple code that performs this calculation is given in the MATLAB® function `langmult` listed below.

```
function dk=lagmult(STF,C,F,Q)
% Function that solves the constrained problem by Lagrange multipliers
% Syntax:
%    langmult(STF,C,F,Q)
%    -STF - stiffness matrix
%    -C - constrain matrix
%    -F - force vector
%    -Q - right-hand side of constrain system
DOF1=size(STF);
DOF=DOF1(1);
const1=size(C);
const=const1(1);
null(1:const,1:const)=0;
STFm=[STF,C';C,null];
dkm=inv(STFm)*[F;Q];
dk=dkm(1:DOF,1);
```

The deformed shape is plotted in Fig. 6.13 with a color bar that depicts the vertical component of the displacements extracted with vector `dy`. The deformed coordinates are computed with the MATLAB® function `deformedsh` provided below.

```
function defc=deformedsh(u,NODS)
% Computing deformed coordinates
DOF1=size(u);
DOF=DOF1(1);
nnod=DOF/2;
for i=1:nnod
    defc(i,1)=NODS(i,1)+u(2*i-1);
    defc(i,2)=NODS(i,2)+u(2*i);
end
```

The proposed code is used to solve a problem with a nearly incompressible material with values of 500 MPa and 0.49999 for the Young's modulus and the Poisson's ratio, respectively. All the upper nodes are loaded with a force equal to

100 N. The nodes on the bottom and left edges of the structure are constrained in both directions. The problem is solved twice, i.e., once with a selective reduced integration and a second time with the irreducible formulation. The latter is calculated by modifying the second cycle (i.e., starting from code line 53) of the MATLAB® function `shpql4nsel.m` by changing the corresponding weight coefficient in line 81 to enable the full Gauss integration. The calculated results are visualized in Fig. 6.13. The problem is here solved over the same mesh with 196 elements and 225 nodes. The listings of the ASCII files containing the mesh are not shown here for brevity, while the MATLAB® codes provided can be used for any quadrilateral meshed arrangement with an arbitrary number of elements and nodes. The results highlight how the irreducible formulation fails to solve the problem due to locking of the mesh and produces a solution with practically no displacements. The selective reduced integration is capable of solving the problem and results in significant displacements in the order of 20% of the size of the structure.

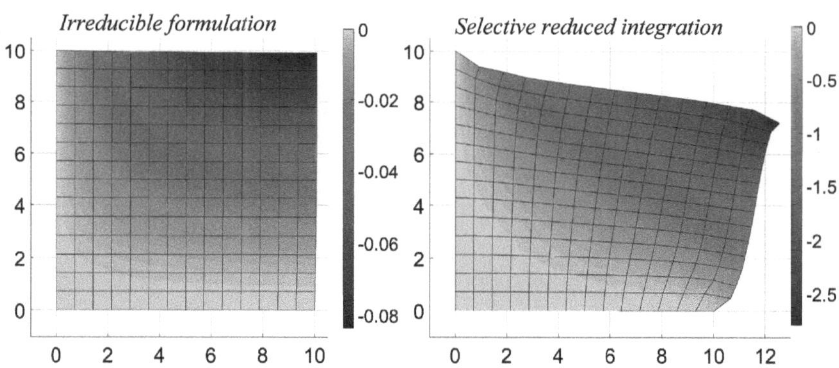

FIGURE 6.13 Solution for nearly incompressible material determined with the irreducible and the selective reduced integration.

6.6.4 Strain energy density function separating the volumetric and deviatoric terms

Based on the numerical difficulties previously encountered for the nearly incompressible materials, it is convenient to formulate the energy density function of hyperelastic materials by separating volumetric and deviatoric parts. From the definition of the invariants of the right Cauchy–Green deformation tensor given in Eq. (6.72), it is evident that the volumetric deformation is fully characterized by the third invariant I_{3C}, while the invariants I_{1C} and I_{2C} depend on both the volumetric and deviatoric parts of the deformation.

In order to separate the strain energy density function into the volumetric part and the deviatoric part, we first need to decompose the deformation gradient as follows:

$$\boldsymbol{F} = \boldsymbol{F}_{vol} \cdot \boldsymbol{F}_{dev} \tag{6.92}$$

where the volumetric part can be written as:

$$\boldsymbol{F}_{vol} = (det\boldsymbol{F})^{\frac{1}{3}}\boldsymbol{I} \tag{6.93}$$

obtained based on the following re-arrangement:

$$det\boldsymbol{F}_{vol} = \left[(det\boldsymbol{F})^{\frac{1}{3}}\right]^3 \cdot \boldsymbol{I} = det\boldsymbol{F} \tag{6.94}$$

From Eq. (6.93), we can also write:

$$\begin{aligned} det\boldsymbol{F} &= det\boldsymbol{F}_{vol} \cdot det\boldsymbol{F}_{dev} \\ \Longrightarrow det\boldsymbol{F}_{dev} &= (det\boldsymbol{F}_{vol})^{-1} \cdot det\boldsymbol{F} = (det\boldsymbol{F})^{-1} \cdot det\boldsymbol{F} = 1 \end{aligned} \tag{6.95}$$

that highlights how $\boldsymbol{F}_{dev}$ is related to the volume preserving deformation.

We can then establish a relationship between the right Cauchy–Green deformation tensor and the modified right Cauchy–Green tensor $\boldsymbol{C}_{dev}$ expressed as a function of the deviatoric deformation as:

$$\begin{aligned} \boldsymbol{C}_{dev} = \boldsymbol{F}_{dev}^T \cdot \boldsymbol{F}_{dev} = (det\boldsymbol{F})^{-\frac{1}{3}}(\boldsymbol{I}\boldsymbol{F})^T \cdot (det\boldsymbol{F})^{-\frac{1}{3}}\boldsymbol{I}\boldsymbol{F} = \left[(det\boldsymbol{F})^2\right]^{\frac{1}{3}} \boldsymbol{F}^T\boldsymbol{F} \\ \Longrightarrow \boldsymbol{C}_{dev} = (I_{3C})^{-\frac{1}{3}}\boldsymbol{C} \end{aligned} \tag{6.96}$$

considering that, by definition, $\boldsymbol{C} = \boldsymbol{F}^T\boldsymbol{F}$, and thus, $(det\boldsymbol{F})^2 = det\boldsymbol{C} = I_{3C}$.

From Eq. (6.96), we can see that the elements of the modified right Cauchy–Green tensor $\boldsymbol{C}_{dev}$ are scaled elements of the tensor $\boldsymbol{C}$. This observation can be used to relate the two sets of invariants. By recalling Eq. (6.72), we can write the first two invariants as follows:

$$\begin{aligned} J_{1C} = (I_{3C})^{-\frac{1}{3}}(C_{11} + C_{22} + C_{33}) = (I_{3C})^{-\frac{1}{3}} I_{1C} \\ J_{2C} = (I_{3C})^{-\frac{2}{3}}(C_{11}C_{22} + C_{22}C_{33} + C_{33}C_{11} - C_{12}C_{21} - C_{23}C_{32} - C_{13}C_{31}) = \\ = (I_{3C})^{-\frac{2}{3}} I_{2C} \end{aligned} \tag{6.97}$$

These two invariants depend only on the deviatoric part of the deformation, while to describe the volumetric part we could use either I_{3C} or $det\boldsymbol{F} = (I_{3C})^{\frac{1}{2}}$. The latter representation is proposed in some literature (see, e.g., [18]) where the term is denoted as J_{3C} to unify the notation. It is important to emphasize that this does not represent the third invariant of the modified right Cauchy–Green tensor $\boldsymbol{C}_{dev}$ that is equal to zero, since the tensor represents the deviatoric part of the deformation.

By using the invariants of the modified right Cauchy–Green tensor, we can formulate the energy density function as the sum of two contributions, i.e., one resulting from deviatoric part of deformation and the other resulting from the volumetric part, as depicted by:

$$U = U^{dev}(J_{1C}, J_{1C}) + U^{vol}(J) \tag{6.98}$$

where $J = det\boldsymbol{F}$, and based on which the modified Mooney–Rivlin can be expressed as:

$$U = C_{10}(J_{1C} - 3) + C_{01}(J_{2C} - 3) + U^{vol}(J) \tag{6.99}$$

where the first two terms are different from those used in Eq. (6.75) because we are now using the invariants of the modified right Cauchy–Green in the place of I_{1C} and I_{2C}. This implies that the meaning and the numerical values of constants C_{10} and C_{01} are not the same.

The remaining part of the elastic potential energy density is expressed by a function depicting the volumetric part of the deformation. A simple form, convenient for numerical implementation, can be obtained by assuming a linear relationship between the hydrostatic pressure p and the change in volume as previously adopted in Eq. (6.82):

$$p = B \cdot e = B \cdot \frac{\Delta V}{V} = B \cdot (J - 1) \tag{6.100}$$

with B being the bulk modulus of elasticity.

For an uniform pressure load, the contribution of the deviatoric part of the strain energy density $U^{dev}(J_{1C}, J_{1C})$ vanishes. By derivation of the potential, the pressure p has to be obtained and this implies:

$$p = \left[U^{vol}(J)\right]' \tag{6.101}$$

with the differentiation meant with respect to J.

By integrating the last equation, we obtain:

$$U^{vol} = \frac{1}{2}B(J - 1)^2 \tag{6.102}$$

which completes the modified Mooney–Rivlin elastic potential that can be rewritten in the following form:

$$U = C_{10}(J_{1C} - 3) + C_{01}(J_{2C} - 3) + \frac{1}{2}B(J - 1)^2 \tag{6.103}$$

Stresses are again computed as the derivative of the elastic potential density with respect to strain components. The derivative with respect to the Green strain gives the second Piola–Kirchhoff stress that can be expressed for the elastic potential given in Eq. (6.103) as:

$$\boldsymbol{S} = \frac{\partial U}{\partial \boldsymbol{E}} = 2\frac{\partial U}{\partial \boldsymbol{C}} = 2\left[C_{10}\frac{\partial J_{1C}}{\boldsymbol{C}} + C_{01}\frac{\partial J_{2C}}{\boldsymbol{C}} + B(J - 1)\frac{\partial J}{\partial \boldsymbol{C}}\right] \tag{6.104}$$

It is convenient to express the derivatives of the invariants of the modified right Cauchy–Green tensor as a function of the invariants of the right Cauchy–Green tensor since, in the course of the implementation, the latter is computed. Considering the relationship of (6.97), we can write:

$$\begin{aligned}
\frac{\partial J_{1C}}{\partial \boldsymbol{C}} &= (I_{3C})^{-\frac{1}{3}}\frac{\partial I_{1C}}{\partial \boldsymbol{C}} - \frac{1}{3}(I_{3C})^{-\frac{4}{3}}\frac{\partial I_{3C}}{\partial \boldsymbol{C}}I_{1C} \\
\frac{\partial J_{2C}}{\partial \boldsymbol{C}} &= (I_{3C})^{-\frac{2}{3}}\frac{\partial I_{2C}}{\partial \boldsymbol{C}} - \frac{2}{3}(I_{3C})^{-\frac{5}{3}}\frac{\partial I_{3C}}{\partial \boldsymbol{C}}I_{2C} \\
\frac{\partial J}{\partial \boldsymbol{C}} &= \frac{\partial}{\partial \boldsymbol{C}}(I_{3C})^{\frac{1}{2}} = \frac{1}{2}(I_{3C})^{-\frac{1}{2}}\frac{\partial I_{3C}}{\partial \boldsymbol{C}}
\end{aligned} \tag{6.105}$$

For the numerical implementation, it is useful to write the derivatives of the invariants I_{iC} in Voigt notation that is adopted throughout this book for the representation of tensors. Considering Eq. (6.72), we can write:

$$\frac{\partial I_{1C}}{\partial \boldsymbol{C}} = \begin{bmatrix} 1 \\ 1 \\ 1 \\ 0 \\ 0 \\ 0 \end{bmatrix}; \quad \frac{\partial I_{2C}}{\partial \boldsymbol{C}} = \begin{bmatrix} C_{22}+C_{33} \\ C_{11}+C_{33} \\ C_{11}+C_{22} \\ -C_{12} \\ -C_{23} \\ -C_{13} \end{bmatrix}; \quad \frac{\partial I_{3C}}{\partial \boldsymbol{C}} = \begin{bmatrix} C_{22}C_{33}-C_{23}^2 \\ C_{11}C_{33}-C_{13}^2 \\ C_{11}C_{22}-C_{12}^2 \\ C_{23}C_{13}-C_{33}C_{12} \\ C_{12}C_{13}-C_{11}C_{23} \\ C_{12}C_{23}-C_{22}C_{13} \end{bmatrix} \tag{6.106}$$

We now provide some remarks regarding the last three components related to the shear stresses. Recalling that the elementary strain energy density can be written as:

$$\begin{aligned}
dU = \boldsymbol{\sigma} : \boldsymbol{\varepsilon} = \sigma_{xx}\varepsilon_{xx} + \sigma_{yy}\varepsilon_{yy} + \sigma_{zz}\varepsilon_{zz} + \sigma_{xy}\varepsilon_{xy} + \sigma_{yx}\varepsilon_{yx} + \sigma_{xz}\varepsilon_{xz} \\
+\sigma_{zx}\varepsilon_{zx} + \sigma_{yz}\varepsilon_{yz} + \sigma_{zy}\varepsilon_{zy}
\end{aligned} \tag{6.107}$$

where $\boldsymbol{\sigma}$ and $\boldsymbol{\varepsilon}$ are here written as second-order tensors (i.e., not in Voigt notation as vectors of six components). As both stress and strain are symmetric tensors, stresses are computed by differentiating the strain energy density function (see Eq. (4.28)). Due to the symmetry of both the tensors (i.e., $\sigma_{ij} = \sigma_{ji}$ and $\varepsilon_{ij} = \varepsilon_{ji}$), the components that result from differentiation with respect to ε_{ij} appear twice. Therefore the components related to the shear stress have to be divided by 2. The particularization of Eq. (4.28) for symmetric tensors may be written as:

$$\sigma_{ii} = \frac{\partial U}{\partial \varepsilon_{ii}}; \quad \sigma_{ij} = \frac{1}{2}\frac{\partial U}{\partial \varepsilon_{ij}} \tag{6.108}$$

In view of Eq. (6.108), the last three components of the vector given in Eq. (6.106), upon the differentiation of (6.72), are divided by 2.

The MATLAB® implementation can be performed by reutilizing the first part for the code used for the calculation of the deformation gradient and of the right Cauchy–Green tensor $\boldsymbol{C}$, included in the MATLAB® script given for the calculation of the compressible Mooney–Rivlin. Different parts of the code are provided in the MATLAB® script given below.

```
...
% Invariants of C tensor:
I1C=trace(C);
I2C=0.5*(trace(C)^2-trace(C*C));
I3C=det(C);
% Derivatives of invariants:
dI1dC=[1;1;1;0;0;0];
dI2dC=[C(2,2)+C(3,3);C(1,1)+C(3,3);C(1,1)+C(2,2);-C(1,2);-C(2,3);-C(1,3)];
dI3dC=[C(2,2)*C(3,3)-C(2,3)^2;C(1,1)*C(3,3)-C(1,3)^2;C(1,1)*C(2,2)-C(1,2)^2;
       C(2,3)*C(1,3)-C(3,3)*C(1,2);C(1,2)*C(1,3)-C(1,1)*C(2,3);C(1,2)*C(2,3)-C(2,2)*C(1,3)];
% Derivatives of modified invariants:
dJ1dC=I3C^(-1/3)*dI1dC-(1/3)*I3C^(-4/3)*dI3dC*I1C;
dJ2dC=I3C^(-2/3)*dI2dC-(2/3)*I3C^(-5/3)*dI3dC*I2C;
dJ3dC=0.5*I3C^(-0.5)*dI3dC;
% Stress tensor (Voight notation):
S=2*(C10*dJ1dC+C01*dJ2dC+B*(det(FGR)-1)*dJ3dC);
```

The tangent stiffness matrix, required for the iterative solution of the problem, is computed based on Eq. (6.81) that can be expressed in the following form:

$$\boldsymbol{D}_t = \frac{\partial \boldsymbol{S}}{\partial \boldsymbol{E}} = 2\frac{\partial \boldsymbol{S}}{\partial \boldsymbol{C}} = 4\left[C_{10}\frac{\partial^2 J_{1C}}{\partial \boldsymbol{C}^2} + C_{01}\frac{\partial^2 J_{2C}}{\partial \boldsymbol{C}^2} + B\frac{\partial J}{\partial \boldsymbol{C}}\frac{\partial J}{\partial \boldsymbol{C}} + B(J-1)\frac{\partial^2 J}{\partial \boldsymbol{C}^2}\right] \tag{6.109}$$

which involves the second derivatives of the modified invariants that are computed by further differentiation of Eq. (6.105). By differentiating Eq. (6.105), the derivatives of modified invariants are expressed as a function of the derivatives of the right Cauchy–Green deformation tensor. Both latter and former are symmetric tensors of fourth order, with the elements:

$$j_{pqrs} = \frac{\partial^2 I_{iC}}{\partial C_{pq}\partial C_{rs}} \tag{6.110}$$

For the purpose of a computer implementation, it is more convenient to express them in Voigt notation (i.e., as matrices).

6.7 Summary

This chapter provided an introduction to geometrical nonlinear problems and to the importance of carefully defining both strain and stress measures when formulating constitutive models for these modeling scenarios.

In the first part of the chapter, particular attention was placed at considering simple finite element models to highlight the influence of specifying different strain measures to describe free-body rotations and deformation. The details of the finite element implementation were presented to gain insight into the specific aspects of the formulations considered.

The derivation of the tangent stiffness matrix, required for the analysis of a component or structure, was then presented by considering one-dimensional and two-dimensional problems. Also in this case, the numerical implementation aimed at highlighting the key aspects involved in the numerical coding.

The last part of the chapter was dedicated to hyperelastic models and to the challenges that arise in their numerical implementation. The advantages in using strain energy density functions in the formulation of constitutive models were then described and simple illustrative examples were provided. The implications of simulating materials with incompressible properties were pointed out and discussed. The occurrence of mesh locking problems that can occur in such scenarios was considered together with some numerical approaches to address it by means of reduced and selective reduced integration strategies.

References

[1] J.N. Reddy, An Introduction to Nonlinear Finite Element Analysis, Oxford University Press, New York, 2004.

[2] K.J. Bathe, Finite Element Procedures, Prentice Hall, Pearson Education Inc., New Jersey, 1982.

[3] M.A. Crisfield, Non-linear Finite Element Analysis of Solids and Structures - vol. 1: Essentials, John Wiley and Sons, New York, 1991.

[4] J. Bonet, R.D. Wood, Non-linear Continuum Mechanics for Finite Element Analysis, Cambridge University Press, New York, 1997.

[5] G.R. Liu, S.S. Quek, The Finite Element Method: A Practical Course, Butterworth Heinemann, Oxford, 2003.

[6] L. Mooney, A theory of large elastic deformation, Journal of Applied Mechanics 11 (1940) 582–592.

[7] R.S. Rivlin, Large elastic deformations of isotropic materials, IV. Further developments of the general theory, Philosophical Transactions of the Royal Society A 241 (1948) 379–397.

[8] L.R.G. Treloar, The Physics of Rubber Elasticity, 3rd edition, Oxford University Press, 1975.

[9] R.W. Ogden, Large deformation isotropic elastic deformation: on the correlation of theory and experiment for incompressible rubberlike solids, Proceedings of Royal Society, London A 326 (1972) 565–584.

[10] R.W. Ogden, Elastic deformation of rubberlike solids, in: H.G. Hopkins, M.J. Sewell (Eds.), Mechanics of Solids, the Rodney Hill 60th Anniversary Volume, Pergamon, Oxford, 1982.

[11] R. De Borst, P.A.J. Van den Bogert, J. Zeilmaker, Modeling and analysis of rubberlike materials, Heron 33 (1) (1988).

[12] M.A. Dobrowlski, A mixed finite element method for approximating incompressible materials, Siam Journal of Numerical Analysis 29 (2) (1992) 365–389.

[13] R.W. Ogden, Non-linear Elastic Deformations, Dover Publications, INC, Mineola, New York, 1984.

[14] M.A. Crisfield, Non-linear Finite Element Analysis of Solids and Structures - vol. 2: Advanced Topics, John Wiley and Sons, New York, 1997.

[15] O. Zienkiewicz, R. Taylor, Finite Element Method, vol. 1, fifth edition, Butterworth-Heinemann, Woburn, 2000.

[16] T. Hughes, Finite Element Method. Linear Static and Dynamic Finite Element Analysis, Dover Publications INC, New York, 2000.

[17] D.S. Malkus, T.J.R. Hughes, Mixed finite element methods - reduced and selective integration techniques: a unification of concept, Computer Methods in Applied Mechanics and Engineering 15 (1) (1978) 63–81.
[18] Rene de Borst, Mike Crisfield, Joris Remmers, Clemens Verhoosel, Non-Linear Finite Element Analysis of Solids and Structures, John Wiley and Sons, West Sussex, UK, 2012.

Chapter 7

Metals

Metallic materials are characterized by pronounced ductility, exhibited through the development of an irreversible plastic deformation, before they rapture. Their structural response is characterized by an initial elastic deformation up to a certain level of applied load. After a certain level of load, a plastic deformation starts to occur while, in general, the material is not damaged as its elastic constants remain unaltered. The degradation of the material, in the form of microcracks or the formation of one or few macroscopically observable cracks, start to develop only after significant amount of plastic deformation has taken place. In order to perform a structural analysis that accounts for the inelastic deformation of metallic materials, a proper elasto-plastic constitutive model has to be employed.

Most metals develop only small elastic deformation. This can be deduced by comparing the value of the Young's modulus with the first yield limit, where the former is usually three orders of magnitude larger than the latter. For example, with a high strength steel alloy such as an Inconel 718, the tensile strength is in the order of 1400 MPa. Considering that its Young's modulus is about 210,000 MPa, the largest elastic strain that the material can develop is:

$$\varepsilon_{max}^{el} = \frac{1400}{210,000} = 0.0067 \tag{7.1}$$

For ordinary steels, or light metals like aluminum and its alloys, this values becomes even smaller. It is therefore common practice and convenient to model metallic materials as linear-elastic up to the onset of plastic deformation. When the stress exceeds the one corresponding to the first yield limit, an irreversible plastic deformation starts to develop. The ultimate strain that metals can develop prior to rapture can reach values of about 0.1 or more. This strain is a total strain (i.e., the sum of the elastic and plastic strains), while its elastic part is still fairly small.

Metals possess a polycrystalline structures containing a very large number of small grains. Each individual grain may, in general, exhibit an anisotropic behavior. The structural components on the other hand, being a couple of orders of magnitude larger than a single grain, contain fairly large number of randomly oriented grains. Owed to this type of structure, macroscopically measured mechanical properties can be determined as average of the properties over different possible orientations of the individual grains, making the global response isotropic. This may not be the case with certain sheet metals which are

Constitutive Modeling of Engineering Materials. https://doi.org/10.1016/B978-0-12-814696-5.00012-5

manufactured by hot or cold rolling. Such manufacturing procedure leads to more uniform grain orientation, resulting in different elastic properties in the rolling and transversal directions. The deviation form globally isotropic material depends mostly on the sheet metal thickness and rolling temperature. In general, if the difference in two perpendicular directions is more than 10%, the material should be modeled as elastically orthotropic.

To account for the mechanical response upon exhausting the elastic capacity, the plastic part of the constitutive model has to be incorporated. We now reconsider the governing equations introduced in Chapter 5 based on elasto-plastic models and place particular attention on their finite element implementation. We also recall that the governing equations of an elasto-plastic model can be expressed in the following form:

$$\begin{aligned} \boldsymbol{\sigma} &= \boldsymbol{D}(\boldsymbol{\varepsilon} - \boldsymbol{\varepsilon}^{pl}) \\ \dot{\boldsymbol{\varepsilon}}^{pl} &= \dot{\lambda}\frac{\partial g}{\partial \boldsymbol{\sigma}} \\ \dot{\lambda} f(\boldsymbol{\sigma}) &= 0 \end{aligned} \tag{7.2}$$

where $\boldsymbol{D}$ is material stiffness tensor (represented as matrix in Voigt notation), f is the adopted yield function, and g is the plastic potential. Metals are generally modeled by associative plasticity, hence there is no need for defining an additional plastic potential, as the plastic flow is perpendicular to the yield surface (i.e., the plastic potential role is associated to the yield surface). When dealing with associative plasticity, function g in the second equation of (7.2) is replaced by the yield function f. In Eqs. (7.2) and throughout this chapter, it is tacitly assumed that $\boldsymbol{\varepsilon}$ without any index represents the total strain, while the plastic and elastic strain components will be specifically identified by a superscript.

The origin of plastic deformation in metals is related to the existence of a plastic slip (see Chapter 5). Consequently, for the modeling of plastic deformation in metals, pressure non-sensitive yield conditions are adopted, as hydrostatic pressure does not contribute to the onset of plastic slip. One of the most commonly used is certainly the von Mises yield condition. The numerical aspects of the implementation of the governing equations (7.2) are presented based on this criterion, although the same approach can be applied to other yield conditions.

The material mechanical behavior given in Eqs. (7.2) constitutes a nonlinear response that generally requires an incremental procedure to trace the solution over a certain range of deformation. In the following, we implement this procedure in the context of a nonlinear finite element analysis, where the nonlinearity is related to the elasto-plastic material response. The implementation of the elasto-plastic constitutive response, as presented in Chapter 5, is performed by means of a trial and correction approach. In the first step of the procedure, a trial stress is computed as if the complete strain increment is elastic. The updated stress is subsequently included in the yield condition to verify if the stress is plastically admissible (i.e., the point is within or on the yield surface). If the calculated point falls outside the yield surface, a correction of the stress state needs to be performed to bring back the point on the yield surface. The incremental solution strategy can be summarized in the following steps.

1. Compute the incremental displacements $\Delta \boldsymbol{d}_i$
2. On the basis of the kinematic equations, compute the corresponding incremental strains $\Delta \boldsymbol{\varepsilon}_i(\Delta \boldsymbol{d}_i)$
3. By employing the constitutive equations, compute trial stress increment and update the stresses: $\Delta \boldsymbol{\sigma}_i = \boldsymbol{D} \cdot \Delta \boldsymbol{\varepsilon}_i$ and $\boldsymbol{\sigma}_i = \boldsymbol{\sigma}_{i-1} + \Delta \boldsymbol{\sigma}_i$
4. Check if the current stress state defined by $\boldsymbol{\sigma}_i$ is plastically admissible and perform the stress corrections if required
5. Calculate the vector of internal forces $\boldsymbol{f}^{\sigma_i}$ and check that the equilibrium condition is satisfied (i.e., compare $\boldsymbol{f}^{\sigma_i}$ with $\boldsymbol{f}_i^{ext}$)

The outlined incremental strategy has highlighted the two most important roles that a plasticity algorithm has within a finite element code. The first one is related to the correction of a stress state. Owing to the adopted trial-correction procedure within the calculation inevitably the stress state at some point will be outside the yield surface (i.e., will be plastically inadmissible). When implementing the elasto-plastic constitutive model within a standard displacement-based FEM program, it is therefore necessary to supply the code that brings back the stress on a yield surface. In computational plasticity, such procedure is referred to as the *stress return algorithm*.

The second important role is related to the calculation of the displacement increment $\Delta \boldsymbol{d}_i$. Since the problem is nonlinear, the solution should be reached by an iterative procedure. It is common, in finite element implementations, to adopt the Newton–Raphson method that requires the evaluation of a tangent stiffness matrix. Following this procedure (as introduced in Chapter 6, see Eq. (6.31)), the iterative displacement increment can be computed as follows:

$$\delta \boldsymbol{d}_i^j = -(\boldsymbol{K}_t^j)^{-1} \cdot \boldsymbol{r}^j \tag{7.3}$$

where $\boldsymbol{r}^j$ is the vector of residuals, here specifically computed as the difference between the internal and external nodal forces:

$$\boldsymbol{r}^j = \boldsymbol{f}_j^{\sigma_i} - \boldsymbol{f}_i^{ext} \tag{7.4}$$

In last two equations, index j indicates the number of iterations needed to compute the equilibrium state for the increment i.

The numerical implementation of an elasto-plastic computational algorithm can be based on the following incremental-iterative procedure.

1. Compute the iterative displacement increment:
   $\delta \boldsymbol{d}_i^j = -(\boldsymbol{K}_t^j)^{-1} \cdot \boldsymbol{r}^j$
2. Update the displacement increment:
   $\Delta \boldsymbol{d}_i = \Delta \boldsymbol{d}_i + \delta \boldsymbol{d}_i^j$
3. Compute the incremental strains as a function of the incremental displacement:
   $\Delta \boldsymbol{\varepsilon}_i(\Delta \boldsymbol{d}_i)$
4. Compute the incremental stress:
   $\Delta \boldsymbol{\sigma}_i = \boldsymbol{D}(\Delta \boldsymbol{\varepsilon}_i - \Delta \boldsymbol{\varepsilon}^{pl^j})$
5. Update the stresses from the previous converged equilibrium state (i.e., denoted by $\boldsymbol{\sigma}_{i-1}$):
   $\boldsymbol{\sigma}_i = \boldsymbol{\sigma}_{i-1} + \Delta \boldsymbol{\sigma}_i$
6. Verify the yield condition (i.e., verify that $f(\boldsymbol{\sigma}_i) \leq 0$) and, if necessary, perform the procedure to bring back the point to the yield surface
7. Calculate the vector of internal forces $\boldsymbol{f}^{\sigma_i}$ and check that equilibrium is satisfied (i.e., compare $\boldsymbol{f}^{\sigma_i}$ with $\boldsymbol{f}_i^{ext}$)

Within the outlined incremental-iterative procedure, each incremental solution (denoted by index i) might require a number of iterations (denoted by index j) when dealing with an elasto-plastic deformation to reach the equilibrium state. When exposed to a purely elastic step (i.e., at each Gauss point the condition $f(\boldsymbol{\sigma}) < 0$ is verified), the equilibrium calculated at step 7 is satisfied after a single iteration. If at certain Gauss points the stress goes outside of the yield surface, an adequate stress return procedure needs to be employed to bring the point back to the yield surface. This then leads to the violation of the equilibrium, and additional iterations have to be performed to reach the equilibrium state.

The iterative procedure requires that the classical stiffness matrix of the FEM formulation has to be replaced by a tangent stiffness matrix. The former is computed as:

$$\boldsymbol{K} = \int \boldsymbol{B}^T \boldsymbol{D} \boldsymbol{B} dV \tag{7.5}$$

where matrix $\boldsymbol{D}$ governs the elastic response and its form depends on the elastic symmetry of the material. The onset of plastic deformation is not changing the elastic constants of the material, i.e., entries of matrix $\boldsymbol{D}$ do not change when plastic deformation occurs. On the other hand, the tangent stiffness matrix should change within each iteration in order to reach the convergence. It is therefore obvious that it cannot be calculated only on the basis of the elastic constants through the matrix $\boldsymbol{D}$ as they are independent of the plastic deformation.

Instead, the tangent stiffness matrix can be formulated through:

$$K_t = \int B^T D_t B dV \tag{7.6}$$

where matrix D_t is called the *standard tangential modular matrix* and is given by:

$$D_t = \frac{\partial \sigma}{\partial \varepsilon} \tag{7.7}$$

which, for a nonlinear stress–strain relationship, does not depend only on the elastic constants of the material but also on current stress state (i.e., the position of the point in stress space where it is calculated). The formulation of this modular matrix, together with the stress return algorithm, are two important aspects that any plasticity algorithm has to address. Before passing to more detailed discussion of these procedure we want to make an additional comment regarding the incremental-iterative algorithm.

Within step 2 of the outlined procedure, the displacement increments are updated in each iteration, and are used to compute the strain increment in step 3. It may be argued that a more logical approach would be to calculate directly the iterative strain as a function of the iterative displacement, $\delta\varepsilon(\delta d)$. This strategy has been employed in certain previous works (see [1]) and it may lead to some spurious unloading (i.e., negative strain increments) during iterations For illustrative purposes, the problem is graphically depicted in the two-dimensional strain space in Fig. 7.1. The figure shows incremental and iterative strains for different iterations when moving from point *A* to point *B*. Due to this problem, which is discussed in [2] and [3], the strategy that first updates the displacement increment and that then determines the strain increment is generally preferable.

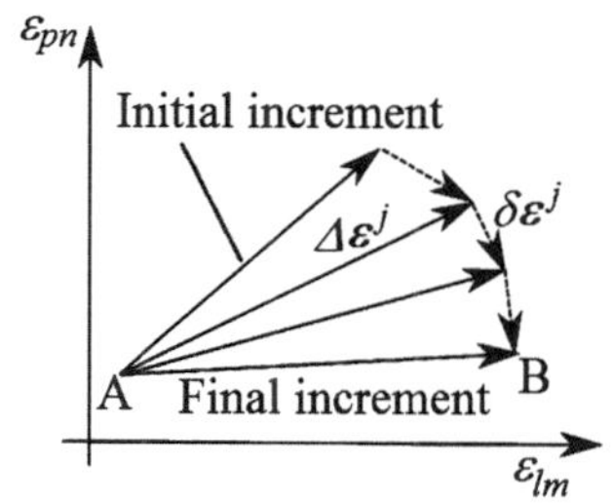

FIGURE 7.1 Strain increments (solid line) and iterative strains (dashed line) for each iteration.

7.1 The standard elasto-plastic modular matrix for the von Mises plasticity model

The most commonly used elasto-plastic constitutive model when modeling metals is the von Mises model (or J_2 plasticity model). In the following, we outline the construction of the modular matrix when implementing this model by considering the plane stress case. This example has been selected for ease of notation and to keep the matrices small. Same procedure is applicable for the general three-dimensional case.

The plane-stress form of the von Mises yield function expressed in terms of stress components is given by the following equation:

$$f = \sqrt{\sigma_x^2 + \sigma_y^2 - \sigma_x\sigma_y + 3\tau_{xy}^2} - \sigma_0 = q - \sigma_0 = 0 \tag{7.8}$$

where q is the equivalent von Mises stress.

For associative plasticity, the flow rule states that the direction of the plastic strain rate is perpendicular to the yield surface:

$$\dot{\boldsymbol{\varepsilon}}^{pl} = \dot{\lambda} \cdot \frac{\partial f}{\partial \boldsymbol{\sigma}} = \dot{\lambda} \cdot \boldsymbol{a} \tag{7.9}$$

where λ is the plastic flow multiplier and the gradient of the yield surface is denoted by $\boldsymbol{a}$.

Considering that the yield function is given by Eq. (7.8), and by performing the differentiation with respect to the stress components, the plastic strain rate becomes:

$$\dot{\boldsymbol{\varepsilon}}^{pl} = \frac{\dot{\lambda}}{2q} \begin{bmatrix} 2\sigma_x - \sigma_y \\ 2\sigma_y - \sigma_x \\ 6\tau_{xy} \end{bmatrix} \tag{7.10}$$

By assuming that elastic deformation is small, we can use a linear relationship between the elastic part of the strain and the corresponding stress. Expressed in a rate form, the stress changes are related to the strain changes as:

$$\dot{\boldsymbol{\sigma}} = \boldsymbol{D}(\dot{\boldsymbol{\varepsilon}} - \dot{\boldsymbol{\varepsilon}}^{pl}) = \boldsymbol{D}(\dot{\boldsymbol{\varepsilon}} - \dot{\lambda}\boldsymbol{a}) \tag{7.11}$$

where $\dot{\boldsymbol{\varepsilon}}$ represents the total strain rate. By further assuming isotropic elasticity, for the plane-stress case, $\boldsymbol{D}$ takes the following form:

$$\boldsymbol{D} = \frac{E}{1-\nu^2} \begin{bmatrix} 1 & \nu & 0 \\ \nu & 1 & 0 \\ 0 & 0 & \frac{1-\nu}{2} \end{bmatrix} \tag{7.12}$$

If, for a trial step, it turns out that the point in the stress space lies inside the yield surface, the step is elastic, and, consequently, the plastic flow multiplier assumes value 0. In an elasto-plastic step (i.e., plastic flow occurs), we need to compute the value of $\dot{\boldsymbol{\varepsilon}}^{pl}$. With associative plasticity, the direction of the plastic flow is known and set by the normal to the yield surface, thus, in such a situation, it is required to quantify only the plastic flow multiplier $\dot{\lambda}$.

During plastic flow, the stress must remain on the yield surface, which means that the stress rate vector and the plastic strain rate vector are orthogonal:

$$(\dot{\boldsymbol{\varepsilon}}^{pl})^T \cdot \dot{\boldsymbol{\sigma}} = \dot{\lambda}\boldsymbol{a}^T \cdot \dot{\boldsymbol{\sigma}} = 0 \quad \rightarrow \quad \boldsymbol{a}^T \cdot \dot{\boldsymbol{\sigma}} = 0 \tag{7.13}$$

We will further use this condition to find the plastic flow multiplier. If we multiply Eq. (7.11) by $\boldsymbol{a}^T$ we get:

$$\boldsymbol{a}^T \dot{\boldsymbol{\sigma}} = \boldsymbol{a}^T \boldsymbol{D}\dot{\boldsymbol{\varepsilon}} - \boldsymbol{a}^T \boldsymbol{D}\dot{\lambda}\boldsymbol{a} \tag{7.14}$$

The left-hand side of the last equation is, due to the orthogonality given by Eq. (7.13), equal to zero. Based on this, we can express the plastic flow multiplier as:

$$\dot{\lambda} = \frac{\boldsymbol{a}^T \boldsymbol{D}\dot{\boldsymbol{\varepsilon}}}{\boldsymbol{a}^T \boldsymbol{D}\boldsymbol{a}} \tag{7.15}$$

We can further substitute this result into Eq. (7.11) to get the relationship between the stress rate and the (total) strain rate as follows:

$$\dot{\boldsymbol{\sigma}} = \boldsymbol{D}\left(\dot{\boldsymbol{\varepsilon}} - \frac{\boldsymbol{a}^T \boldsymbol{D}\dot{\boldsymbol{\varepsilon}}}{\boldsymbol{a}^T \boldsymbol{D}\boldsymbol{a}}\boldsymbol{a}\right) = \left(\boldsymbol{D} - \frac{\boldsymbol{D}\boldsymbol{a}\boldsymbol{a}^T \boldsymbol{D}}{\boldsymbol{a}^T \boldsymbol{D}\boldsymbol{a}}\right)\dot{\boldsymbol{\varepsilon}} \quad \rightarrow \dot{\boldsymbol{\sigma}} = \boldsymbol{D}_t\dot{\boldsymbol{\varepsilon}} \tag{7.16}$$

where:

$$\boldsymbol{D}_t = \boldsymbol{D} - \frac{\boldsymbol{D}\boldsymbol{a}\boldsymbol{a}^T \boldsymbol{D}}{\boldsymbol{a}^T \boldsymbol{D}\boldsymbol{a}} \tag{7.17}$$

and $\boldsymbol{D}_t$ depicts the tangent modular matrix that is expressed as function of the elastic constants E and ν as well as the gradient $\boldsymbol{a}$ that depends on the current stress state. This matrix can be used in the integral of Eq. (7.6) to evaluate the structural tangent stiffness matrix $\boldsymbol{K}_t$, required for the iterative solution strategy. The presented procedure is suitable for a numerical implementation within a FEM code. It uses the reduced Voigt notation so the tangent modular matrix is indeed a matrix. In a complete notation, it would be represented by a fourth-order tensor.

7.2 Stress return algorithm

Even though within the incremental solution strategy the increments are in general small, they are never infinitesimally small, which has as a consequence that, inevitably, during the calculations performed at some points, the stress can end up outside the yield surface. In order to be plastically admissible, such a stress has to be brought back to the yield surface. The problem is solved by employing a stress return algorithm.

The approach may be viewed as successive application of stress "predictor" and "corrector" steps, in which the corrector should be applied if the predictor brings the stress outside the yield surface (see Fig. 7.2). The predictor (i.e., the trial step) is always computed by assuming the whole total strain is elastic. By using Eq. (7.11) and expressing rates in terms of changes, we may write:

$$\boldsymbol{\sigma}_C = \boldsymbol{\sigma}_B - \Delta\lambda \boldsymbol{D}\boldsymbol{a} \tag{7.18}$$

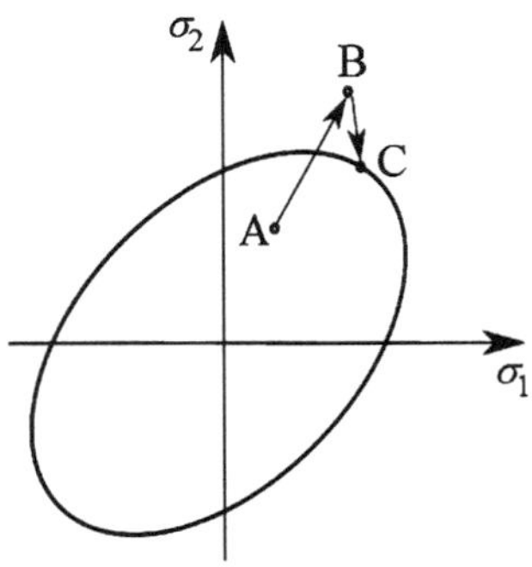

FIGURE 7.2 Stress predictor step (A–B) and stress corrector step (B–C) for a plastically admissible stress.

Within this strategy, the corrector is determined by keeping the total strain fixed, while adding an additional plastic strain to "relax" the stresses to get them back to the yield surface. Different stress return algorithms differ in the way the corrector step is calculated. As the focus of this chapter is the modeling of the elasto-plastic behavior, a stress return algorithm is presented in the following that works well with the von Mises yield surface. For alternative approaches, reference could be made, for example, to [4], [5], and [6].

Finding the stress corrector represents a nonlinear problem. This can be deduced by inspection of Eq. (7.18) in which $\boldsymbol{a}$ depicts the normal to the yield surface. If the trial step (i.e., predictor) brings the stress outside the yield surface, such a step connects the following two points: the initial point that is inside the yield surface, and the final point that is outside of the yield surface. There are now different possibilities for the choice where the normal should be calculated. If the normal is calculated at the intersection point of the step increment $A - B$ and the yield surface, a forward-Euler procedure is adopted. With the use of a backward-Euler procedure, the normal is calculated at the final point C. In the following, we consider the implementation of the backward-Euler return procedure.

7.2.1 Backward-Euler return algorithm

The backward-Euler return is based on the following equation:

$$\boldsymbol{\sigma}_C = \boldsymbol{\sigma}_B - \Delta\lambda \boldsymbol{D}\boldsymbol{a}_C \tag{7.19}$$

Point C is located on the yield surface and, therefore, the yield criterion needs to be satisfied at this point:

$$f(\boldsymbol{\sigma}_C) = f(\boldsymbol{\sigma}_B - \Delta\lambda \boldsymbol{D}\boldsymbol{a}_C) = 0 \tag{7.20}$$

Since at the beginning, the location of point C is unknown, it is not possible to compute the normal to the yield surface at that point and the problem becomes nonlinear, therefore requiring an iterative scheme to reach the solution. An initial estimate for $\boldsymbol{\sigma}_C$ can be obtained by computing the normal at point B with the forward-Euler return (this is an advantage as it avoids the computation of the intersection point), therefore we can write:

$$\boldsymbol{\sigma}_C^{initial} = \boldsymbol{\sigma}_B - \Delta\lambda \boldsymbol{D}\boldsymbol{a}_B \tag{7.21}$$

The value of the yield function at point B is larger than zero as the point is outside the yield surface. The returning step then quantifies the stress increment to correct stress at point B to reduce the value of the objective function to zero. This condition is used to determine the initial value of $\Delta\lambda$ for Eq. (7.21). The truncated first-order Taylor expansion around the point B can be written as:

$$f = f_B + \left(\frac{\partial f}{\partial \boldsymbol{\sigma}}\right)^T \Delta\boldsymbol{\sigma} = f_B + \boldsymbol{a}_B^T \Delta\boldsymbol{\sigma} \tag{7.22}$$

The stress increment that corrects the initial stress $\boldsymbol{\sigma}_B$ from Eq. (7.21) is equal to $-\Delta\lambda \boldsymbol{D}\boldsymbol{a}_B$. Hence, Eq. (7.22) can be rewritten as:

$$f = f_B - \Delta\lambda \boldsymbol{a}_B^T \boldsymbol{D}\boldsymbol{a}_B \tag{7.23}$$

By setting the yield function to zero (as the corrected stress should be on the yield surface), the expression which can be used for the initial plastic flow multiplier is obtained as follows:

$$\Delta\lambda^{initial} = \frac{f_B}{\boldsymbol{a}_B^T \boldsymbol{D}\boldsymbol{a}_B} \tag{7.24}$$

The normal at the trial point B in general differs from the normal passing through the final point C. As a consequence, the yield function at the point defined by the stress computed through Eq. (7.21) is not satisfied, and the use of the iterative procedure is required to reach the solution.

In order to setup the iterative loop, we construct a residual vector $\boldsymbol{r}$ to represent the difference between the current stress and the backward-Euler stress. By denoting the iterative quantities with the index i, the residual takes the following form:

$$\boldsymbol{r}_i = \boldsymbol{\sigma}_i - (\boldsymbol{\sigma}_B - \Delta\lambda_i \boldsymbol{D}\boldsymbol{a}_i) \tag{7.25}$$

The iterations are terminated when simultaneously both the residual given by Eq. (7.25) and the yield function equal zero (i.e., $f(\boldsymbol{\sigma}_i) = 0$). The iterative scheme is built from the truncated Taylor expansion of these two equations.

By keeping the first derivative only, a truncated Taylor expansion of the yield function takes the following form:

$$f_n = f_0 + \left(\frac{\partial f}{\partial \boldsymbol{\sigma}}\right)^T \dot{\boldsymbol{\sigma}} = f_0 + \boldsymbol{a}^T \dot{\boldsymbol{\sigma}} \tag{7.26}$$

where $\dot{\boldsymbol{\sigma}}$ is the change in stress. Similarly, a truncated Taylor expansion up to the first derivatives of the residual (i.e., given by Eq. (7.25)) can be written as (while exploiting the fact that $\boldsymbol{\sigma}_B$ remains constant):

$$\boldsymbol{r}_n = \boldsymbol{r}_0 + \dot{\boldsymbol{\sigma}} + \dot{\lambda}\boldsymbol{D}\boldsymbol{a} + \Delta\lambda \boldsymbol{D}\frac{\partial \boldsymbol{a}}{\partial \boldsymbol{\sigma}}\dot{\boldsymbol{\sigma}} \tag{7.27}$$

where, aside from the change in stress $\dot{\boldsymbol{\sigma}}$, also the change in $\Delta\lambda$ is present, denoted by $\dot{\lambda}$.

By setting r_n to zero, the change in stress results equal to:

$$\dot{\boldsymbol{\sigma}} = \left(\boldsymbol{I} + \Delta\lambda \boldsymbol{D}\frac{\partial \boldsymbol{a}}{\partial \boldsymbol{\sigma}}\right)^{-1} \left(-\boldsymbol{r}_0 - \dot{\lambda}\boldsymbol{D}\boldsymbol{a}\right) = \boldsymbol{Q}^{-1}\left(-\boldsymbol{r}_0 - \dot{\lambda}\boldsymbol{D}\boldsymbol{a}\right) \tag{7.28}$$

where, for ease of notation, the term in brackets is denoted by $\boldsymbol{Q}$ and it represent a matrix in reduced Voigt notation.

Similarly, by setting f_n equal to zero, and by substituting the value of $\dot{\boldsymbol{\sigma}}$ obtained from Eq. (7.28) into it, the following relation is obtained:

$$\dot{\lambda} = \frac{f_0 - \boldsymbol{a}^T \boldsymbol{Q}^{-1}\boldsymbol{r}_0}{\boldsymbol{a}^T \boldsymbol{Q}^{-1}\boldsymbol{D}\boldsymbol{a}} \tag{7.29}$$

The value for $\dot{\lambda}$ is then used in Eq. (7.28) to get the iterative stresses. The summary of the backward-Euler return algorithm is presented in the following.

1. For a given strain increment $\Delta\boldsymbol{\varepsilon}$, compute the trial stress increment as if all the strain is elastic: $\Delta\boldsymbol{\sigma} = \boldsymbol{D}\Delta\boldsymbol{\varepsilon}$
2. Update the stress tensor from the previous converged increment: $\boldsymbol{\sigma}_B = \boldsymbol{\sigma}_A + \Delta\boldsymbol{\sigma}$
3. Compute the gradient to the yield surface at point B: $\frac{\partial f(\boldsymbol{\sigma}_B)}{\partial \boldsymbol{\sigma}}$
4. Compute the initial value of the plastic flow multiplier: $\Delta\lambda^0 = \frac{f_B}{\boldsymbol{a}_B^T \boldsymbol{D}\boldsymbol{a}_B}$
5. Compute the initial value of the stress at point C: $\boldsymbol{\sigma}_C^0 = \boldsymbol{\sigma}_B - \Delta\lambda^0 \boldsymbol{D}\boldsymbol{a}_B$
6. Calculate the yield function value for the newly evaluated stress at point C: $f(\boldsymbol{\sigma}_C^i)$
7. Compute the gradient to the yield surface at the current point C: $\boldsymbol{a}_C^i = \frac{\partial f(\boldsymbol{\sigma}_C^i)}{\partial \boldsymbol{\sigma}}$

8. Compute the current residual: $\boldsymbol{r}_i = \boldsymbol{\sigma}_C^i - (\boldsymbol{\sigma}_B - \Delta\lambda_i \boldsymbol{D}\boldsymbol{a}_C^i)$
9. Compute matrix $\boldsymbol{Q}$ by using the current iterative values: $\boldsymbol{Q}_i = \boldsymbol{I} + \Delta\lambda_i \boldsymbol{D}\left(\frac{\partial \boldsymbol{a}}{\partial \boldsymbol{\sigma}}\right)_i$
10. Compute the change in the plastic flow multiplier by using the current iterative values: $\dot{\lambda} = \frac{f(\boldsymbol{\sigma}_C^i) - (\boldsymbol{a}_C^i)^T \boldsymbol{Q}_i^{-1} \boldsymbol{r}_0}{(\boldsymbol{a}_C^i)^T \boldsymbol{Q}_i^{-1} \boldsymbol{D}\boldsymbol{a}_C^i}$
11. Compute the change in stress by using the current iterative values: $\dot{\boldsymbol{\sigma}} = \boldsymbol{Q}_i^{-1}\left(-\boldsymbol{r}_i - \dot{\lambda}\boldsymbol{D}\boldsymbol{a}_C^i\right)$
12. Update the plastic flow multiplier rate and stress: $\Delta\lambda_{i+1} = \Delta\lambda_i + \dot{\lambda}$ $\boldsymbol{\sigma}_C^{i+1} = \boldsymbol{\sigma}_C^i + \dot{\boldsymbol{\sigma}}$
13. Proceed to the next iteration and repeat from step 6 until the convergence criteria is satisfied.

In the above algorithm, the trial step calculation has also been included for completeness, while steps 3–13 deal with the backward-Euler return procedure. The initial values are computed within steps 3–5 and the cyclic loop of the return is described in steps 6–13. The convergence criteria are commonly imposed on the value of the yield function as it should be reduced to zero, or an acceptably small value.

For the numerical implementation of the presented algorithm it is useful to evaluate the form of $\frac{\partial \boldsymbol{a}}{\partial \boldsymbol{\sigma}}$. In Voight notation, this is expressed as a matrix. In the following, we consider the von Mises yield condition applied to the plane-stress case for ease of notation, in order to keep matrices relatively small.

The von Mises yield criterion for the two-dimensional stress state, expressed in terms of the principal stress components, has the following form:

$$f(\boldsymbol{\sigma}) = \sqrt{\sigma_1^2 + \sigma_2^2 - \sigma_1\sigma_2} - \sigma_0 = q - \sigma_0 \tag{7.30}$$

The gradient vector is obtained by differentiating the last equation with respect to the stress components:

$$\boldsymbol{a} = \frac{1}{2q}\begin{bmatrix} 2\sigma_1 - \sigma_2 \\ 2\sigma_2 - \sigma_1 \end{bmatrix} \tag{7.31}$$

The derivative of the gradient vector with respect to the stress tensor (written as a vector in Voigt notation) is required for the implementation of the backward-Euler return, and in Voigt notation can be written as a 2×2 matrix as:

$$\frac{\partial \boldsymbol{a}}{\partial \boldsymbol{\sigma}} = \begin{bmatrix} \frac{\partial a_1}{\partial \sigma_1} & \frac{\partial a_1}{\partial \sigma_2} \\ \frac{\partial a_2}{\partial \sigma_1} & \frac{\partial a_2}{\partial \sigma_2} \end{bmatrix} \tag{7.32}$$

Considering that the gradient is given by Eq. (7.31), we can write:

$$\frac{\partial \boldsymbol{a}}{\partial \boldsymbol{\sigma}} = \frac{1}{2q}\frac{\partial}{\partial \boldsymbol{\sigma}}\begin{bmatrix} 2\sigma_1 - \sigma_2 \\ 2\sigma_2 - \sigma_1 \end{bmatrix} + \frac{\partial}{\partial \boldsymbol{\sigma}}\left(\frac{1}{2q}\right)\begin{bmatrix} 2\sigma_1 - \sigma_2 \\ 2\sigma_2 - \sigma_1 \end{bmatrix} \tag{7.33}$$

where the first member on the right-hand side results in the following matrix:

$$\frac{1}{2q}\frac{\partial}{\partial \boldsymbol{\sigma}}\begin{bmatrix} 2\sigma_1 - \sigma_2 \\ 2\sigma_2 - \sigma_1 \end{bmatrix} = \frac{1}{2q}\begin{bmatrix} 2 & -1 \\ -1 & 2 \end{bmatrix} \tag{7.34}$$

and the second member can be written as:

$$\begin{aligned} &\frac{\partial}{\partial \boldsymbol{\sigma}}\left(\frac{1}{2q}\right)\begin{bmatrix} 2\sigma_1 - \sigma_2 \\ 2\sigma_2 - \sigma_1 \end{bmatrix} = \\ &-\frac{1}{4q^3}\begin{bmatrix} (2\sigma_1 - \sigma_2)^2 & (2\sigma_2 - \sigma_1)(2\sigma_1 - \sigma_2) \\ (2\sigma_1 - \sigma_2)(2\sigma_2 - \sigma_1) & (2\sigma_2 - \sigma_1)^2 \end{bmatrix} = \\ &-\frac{1}{q}\cdot\frac{1}{2q}\begin{bmatrix} 2\sigma_1 - \sigma_2 \\ 2\sigma_2 - \sigma_1 \end{bmatrix}\frac{1}{2q}\begin{bmatrix} 2\sigma_1 - \sigma_2 & 2\sigma_2 - \sigma_1 \end{bmatrix} = -\frac{1}{q}\boldsymbol{a}\cdot\boldsymbol{a}^T \end{aligned} \tag{7.35}$$

Based on the above, we can write:

$$\frac{\partial \boldsymbol{a}}{\partial \boldsymbol{\sigma}} = \frac{1}{2q}\begin{bmatrix} 2 & -1 \\ -1 & 2 \end{bmatrix} - \frac{1}{q}\boldsymbol{a}\cdot\boldsymbol{a}^T = \frac{1}{2q}\boldsymbol{A} - \frac{1}{q}\boldsymbol{a}\cdot\boldsymbol{a}^T \tag{7.36}$$

The last equation completes the set of matrices required for the numerical implementation of the backward-Euler return algorithm. The following MATLAB® code provides the implementation in the form of a function that, for a given stress state outside of the yield surface, performs a backward-Euler return, and computes the final stress state on the yield surface, and the corresponding normal to the yield surface at that point. The function is written for the von Mises surface without hardening, and it takes as arguments, aside of the stress state, the Young's modulus and the yield limit. To minimize the complexity of the example, the Poisson coefficient is assumed to be equal to zero.

```
function [stressC,aC]=bckw_euler_ret(stressB,E,sig0)
% Backward-Euler return to the von Mises yeald surface
A=[2,-1;-1,2]; % Matrix for da/dsigma calculation
D=E*eye(2); % Matrix D
fC=sqrt(stressB(1)^2+stressB(2)^2-stressB(1)*stressB(2))-sig0;
se0=sqrt(stressB(1)^2+stressB(2)^2-stressB(1)*stressB(2));
a0=1/(2*se0)*[2*stressB(1)-stressB(2);2*stressB(2)-stressB(1)];
dL0=fC/(a0'*D*a0);
stressC=stressB-dL0*D*a0; % Stress estimate
while fC>1e-5
    s1=stressC(1);
```

```
    s2=stressC(2);
    seC=sqrt(s1^2+s2^2-s1*s2);
    fC=sqrt(s1^2+s2^2-s1*s2)-sig0;
    aC=1/(2*seC)*[2*s1-s2;2*s2-s1];
    resid=stressC-(stressB-dL0*D*aC);
    dads=1/(2*seC)*A-1/seC*aC*aC';
    Q=eye(2)+dL0*D*dads;
    Ldot=(fC-aC'*inv(Q)*resid)/(aC'*inv(Q)*D*aC);
    strDOT=-inv(Q)*resid-Ldot*inv(Q)*D*aC;
    stressC=stressC+strDOT; % New updated stress;
    dL0=dL0+Ldot; % New updated Delta_Lambda
end
```

A few remarks are provided below on how the function works by considering the following numerical example: plane-stress case without hardening, $\sigma_0 = 200$ MPa, and $E = 200{,}000$ MPa. We further assume that the previous equilibrium solution corresponds to the stress state defined by $\boldsymbol{\sigma}_A = [100, 130]^T$ MPa, elastic since inside the yield surface. To this state a further step increment of $\Delta\boldsymbol{\varepsilon} = [0.0012, 0.0007]^T$ is applied. The trial (elastic step) results to be equal to (in MPa):

$$\boldsymbol{\sigma}_B = \boldsymbol{\sigma}_A + E \cdot \Delta\boldsymbol{\varepsilon} = \begin{bmatrix} 340 \\ 270 \end{bmatrix} \tag{7.37}$$

As this stress state is outside the yield surface, the MATLAB® function `bckw_euler_ret` can be used to bring the stress back on the yield surface. In the following, a short MATLAB® script is provided to solve this problem:

```
clear
strain_inc=[0.0012,0.0007]';
E=200000;
C=E*eye(2);
sig0=200;
sigA=[100,130]';
sigB=sigA+E*strain_inc;
[stress,a0]=bckw_euler_ret(sigB,E,sig0);
```

The backward-Euler return involves few iterations to reach the solution that satisfies the prescribed tolerance (i.e., the yield function is smaller than $1e-5$). The graphical representation of the trial stress (i.e., point B) and all the iterative stresses are illustrated in Fig. 7.3.

It is worth noting that the material tangent stiffness matrix given by Eq. (7.17) is denoted as the *standard* material tangent stiffness matrix. The incremental-iterative algorithm for the implementation of the elasto-plastic constitutive model outlined above indicates that, within the procedure, there are two iterative loops: the inner one representing the backward Euler return (or any other return to the yield surface), and the outer, overall equilibrium iteration, used to solve for the displacement increment. The material tangent stiffness matrix given by Eq. (7.17) is derived by directly combining the plasticity flow

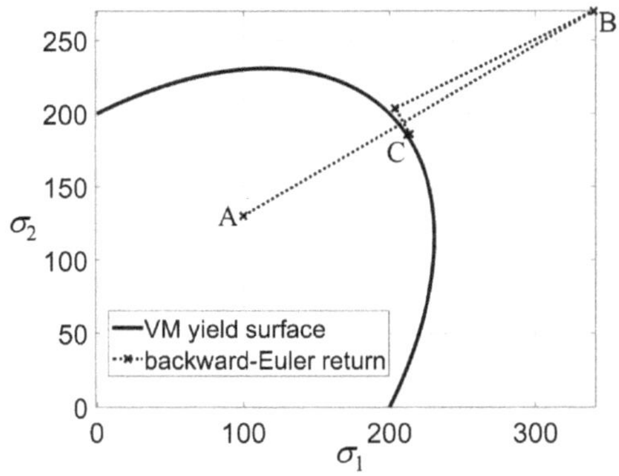

FIGURE 7.3 Backward-Euler return iterations from the trial stress.

rule with the stress–strain relation valid when plastic deformation occurs (i.e., the stress is related only to the elastic part of the strain). On the other hand, the plastic admissibility is guaranteed by employing an adequate return stress algorithm. When combined with a specific stress return algorithm, the material tangent stiffness matrix is, in general, not consistent, and does not provide a quadratic convergence with the overall Newton–Raphson procedure. Based on this, the standard material tangent stiffness matrix is also denoted as *inconsistent*. To address this issue, it is possible, for a specific stress return algorithm, to derive a modified, *consistent*, matrix that can guarantee the full quadratic convergence of the overall Newton–Raphson procedure. The material tangent stiffness matrix that is fully consistent with the backward Euler return (to exhibit a quadratic convergence) was first derived by Simo and Taylor [7]. The consistent material tangent takes the following form:

$$\boldsymbol{D}_t^c = \left(\boldsymbol{R} - \frac{\boldsymbol{R}\boldsymbol{a}\boldsymbol{a}^T\boldsymbol{R}}{\boldsymbol{a}^T\boldsymbol{R}\boldsymbol{a}} \right) \tag{7.38}$$

where matrix $\boldsymbol{R}$ is given by:

$$\boldsymbol{R} = \boldsymbol{Q}^{-1}\boldsymbol{D} \tag{7.39}$$

with matrix $\boldsymbol{Q}$ being already defined (see Eq. (7.28)).

For the details on the derivation of the consistent material tangent stiffness matrix, the reader should refer to the original paper [7] or [1].

7.3 Numerical implementation of the von Mises plasticity without hardening

Writing a computer program "from the scratch" that uses finite element method and that can account for the elasto-plastic material behavior is a task that is beyond the scope of this book. Despite this, examples of working computer codes related to the implementation of particular aspects of the constitutive model within the FEM procedure are provided in the following by considering the case of a single finite element. It is worth mentioning that for a numerical implementation of any constitutive model it is highly recommended to first perform tests on a single element before incorporating it into a larger code.

In the following numerical implementation, we consider the von Mises plasticity model without hardening. In the previous examples, we relied on the backward Euler return and considered the plane stress case merely for easier visualization and for reducing matrix sizes. At this point, it is worth to emphasize that it has been pointed out that the plane stress case within the context of elasto-plastic deformation, when combined with the backward Euler return method, may lead to numerical instabilities and may require special treatment (details on this could be found in [8], [9]). We now leave aside the plane stress case and move to the general procedure applicable to the three-dimensional stress state.

Let us consider a single hexagonal, 8-node finite element, with the boundary conditions depicted in Fig. 7.4. Let us further consider that the material elastic behavior is isotropic with $E = 200{,}000$ MPa and $\nu = 0.3$, up to the yield limit of $\sigma_0 = 100$ MPa. After this point, the material follows perfect plasticity. The onset of plastic deformation is modeled with the von Mises criterion. The main MATLAB® code that can be used to solve this problem is given in the following.

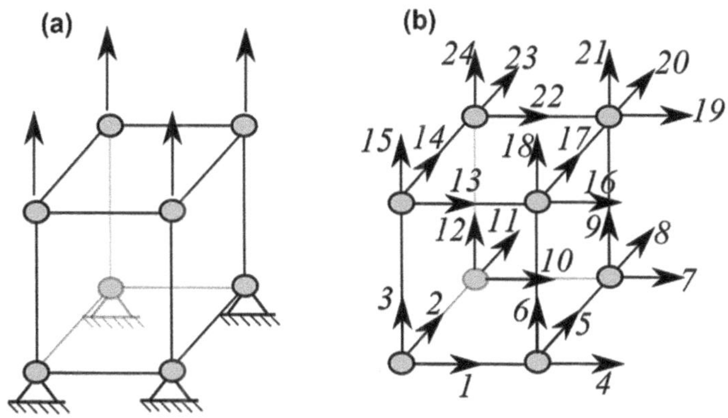

FIGURE 7.4 Single 8-node element considered in the elasto-plastic calculations: (a) boundary conditions and (b) degrees of freedom at the nodes.

```
% Main code for single element 3D plasticity
clear
E=200000; nu=0.3; % Elastic constants
% Specifying yield function
yield=inline('sqrt(s1^2+s2^2+s3^2-s1*s2-s1*s3-s2*s3+3*s4^2+3*s5^2+3*s6^2)-100');
% Material elastic stiffness matrix
D=(E/((1+nu)*(1-2*nu)))*[1-nu,nu,nu,0,0,0;
                          nu,1-nu,nu,0,0,0;
                          nu,nu,1-nu,0,0,0;
```

```
                         0,0,0,0.5*(1-2*nu),0,0;
                         0,0,0,0,0.5*(1-2*nu),0;
                         0,0,0,0,0,0.5*(1-2*nu)];
% Coordinates
coo=[0,0,0;
    1,0,0;
    1,1,0;
    0,1,0;
    0,0,1;
    1,0,1;
    1,1,1;
    0,1,1];
% Building constrain matrix CC and QQ for Lagrange multipliers
CC(1:12,1:24)=0;
for i=1:12
  CC(i,i)=1;
end
QQ(1:12,1)=0;
% Building force vector
FF(1:24,1)=0;
for i=5:8
    FF(3*i)=30;
end
NInc=10; % Number of increments
dF=1/NInc*FF;
% Setting initial values:
disp(24,1)=0;
stress(1:8,1:6)=0;
strain(1:8,1:6)=0;
pl_strain(1:8,1:6)=0;
%~~~~~~~~~~~~~~~~~~~~~~~~~~~~~~~~~~~~~~~~~~~~~~~~~~~~~~~~~~~~~~~~~~
% Incremental analysis
%%%%%%%%%%%%%%%%%%%%%%%%%%%%%%%%%%%%%%%%%%%%%%%%%%%%%%%%%%
for inc=1:NInc
% Vector of normals to the yield surface at each Gauss point (one line
% corresponds to one Gauss point)
AA(1:8,1:6)=0;
residual=dF;
dstrain_pl(1:8,1:6)=0; % This should be zero before each increment
Iter=0; % Counter of iterations
d_temp(1:24,1)=0;
%~~~~~~~~~~~~~~~~~~~~~~~~~~~~~~~~~~~~~~~~~~~~~~~
% Cyclic calculations
%%%%%%%%%%%%%%%%%%%%%%%%%%%%%%%%%%%%%%%%%%%%%%%%%
while norm(residual)>1e-7
Iter=Iter+1;
% 1. Building (tangent) stiffness matrix
tstf=stiff_elm8n(coo,E,nu,AA);
AA = zeros(8,6); %clear AA
% 2. Solving for displacement increment
dk=lagmult(tstf,CC,residual,QQ); % function provided in Section 6.5
d_temp=d_temp+dk; % Updating displacements
```

```
62 % 3. Computing strain increment form updated displacement increment
63 [dstrain]=strain_elm8n(coo,d_temp);
64 % 4. Calculate stresses on the basis of constitutive equation
65 dstress=(D*(dstrain'-dstrain_pl'))';
66 % 5. Updating stress
67 stress_temp=stress+dstress;
68 % 6. Yield condition check and returning to yield surface where necessary
69 for i=1:8
70     fc=yield(stress_temp(i,1),stress_temp(i,2),stress_temp(i,3),...
71         stress_temp(i,4),stress_temp(i,5),stress_temp(i,6));
72     if fc>0 % Plastic deformation occured
73         [stressC,aC,dlam,Q]=bckw_euler_ret_3dc(stress_temp(i,:)',D,100);
74         stress_temp(i,:)=stressC'; % Modify stresses
75         dstrain_pl(i,:)=dstrain_pl(i,:)+dlam*aC';
76         AA(i,:)=aC';
77     else
78         AA(i,:)=0;
79     end
80 end
81 % 7. Calculating (incremental) internal nodal forces
82 dstress_mod=stress_temp-stress;
83 EqNF=node_force_3d(coo,dstress_mod);
84 % Excluding forces corresponding to blocked DOF
85 for i=1:size(CC,1)
86     pos=find(CC(i,:));
87     EqNF(pos)=0;
88 end
89 % 8. Calculating new residual
90 residual=dF-EqNF;
91 end % End of iteration
92 % Updating values:
93 stress=stress+dstress_mod;
94 strain=strain+dstrain;
95 pl_strain=pl_strain+dstrain_pl;
96 disp=disp+d_temp;
97 end % End of increment
```

The initial part of the code (up to the line number 39) is dedicated to the definition of the problem where coordinates, material properties, constraining and all updating vectors, and matrices are set. Specifically, four main variables are prepared: `disp` – a vector of nodal displacements, `stress` – a matrix of stresses where each line contains the 6 stress components corresponding to one Gauss point, matrix `strain` – a matrix with rows collecting strain components, and `pl_strain` – a matrix collecting the plastic strain components at the different Gauss points.

The main, equilibrium incremental analysis starts from line number 43. As the incremental strains are used (rather than the iterative ones), an additional variable `d_temp` is introduced, and it is set to zero before each increment. This variable is updated through the main loop cycle. For the calculation of the material tangent stiffness matrix, when the material undergoes plastic deformation, a vector of normal to the yield surface needs to be evaluated at each Gauss point. For this purpose, a matrix AA is introduced that collects normals (placed as rows of the matrix) for each of the 8 Gauss points. For the initial calculation at each increment, this matrix is set to zero.

The loop cycle of the main Newton–Raphson procedure starts at line 54, and proceeds until the norm of the residual, represented here by the out-of-balance force vector, is below a small number, here set equal to 10^{-7}. Within the first step of this loop, a material tangent stiffness matrix is calculated, by invoking the function `stiff_elm8n`.

```
function [STF]=stiff_elm8n(coo,E,nu,AA)
% Materials (tangent) stiffness matrix calculation
c11=E*(1-nu)/((1-2*nu)*(1+nu));
c12=E*nu/((1-2*nu)*(1+nu));
D=[c11,c12,c12,0,0,0;c12,c11,c12,0,0,0;c12,c12,c11,0,0,0;0,0,0,...
    (c11-c12)/2,0,0;0,0,0,0,(c11-c12)/2,0;0,0,0,0,0,(c11-c12)/2];
gauss_coor=[-1/sqrt(3),-1/sqrt(3),-1/sqrt(3);
            1/sqrt(3),-1/sqrt(3),-1/sqrt(3);
            1/sqrt(3),1/sqrt(3),-1/sqrt(3);
            -1/sqrt(3),1/sqrt(3),-1/sqrt(3);
            -1/sqrt(3),-1/sqrt(3),1/sqrt(3);
            1/sqrt(3),-1/sqrt(3),1/sqrt(3);
            1/sqrt(3),1/sqrt(3),1/sqrt(3);
            -1/sqrt(3),1/sqrt(3),1/sqrt(3)];
Nmb=0;
for i=1:8 % Cycle over Gauss points
    a=AA(i,:)'; % Taking the normal from matrix
    if a'*a> 0 % Means plastic step
        Dt=(D-(D*a*a'*D)/(a'*D*a));
    else
        Dt=D;
    end
ksi=gauss_coor(i,1);
eta=gauss_coor(i,2);
zeta=gauss_coor(i,3);
% Nodal coordiantes
x1=coo(1,1); y1=coo(1,2); z1=coo(1,3);
x2=coo(2,1); y2=coo(2,2); z2=coo(2,3);
x3=coo(3,1); y3=coo(3,2); z3=coo(3,3);
x4=coo(4,1); y4=coo(4,2); z4=coo(4,3);
x5=coo(5,1); y5=coo(5,2); z5=coo(5,3);
x6=coo(6,1); y6=coo(6,2); z6=coo(6,3);
x7=coo(7,1); y7=coo(7,2); z7=coo(7,3);
x8=coo(8,1); y8=coo(8,2); z8=coo(8,3);
% Derivatives of shape function
dN1ksi=-0.125*(1-eta)*(1-zeta);
dN1eta=-0.125*(1-ksi)*(1-zeta);
dN1zeta=-0.125*(1-ksi)*(1-eta);
dN2ksi=0.125*(1-eta)*(1-zeta);
dN2eta=-0.125*(1+ksi)*(1-zeta);
dN2zeta=-0.125*(1+ksi)*(1-eta);
```

```
dN3ksi=0.125*(1+eta)*(1-zeta);
dN3eta=0.125*(1+ksi)*(1-zeta);
dN3zeta=-0.125*(1+ksi)*(1+eta);
dN4ksi=-0.125*(1+eta)*(1-zeta);
dN4eta=0.125*(1-ksi)*(1-zeta);
dN4zeta=-0.125*(1-ksi)*(1+eta);
dN5ksi=-0.125*(1-eta)*(1+zeta);
dN5eta=-0.125*(1-ksi)*(1+zeta);
dN5zeta=0.125*(1-ksi)*(1-eta);
dN6ksi=0.125*(1-eta)*(1+zeta);
dN6eta=-0.125*(1+ksi)*(1+zeta);
dN6zeta=0.125*(1+ksi)*(1-eta);
dN7ksi=0.125*(1+eta)*(1+zeta);
dN7eta=0.125*(1+ksi)*(1+zeta);
dN7zeta=0.125*(1+ksi)*(1+eta);
dN8ksi=-0.125*(1+eta)*(1+zeta);
dN8eta=0.125*(1-ksi)*(1+zeta);
dN8zeta=0.125*(1-ksi)*(1+eta);
% Defining Jacobean and inverting it
JAK=[dN1ksi,dN2ksi,dN3ksi,dN4ksi,dN5ksi,dN6ksi,dN7ksi,dN8ksi;
    dN1eta,dN2eta,dN3eta,dN4eta,dN5eta,dN6eta,dN7eta,dN8eta;
    dN1zeta,dN2zeta,dN3zeta,dN4zeta,dN5zeta,dN6zeta,dN7zeta,dN8zeta]...
    *[x1,y1,z1;x2,y2,z2;x3,y3,z3;x4,y4,z4;
    x5,y5,z5;x6,y6,z6;x7,y7,z7;x8,y8,z8];
JAKin=inv(JAK);
% Derivatives with respect to material coordinates
dN1dx=dN1ksi*JAKin(1,1)+dN1eta*JAKin(1,2)+dN1zeta*JAKin(1,3);
dN2dx=dN2ksi*JAKin(1,1)+dN2eta*JAKin(1,2)+dN2zeta*JAKin(1,3);
dN3dx=dN3ksi*JAKin(1,1)+dN3eta*JAKin(1,2)+dN3zeta*JAKin(1,3);
dN4dx=dN4ksi*JAKin(1,1)+dN4eta*JAKin(1,2)+dN4zeta*JAKin(1,3);
dN5dx=dN5ksi*JAKin(1,1)+dN5eta*JAKin(1,2)+dN5zeta*JAKin(1,3);
dN6dx=dN6ksi*JAKin(1,1)+dN6eta*JAKin(1,2)+dN6zeta*JAKin(1,3);
dN7dx=dN7ksi*JAKin(1,1)+dN7eta*JAKin(1,2)+dN7zeta*JAKin(1,3);
dN8dx=dN8ksi*JAKin(1,1)+dN8eta*JAKin(1,2)+dN8zeta*JAKin(1,3);
dN1dy=dN1ksi*JAKin(2,1)+dN1eta*JAKin(2,2)+dN1zeta*JAKin(2,3);
dN2dy=dN2ksi*JAKin(2,1)+dN2eta*JAKin(2,2)+dN2zeta*JAKin(2,3);
dN3dy=dN3ksi*JAKin(2,1)+dN3eta*JAKin(2,2)+dN3zeta*JAKin(2,3);
dN4dy=dN4ksi*JAKin(2,1)+dN4eta*JAKin(2,2)+dN4zeta*JAKin(2,3);
dN5dy=dN5ksi*JAKin(2,1)+dN5eta*JAKin(2,2)+dN5zeta*JAKin(2,3);
dN6dy=dN6ksi*JAKin(2,1)+dN6eta*JAKin(2,2)+dN6zeta*JAKin(2,3);
dN7dy=dN7ksi*JAKin(2,1)+dN7eta*JAKin(2,2)+dN7zeta*JAKin(2,3);
dN8dy=dN8ksi*JAKin(2,1)+dN8eta*JAKin(2,2)+dN8zeta*JAKin(2,3);
dN1dz=dN1ksi*JAKin(3,1)+dN1eta*JAKin(3,2)+dN1zeta*JAKin(3,3);
dN2dz=dN2ksi*JAKin(3,1)+dN2eta*JAKin(3,2)+dN2zeta*JAKin(3,3);
dN3dz=dN3ksi*JAKin(3,1)+dN3eta*JAKin(3,2)+dN3zeta*JAKin(3,3);
dN4dz=dN4ksi*JAKin(3,1)+dN4eta*JAKin(3,2)+dN4zeta*JAKin(3,3);
dN5dz=dN5ksi*JAKin(3,1)+dN5eta*JAKin(3,2)+dN5zeta*JAKin(3,3);
dN6dz=dN6ksi*JAKin(3,1)+dN6eta*JAKin(3,2)+dN6zeta*JAKin(3,3);
dN7dz=dN7ksi*JAKin(3,1)+dN7eta*JAKin(3,2)+dN7zeta*JAKin(3,3);
dN8dz=dN8ksi*JAKin(3,1)+dN8eta*JAKin(3,2)+dN8zeta*JAKin(3,3);
% Strain matrix:
B1=[dN1dx,0,0;
```

```
    0,dN1dy,0;
    0,0,dN1dz;
    0,dN1dz,dN1dy;
    dN1dz,0,dN1dx;
    dN1dy,dN1dx,0];
B2=[dN2dx,0,0;
    0,dN2dy,0;
    0,0,dN2dz;
    0,dN2dz,dN2dy;
    dN2dz,0,dN2dx;
    dN2dy,dN2dx,0];
B3=[dN3dx,0,0;
    0,dN3dy,0;
    0,0,dN3dz;
    0,dN3dz,dN3dy;
    dN3dz,0,dN3dx;
    dN3dy,dN3dx,0];
B4=[dN4dx,0,0;
    0,dN4dy,0;
    0,0,dN4dz;
    0,dN4dz,dN4dy;
    dN4dz,0,dN4dx;
    dN4dy,dN4dx,0];
B5=[dN5dx,0,0;
    0,dN5dy,0;
    0,0,dN5dz;
    0,dN5dz,dN5dy;
    dN5dz,0,dN5dx;
    dN5dy,dN5dx,0];
B6=[dN6dx,0,0;
    0,dN6dy,0;
    0,0,dN6dz;
    0,dN6dz,dN6dy;
    dN6dz,0,dN6dx;
    dN6dy,dN6dx,0];
B7=[dN7dx,0,0;
    0,dN7dy,0;
    0,0,dN7dz;
    0,dN7dz,dN7dy;
    dN7dz,0,dN7dx;
    dN7dy,dN7dx,0];
B8=[dN8dx,0,0;
    0,dN8dy,0;
    0,0,dN8dz;
    0,dN8dz,dN8dy;
    dN8dz,0,dN8dx;
    dN8dy,dN8dx,0];
B=[B1,B2,B3,B4,B5,B6,B7,B8]; % Forming matrix B
Nmb=Nmb+1;
kk(:,:,Nmb)=B'*Dt*B*det(JAK)*1*1*1;
end
STF(1:24,1:24)=0;
```

```
146 for i=1:8
147     STF=STF+kk(:,:,i);
148 end
```

This rather long routine represents a classic way of forming the stiffness matrix of an 8-node hexahedron element, and it therefore will not be commented here in details. For details on this routine, reference should be made to elementary finite element textbooks, e.g., [10], [11]. Here, we intend to point out only the main differences introduced to the function to build a tangent stiffness matrix when dealing with plastic deformation. The principal difference consists in the calculation of the material stiffness matrix $\boldsymbol{D}$, that is done between lines 19 and 21. For each Gauss point, a normal is taken from the matrix `AA` that is passed as input. If the norm of the normal is larger than zero, the step is plastic so the inconsistent material tangent stiffness is computed (line 19). If this is not the case, the response is build based on an elastic behavior. It is worth noting that the presented implementation works with the inconsistent material tangent stiffness matrix, therefore a quadratic convergence should not be expected.

Within step 2 of the main routine, the incremental displacements are computed. For this purpose, a Lagrange multipliers' technique is used by calling the MATLAB® function with the listing previously provided in Chapter 6. The calculated iterative displacement is used to update the incremental displacements, that are further used within step 3 to compute the strain increment by calling the function `strain_elm8n`. In order to emphasize the main steps required for the implementation of the elasto-plastic constitutive model while minimizing the complexity of the scripts, only a small strain regime is considered in the following. The listing of the function computing infinitesimal strains is provided below.

```
function [E]=strain_elm8n(coo,de)
% Function that computes strain in 8-node element
gauss_coor=[-1/sqrt(3),-1/sqrt(3),-1/sqrt(3);
            1/sqrt(3),-1/sqrt(3),-1/sqrt(3);
            1/sqrt(3),1/sqrt(3),-1/sqrt(3);
            -1/sqrt(3),1/sqrt(3),-1/sqrt(3);
            -1/sqrt(3),-1/sqrt(3),1/sqrt(3);
            1/sqrt(3),-1/sqrt(3),1/sqrt(3);
            1/sqrt(3),1/sqrt(3),1/sqrt(3);
            -1/sqrt(3),1/sqrt(3),1/sqrt(3)];
for i=1:8 % Cycle over Gauss points
ksi=gauss_coor(i,1);
eta=gauss_coor(i,2);
zeta=gauss_coor(i,3);
% Decomposing vector of nodal displacements to x- y- and z-part
dx=[de(1);de(4);de(7);de(10);de(13);de(16);de(19);de(22)];
dy=[de(2);de(5);de(8);de(11);de(14);de(17);de(20);de(23)];
dz=[de(3);de(6);de(9);de(12);de(15);de(18);de(21);de(24)];
% Nodal coordinates
x1=coo(1,1); y1=coo(1,2); z1=coo(1,3);
x2=coo(2,1); y2=coo(2,2); z2=coo(2,3);
```

```
x3=coo(3,1); y3=coo(3,2); z3=coo(3,3);
x4=coo(4,1); y4=coo(4,2); z4=coo(4,3);
x5=coo(5,1); y5=coo(5,2); z5=coo(5,3);
x6=coo(6,1); y6=coo(6,2); z6=coo(6,3);
x7=coo(7,1); y7=coo(7,2); z7=coo(7,3);
x8=coo(8,1); y8=coo(8,2); z8=coo(8,3);
% Derivatives of shape function
dN1ksi=-0.125*(1-eta)*(1-zeta);
dN1eta=-0.125*(1-ksi)*(1-zeta);
dN1zeta=-0.125*(1-ksi)*(1-eta);
dN2ksi=0.125*(1-eta)*(1-zeta);
dN2eta=-0.125*(1+ksi)*(1-zeta);
dN2zeta=-0.125*(1+ksi)*(1-eta);
dN3ksi=0.125*(1+eta)*(1-zeta);
dN3eta=0.125*(1+ksi)*(1-zeta);
dN3zeta=-0.125*(1+ksi)*(1+eta);
dN4ksi=-0.125*(1+eta)*(1-zeta);
dN4eta=0.125*(1-ksi)*(1-zeta);
dN4zeta=-0.125*(1-ksi)*(1+eta);
dN5ksi=-0.125*(1-eta)*(1+zeta);
dN5eta=-0.125*(1-ksi)*(1+zeta);
dN5zeta=0.125*(1-ksi)*(1-eta);
dN6ksi=0.125*(1-eta)*(1+zeta);
dN6eta=-0.125*(1+ksi)*(1+zeta);
dN6zeta=0.125*(1+ksi)*(1-eta);
dN7ksi=0.125*(1+eta)*(1+zeta);
dN7eta=0.125*(1+ksi)*(1+zeta);
dN7zeta=0.125*(1+ksi)*(1+eta);
dN8ksi=-0.125*(1+eta)*(1+zeta);
dN8eta=0.125*(1-ksi)*(1+zeta);
dN8zeta=0.125*(1-ksi)*(1+eta);
% Defining Jacobean and inverting it
JAK=[dN1ksi,dN2ksi,dN3ksi,dN4ksi,dN5ksi,dN6ksi,dN7ksi,dN8ksi;
    dN1eta,dN2eta,dN3eta,dN4eta,dN5eta,dN6eta,dN7eta,dN8eta;
    dN1zeta,dN2zeta,dN3zeta,dN4zeta,dN5zeta,dN6zeta,dN7zeta,dN8zeta]...
    *[x1,y1,z1;x2,y2,z2;x3,y3,z3;x4,y4,z4;
    x5,y5,z5;x6,y6,z6;x7,y7,z7;x8,y8,z8];
JAKin=inv(JAK);
% Derivatives with respect to material coordinates
dN1dx=dN1ksi*JAKin(1,1)+dN1eta*JAKin(1,2)+dN1zeta*JAKin(1,3);
dN2dx=dN2ksi*JAKin(1,1)+dN2eta*JAKin(1,2)+dN2zeta*JAKin(1,3);
dN3dx=dN3ksi*JAKin(1,1)+dN3eta*JAKin(1,2)+dN3zeta*JAKin(1,3);
dN4dx=dN4ksi*JAKin(1,1)+dN4eta*JAKin(1,2)+dN4zeta*JAKin(1,3);
dN5dx=dN5ksi*JAKin(1,1)+dN5eta*JAKin(1,2)+dN5zeta*JAKin(1,3);
dN6dx=dN6ksi*JAKin(1,1)+dN6eta*JAKin(1,2)+dN6zeta*JAKin(1,3);
dN7dx=dN7ksi*JAKin(1,1)+dN7eta*JAKin(1,2)+dN7zeta*JAKin(1,3);
dN8dx=dN8ksi*JAKin(1,1)+dN8eta*JAKin(1,2)+dN8zeta*JAKin(1,3);
dN1dy=dN1ksi*JAKin(2,1)+dN1eta*JAKin(2,2)+dN1zeta*JAKin(2,3);
dN2dy=dN2ksi*JAKin(2,1)+dN2eta*JAKin(2,2)+dN2zeta*JAKin(2,3);
dN3dy=dN3ksi*JAKin(2,1)+dN3eta*JAKin(2,2)+dN3zeta*JAKin(2,3);
dN4dy=dN4ksi*JAKin(2,1)+dN4eta*JAKin(2,2)+dN4zeta*JAKin(2,3);
dN5dy=dN5ksi*JAKin(2,1)+dN5eta*JAKin(2,2)+dN5zeta*JAKin(2,3);
```

```
dN6dy=dN6ksi*JAKin(2,1)+dN6eta*JAKin(2,2)+dN6zeta*JAKin(2,3);
dN7dy=dN7ksi*JAKin(2,1)+dN7eta*JAKin(2,2)+dN7zeta*JAKin(2,3);
dN8dy=dN8ksi*JAKin(2,1)+dN8eta*JAKin(2,2)+dN8zeta*JAKin(2,3);
dN1dz=dN1ksi*JAKin(3,1)+dN1eta*JAKin(3,2)+dN1zeta*JAKin(3,3);
dN2dz=dN2ksi*JAKin(3,1)+dN2eta*JAKin(3,2)+dN2zeta*JAKin(3,3);
dN3dz=dN3ksi*JAKin(3,1)+dN3eta*JAKin(3,2)+dN3zeta*JAKin(3,3);
dN4dz=dN4ksi*JAKin(3,1)+dN4eta*JAKin(3,2)+dN4zeta*JAKin(3,3);
dN5dz=dN5ksi*JAKin(3,1)+dN5eta*JAKin(3,2)+dN5zeta*JAKin(3,3);
dN6dz=dN6ksi*JAKin(3,1)+dN6eta*JAKin(3,2)+dN6zeta*JAKin(3,3);
dN7dz=dN7ksi*JAKin(3,1)+dN7eta*JAKin(3,2)+dN7zeta*JAKin(3,3);
dN8dz=dN8ksi*JAKin(3,1)+dN8eta*JAKin(3,2)+dN8zeta*JAKin(3,3);
% Strain matrix:
B1=[dN1dx,0,0;
    0,dN1dy,0;
    0,0,dN1dz;
    0,dN1dz,dN1dy;
    dN1dz,0,dN1dx;
    dN1dy,dN1dx,0];
B2=[dN2dx,0,0;
    0,dN2dy,0;
    0,0,dN2dz;
    0,dN2dz,dN2dy;
    dN2dz,0,dN2dx;
    dN2dy,dN2dx,0];
B3=[dN3dx,0,0;
    0,dN3dy,0;
    0,0,dN3dz;
    0,dN3dz,dN3dy;
    dN3dz,0,dN3dx;
    dN3dy,dN3dx,0];
B4=[dN4dx,0,0;
    0,dN4dy,0;
    0,0,dN4dz;
    0,dN4dz,dN4dy;
    dN4dz,0,dN4dx;
    dN4dy,dN4dx,0];
B5=[dN5dx,0,0;
    0,dN5dy,0;
    0,0,dN5dz;
    0,dN5dz,dN5dy;
    dN5dz,0,dN5dx;
    dN5dy,dN5dx,0];
B6=[dN6dx,0,0;
    0,dN6dy,0;
    0,0,dN6dz;
    0,dN6dz,dN6dy;
    dN6dz,0,dN6dx;
    dN6dy,dN6dx,0];
B7=[dN7dx,0,0;
    0,dN7dy,0;
    0,0,dN7dz;
    0,dN7dz,dN7dy;
```

```
    dN7dz,0,dN7dx;
    dN7dy,dN7dx,0];
B8=[dN8dx,0,0;
    0,dN8dy,0;
    0,0,dN8dz;
    0,dN8dz,dN8dy;
    dN8dz,0,dN8dx;
    dN8dy,dN8dx,0];
B=[B1,B2,B3,B4,B5,B6,B7,B8]; % Forming matrix B
E(i,:)=B*de; % Calculating strain
end
```

Within step 4, the incremental stresses are computed on the basis of the calculated strain increment that is treated as the total strain, from which the plastic part is subtracted. As the problem is solved incrementally, here only the incremental plastic strain is subtracted (stored as variable `dstrain_pl`), while the absolute plastic strain is updated at the end of each increment with variable `pl_strain`.

Incremental stresses are added to the absolute stress values obtained from previous converged iterations in step 5 of the main routine. The result is stored as variable `stress_temp` that, at this stage, should be treated as a temporary value in view of the potential need to perform the stress return procedure in the case of violation of the yield condition. The yield condition check, followed by the subsequent return to the yield surface in the case in which the value of yield function is positive, is performed at each Gauss point within step 6. The return is performed by calling the function `bckw_euler_ret_3dc`. The function is similar to the one already given in the previous section, except that here it is written for the stress state expressed by 6 components. The listing of the function is given below.

```
function [stressC,aC,dL0,Q]=bckw_euler_ret_3dc(stressB,D,sig0)
% Backward-Euler return to the von Mises yeald surface
A=[2,-1,-1,0,0,0;
   -1,2,-1,0,0,0;
   -1,-1,2,0,0,0;
    0,0,0,6,0,0;
    0,0,0,0,6,0;
    0,0,0,0,0,6]; % Matrix for da/dsigma calculation
se0=sqrt(stressB(1)^2+stressB(2)^2+stressB(3)^2-stressB(1)*stressB(2)- ...
    stressB(1)*stressB(3)-stressB(2)*stressB(3)+3*stressB(4)^2+ ...
    3*stressB(5)^2+3*stressB(6)^2);
fC=se0-sig0;
aC=1/(2*se0)*[2*stressB(1)-stressB(2)-stressB(3);
              2*stressB(2)-stressB(1)-stressB(3);
              2*stressB(3)-stressB(1)-stressB(2);
              6*stressB(4);
              6*stressB(5);
              6*stressB(6)];
dL0=fC/(aC'*D*aC);
stressC=stressB-dL0*D*aC; % Stress estimate
```

```
while fC>1e-5
 seC=sqrt(stressC(1)^2+stressC(2)^2+stressC(3)^2-stressC(1)*stressC(2)- ...
    stressC(1)*stressC(3)-stressC(2)*stressC(3)+3*stressC(4)^2+ ...
    3*stressC(5)^2+3*stressC(6)^2);
 fC=seC-sig0;
aC=1/(2*seC)*[2*stressC(1)-stressC(2)-stressC(3);
              2*stressC(2)-stressC(1)-stressC(3);
              2*stressC(3)-stressC(1)-stressC(2);
              6*stressC(4);
              6*stressC(5);
              6*stressC(6)];
 resid=stressC-(stressB-dL0*D*aC);
 dads=1/(2*seC)*A-1/seC*aC*aC';
 Q=eye(6)+dL0*D*dads;
 Ldot=(fC-aC'*inv(Q)*resid)/(aC'*inv(Q)*D*aC);
 strDOT=-inv(Q)*resid-Ldot*inv(Q)*D*aC;
 stressC=stressC+strDOT; % New updated stress;
 dL0=dL0+Ldot; % New updated Delta_Lambda
 end
```

The function takes, as inputs, the value of stress state that is outside the yield surface (i.e., stress predictor) and returns the value of stress state that is on the yield surface, together with a corresponding normal to the yield surface, required for further calculation of the material tangent stiffness matrix, and of the plastic load multiplier.

The stress return algorithm works on the absolute value of stress (not on the stress increment). On the other hand, the iterative procedure is incremental, therefore the internal forces for the equilibrium check have to be expressed in an incremental form. In step 7 of the main routine, the modified incremental stress is computed (considering the stress value computed with the backward Euler return and the absolute stress from the previous converged increment) and the incremental forces are determined with function `node_force_3d`. Listing of this MATLAB® function is given below.

```
function FE=node_force_3d(coo,stress)
% Computing nodal forces in 8-node element
gauss_coor=[-1/sqrt(3),-1/sqrt(3),-1/sqrt(3);
            1/sqrt(3),-1/sqrt(3),-1/sqrt(3);
            1/sqrt(3),1/sqrt(3),-1/sqrt(3);
            -1/sqrt(3),1/sqrt(3),-1/sqrt(3);
            -1/sqrt(3),-1/sqrt(3),1/sqrt(3);
            1/sqrt(3),-1/sqrt(3),1/sqrt(3);
            1/sqrt(3),1/sqrt(3),1/sqrt(3);
            -1/sqrt(3),1/sqrt(3),1/sqrt(3)];
Nmb=0;
for i=1:8 % Cycle over Gauss points
ksi=gauss_coor(i,1);
eta=gauss_coor(i,2);
zeta=gauss_coor(i,3);
% Nodal coordiantes
x1=coo(1,1); y1=coo(1,2); z1=coo(1,3);
```

```
x2=coo(2,1); y2=coo(2,2); z2=coo(2,3);
x3=coo(3,1); y3=coo(3,2); z3=coo(3,3);
x4=coo(4,1); y4=coo(4,2); z4=coo(4,3);
x5=coo(5,1); y5=coo(5,2); z5=coo(5,3);
x6=coo(6,1); y6=coo(6,2); z6=coo(6,3);
x7=coo(7,1); y7=coo(7,2); z7=coo(7,3);
x8=coo(8,1); y8=coo(8,2); z8=coo(8,3);
% Derivatives of shape function
dN1ksi=-0.125*(1-eta)*(1-zeta);
dN1eta=-0.125*(1-ksi)*(1-zeta);
dN1zeta=-0.125*(1-ksi)*(1-eta);
dN2ksi=0.125*(1-eta)*(1-zeta);
dN2eta=-0.125*(1+ksi)*(1-zeta);
dN2zeta=-0.125*(1+ksi)*(1-eta);
dN3ksi=0.125*(1+eta)*(1-zeta);
dN3eta=0.125*(1+ksi)*(1-zeta);
dN3zeta=-0.125*(1+ksi)*(1+eta);
dN4ksi=-0.125*(1+eta)*(1-zeta);
dN4eta=0.125*(1-ksi)*(1-zeta);
dN4zeta=-0.125*(1-ksi)*(1+eta);
dN5ksi=-0.125*(1-eta)*(1+zeta);
dN5eta=-0.125*(1-ksi)*(1+zeta);
dN5zeta=0.125*(1-ksi)*(1-eta);
dN6ksi=0.125*(1-eta)*(1+zeta);
dN6eta=-0.125*(1+ksi)*(1+zeta);
dN6zeta=0.125*(1+ksi)*(1-eta);
dN7ksi=0.125*(1+eta)*(1+zeta);
dN7eta=0.125*(1+ksi)*(1+zeta);
dN7zeta=0.125*(1+ksi)*(1+eta);
dN8ksi=-0.125*(1+eta)*(1+zeta);
dN8eta=0.125*(1-ksi)*(1+zeta);
dN8zeta=0.125*(1-ksi)*(1+eta);
% Defining Jacobean and inverting it
JAK=[dN1ksi,dN2ksi,dN3ksi,dN4ksi,dN5ksi,dN6ksi,dN7ksi,dN8ksi;
    dN1eta,dN2eta,dN3eta,dN4eta,dN5eta,dN6eta,dN7eta,dN8eta;
    dN1zeta,dN2zeta,dN3zeta,dN4zeta,dN5zeta,dN6zeta,dN7zeta,dN8zeta]...
    *[x1,y1,z1;x2,y2,z2;x3,y3,z3;x4,y4,z4;
    x5,y5,z5;x6,y6,z6;x7,y7,z7;x8,y8,z8];
JAKin=inv(JAK);
% Derivatives with respect to material coordinates
dN1dx=dN1ksi*JAKin(1,1)+dN1eta*JAKin(1,2)+dN1zeta*JAKin(1,3);
dN2dx=dN2ksi*JAKin(1,1)+dN2eta*JAKin(1,2)+dN2zeta*JAKin(1,3);
dN3dx=dN3ksi*JAKin(1,1)+dN3eta*JAKin(1,2)+dN3zeta*JAKin(1,3);
dN4dx=dN4ksi*JAKin(1,1)+dN4eta*JAKin(1,2)+dN4zeta*JAKin(1,3);
dN5dx=dN5ksi*JAKin(1,1)+dN5eta*JAKin(1,2)+dN5zeta*JAKin(1,3);
dN6dx=dN6ksi*JAKin(1,1)+dN6eta*JAKin(1,2)+dN6zeta*JAKin(1,3);
dN7dx=dN7ksi*JAKin(1,1)+dN7eta*JAKin(1,2)+dN7zeta*JAKin(1,3);
dN8dx=dN8ksi*JAKin(1,1)+dN8eta*JAKin(1,2)+dN8zeta*JAKin(1,3);
dN1dy=dN1ksi*JAKin(2,1)+dN1eta*JAKin(2,2)+dN1zeta*JAKin(2,3);
dN2dy=dN2ksi*JAKin(2,1)+dN2eta*JAKin(2,2)+dN2zeta*JAKin(2,3);
dN3dy=dN3ksi*JAKin(2,1)+dN3eta*JAKin(2,2)+dN3zeta*JAKin(2,3);
dN4dy=dN4ksi*JAKin(2,1)+dN4eta*JAKin(2,2)+dN4zeta*JAKin(2,3);
```

```
dN5dy=dN5ksi*JAKin(2,1)+dN5eta*JAKin(2,2)+dN5zeta*JAKin(2,3);
dN6dy=dN6ksi*JAKin(2,1)+dN6eta*JAKin(2,2)+dN6zeta*JAKin(2,3);
dN7dy=dN7ksi*JAKin(2,1)+dN7eta*JAKin(2,2)+dN7zeta*JAKin(2,3);
dN8dy=dN8ksi*JAKin(2,1)+dN8eta*JAKin(2,2)+dN8zeta*JAKin(2,3);
dN1dz=dN1ksi*JAKin(3,1)+dN1eta*JAKin(3,2)+dN1zeta*JAKin(3,3);
dN2dz=dN2ksi*JAKin(3,1)+dN2eta*JAKin(3,2)+dN2zeta*JAKin(3,3);
dN3dz=dN3ksi*JAKin(3,1)+dN3eta*JAKin(3,2)+dN3zeta*JAKin(3,3);
dN4dz=dN4ksi*JAKin(3,1)+dN4eta*JAKin(3,2)+dN4zeta*JAKin(3,3);
dN5dz=dN5ksi*JAKin(3,1)+dN5eta*JAKin(3,2)+dN5zeta*JAKin(3,3);
dN6dz=dN6ksi*JAKin(3,1)+dN6eta*JAKin(3,2)+dN6zeta*JAKin(3,3);
dN7dz=dN7ksi*JAKin(3,1)+dN7eta*JAKin(3,2)+dN7zeta*JAKin(3,3);
dN8dz=dN8ksi*JAKin(3,1)+dN8eta*JAKin(3,2)+dN8zeta*JAKin(3,3);
% Strain matrix:
B1=[dN1dx,0,0;
    0,dN1dy,0;
    0,0,dN1dz;
    0,dN1dz,dN1dy;
    dN1dz,0,dN1dx;
    dN1dy,dN1dx,0];
B2=[dN2dx,0,0;
    0,dN2dy,0;
    0,0,dN2dz;
    0,dN2dz,dN2dy;
    dN2dz,0,dN2dx;
    dN2dy,dN2dx,0];
B3=[dN3dx,0,0;
    0,dN3dy,0;
    0,0,dN3dz;
    0,dN3dz,dN3dy;
    dN3dz,0,dN3dx;
    dN3dy,dN3dx,0];
B4=[dN4dx,0,0;
    0,dN4dy,0;
    0,0,dN4dz;
    0,dN4dz,dN4dy;
    dN4dz,0,dN4dx;
    dN4dy,dN4dx,0];
B5=[dN5dx,0,0;
    0,dN5dy,0;
    0,0,dN5dz;
    0,dN5dz,dN5dy;
    dN5dz,0,dN5dx;
    dN5dy,dN5dx,0];
B6=[dN6dx,0,0;
    0,dN6dy,0;
    0,0,dN6dz;
    0,dN6dz,dN6dy;
    dN6dz,0,dN6dx;
    dN6dy,dN6dx,0];
B7=[dN7dx,0,0;
    0,dN7dy,0;
    0,0,dN7dz;
```

```
    0,dN7dz,dN7dy;
    dN7dz,0,dN7dx;
    dN7dy,dN7dx,0];
B8=[dN8dx,0,0;
    0,dN8dy,0;
    0,0,dN8dz;
    0,dN8dz,dN8dy;
    dN8dz,0,dN8dx;
    dN8dy,dN8dx,0];
B=[B1,B2,B3,B4,B5,B6,B7,B8]; % Forming matrix B:
Nmb=Nmb+1;
FF(:,:,Nmb)=B'*stress(Nmb,:)'*det(JAK)*1*1*1;
end
FE=FF(:,:,1);
for i=2:Nmb
    FE=FE+FF(:,:,i);
end
```

The presented MATLAB® codes can be used to predict the elasto-plastic response of the problem depicted in Fig. 7.4. The implementation is generally the same also for large scale codes, where the assembling of the global stiffness matrix corresponding to the overall mesh of the model invokes the codes describing the contribution of the individual elements.

7.4 Mechanical characterization of metals

Metals are characterized by a pronounced capability to develop a relatively large plastic deformation prior to rapture. It is therefore most convenient to use for their modeling an elasto-plastic constitutive model. If the targeted material should be modeled within the deformation range that is not close to the ultimate deformation, damage effects can be ignored, therefore resulting in a significant simplification of the mechanical characterization. While most of the metals are modeled by employing the same constitutive models (traditionally the von Mises plasticity model), particularization of the response to a specific metal being modeled is achieved by the quantification of the parameters that are entering into the governing equations. The same constitutive model can be used to model, for example, steel and aluminum, even if fairly different quantitative values of the governing parameters are specified in the two cases.

The mechanical characterization is therefore expressed in terms of the calibration of the constitutive parameters (i.e., constants) that enter into the selected constitutive model. The constants can be divided into two groups: those related to the elastic response, and those that are governing the plastic response. The values of the elastic constants can, in general, be assessed by means of non-destructive tests (e.g., measuring sound propagation velocity). In order to quantify the plastic parameters, it is necessary to introduce a plastic deformation into the material.

The most commonly used test, and probably the easiest to perform, is a tensile test that gives insight into both elastic and inelastic response of the material. While it provides a global picture of the material response, the detailed interpretation of the measurements is not that easy when complex phenomena are taking place (e.g., combined damage with plasticity, local concentration of plastic deformation, different types of hardening, etc.). If available, a tensile test can be used mainly for a qualitative characterization of the material, particularly related to the identification of the type of hardening it exhibits. It is worth noting that, in order to distinguish between the isotropic and kinematic hardening, it is required to have the measurements of tensile tests of specimens subjected to cyclic loads.

After analyzing the uniaxial stress–strain curve, an appropriate elasto-plastic model should be adopted (with or without hardening). The mechanical characterization of the material is now expressed in terms of the quantification of the constants that are entering into the governing equations of the adopted model. Most of these constants cannot be directly measured from the experiment, thus an inverse analysis procedure is required to bridge the gap between the experimentally measured quantities and the sought material parameters. We now consider an example to outline the operative procedure required in the material parameter calibration.

7.4.1 Elasto-plastic material calibration based on an indentation test

The instrumented indentation test, introduced in Chapter 2, represents a good experiment for the characterization of the elasto-plastic material parameters. The experiment gives rise to quite localized plastic deformation, thus enabling the collection of the information on the material inelastic response. This test is usually regarded as almost non-destructive as the introduced irreversible deformation is generally restrained within the zone measured by a fraction of a millimeter. The main experimental information collected from an indentation test is an indentation curve, i.e., the curve which relates the force applied by an indenter to the penetration depth measured at its tip, due to the elasto-plastic deformation of the tested specimen. For the assessment of the direction dependent properties (i.e., mechanical characterization of an orthotropic material), it was demonstrated (see, e.g., [12] [13]) that the shape of residual imprint contains valuable information and should be included in the experimental data used in input of the inverse analysis procedure to enhance the robustness of the identification process.

In order to illustrate how a practical procedure for the material characterization based on an inverse analysis is constructed, we assume that the material behaves isotropically both in elasticity and in plasticity. To maintain the proposed procedure as simple as possible, we consider only an indentation curve in input of the inverse analysis for the material characterization. In reality, additional experimental information needs to be included for the parameter identification to define a well-posed problem and to distinguish between the sets of possible material parameters that produce equivalent indentation curves (these sets of materials have been referred to as "mystical" materials in the literature and have already been discussed in Chapter 2). We then select the geometry of the indenter to be a standard Rockwell with a conical shape [14]. This indenter is popular and widely used, and, for an isotropic material, it gives rise to an axially symmetric deformation, therefore offering the advantage of a simplified, two-dimensional numerical model.

Once the experiment is completed and the indentation curve is recorded, the subsequent step consists in the building of a numerical model that can be used to simulate the experiment (we assumed to have already investigated the well-posed condition of the inverse problem based on pseudo-experimental results). As the present example deals with a metal, we adopted the von Mises elasto-plastic model for the material response. This constitutive model is widely used, and it is implemented in most commercial FEM codes. At this point, we need to decide whether or not we need to perform a particularization regarding material hardening or whether we should just assume an elastic-perfectly plastic response. If no previous knowledge about the material behavior is available, it is advisable to include an isotropic, exponential hardening, being simpler than the kinematic one. If the assessed exponent of hardening turns out to be rather low, it may be reasonable to adopt a simpler form of the elastic-perfectly plastic behavior in the subsequent numerical simulations.

For numerical modeling, in principle, any FEM software has to be used. It is worth noting that, in order to be incorporated within the inverse analysis procedure, it is required for the numerical model to satisfy the following conditions:

- The analysis should be prepared within a single file (i.e., the analysis *input file*), possibly in ASCII format for ease of handling by an external software. It should be possible to execute the analysis from an external (i.e., command window) just by supplying the information about the name and the location of the input file.
- The sought material parameters should be easily modified within the analysis input file so that these can be controlled by the optimization procedure, in context of which they can be regarded as the "design variables").
- The main analysis results (in the present case the indentation curve) should be written in a separate file for further handling by an inverse analysis procedure.

The computed counterpart to the experimentally measured indentation curve serves to build the discrepancy function which quantifies the difference between the two data sets. It is common to construct this function in the least square form:

$$\omega = \frac{1}{2}\Sigma_{i=1}^{i=N}(d_{exp}^{i} - d_{com}^{i})^{2} \tag{7.40}$$

where d_{exp}^{i} and d_{com}^{i} are the experimental quantity and its computed counterpart, respectively. In the present case, we may perform the experiment simulation under displacement control as, in general, the nonlinear analyses run faster under displacement control than in a force control regime. Subsequently, we should compare the two curves over a fixed grid of N points that correspond to the different values of the penetration depth (see Fig. 7.5).

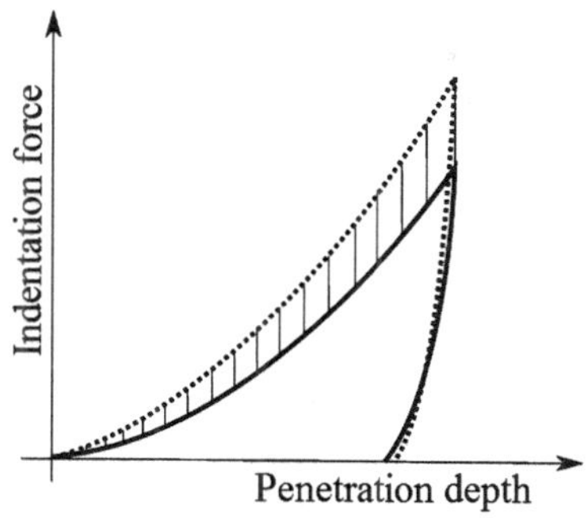

FIGURE 7.5 Comparison of the experimental and computed indentation curves by considering the same maximum penetration depth.

The solution of the minimization problem, central within the inverse analysis procedure, is generally solved in a program external to the one used for the FEM simulations. A fully automatic procedure is generated by the external software, for which it is required to establish the communication between the two programs. The algorithm for the inverse analysis procedure of the considered problem may take the following form:

1. Take the initial "guess" of the material parameters and introduce them into the analysis input file.
2. Run the FEM analysis and extract the main results (in the present example, the indentation curve).
3. Load the experimental indentation curve and use it to calculate the value of the objective function given by Eq. (7.40).
4. Employ the appropriate optimization algorithm to compute the material parameters for the second iteration that improves (i.e., reduces) the value of the objective function.
5. Update the analysis input file and get back to step number 2.

In general, the outlined iterative loop (from step 2 to step 5) is repeated until either the change in the objective function or the parameters' values between two subsequent iterations is smaller than the prescribed tolerance.

In order to solve the present problem, we adopt the ABAQUS software [15] for the FE simulations and implement the optimization tasks in MATLAB® [16]. The two programs are easy to combine, and both offer certain built-in functions that can be used for the solution of the problem. We first look into the numerical model of the indentation test.

Exploiting the axial-symmetry, we can build a two-dimensional model. The first decision we have to make is about the size of the simulation domain and the boundary conditions. The indented specimen usually has dimensions by one or more orders of magnitude larger than the area which undergoes plastic deformation under the indenter. What is of our primary interest is the zone around the indenter, while the remaining part, after certain distance from the indenter can be modeled, e.g., by semiinfinite elements (see Fig. 7.6b). A simpler alternative can be achieved if the zone is taken large enough by simply applying the boundary conditions at the end of the domain (see Fig. 7.6a). The correct selection of the size can be verified a posteriori, upon performing the simulation and by controlling the value of stress induced at the simulation domain boundaries that should be ideally be nil. If required, the domain may be further extended to satisfy this condition.

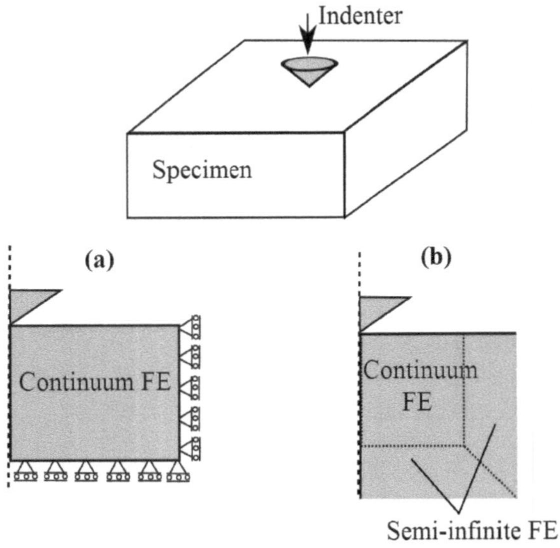

FIGURE 7.6 Modeling of the indenter and sample for an indentation test.

The preparation of the analysis input file in ABAQUS can be done within its graphical user interface (i.e., ABAQUS CAE), while it is recommended to perform additional manipulation of the file manually by operating on the ASCII file. First, it is required to write the necessary output, in this case the indentation curve, in a separate ASCII file. We first generate the node set related to the reference point of the indenter (with the name `REFPNT`), and then read the reaction force corresponding to that node set. It is convenient to keep this data in ASCII format, which should be explicitly stated in the input file, otherwise the output file will be in binary format. The commands should be placed within the step section of the input file (right before `End Step` command) and the syntax should have the following form:

```
*Output, history
*Node file, nset=REFPNT
RF
*File format, ASCII
*End Step
```

The second change to be carried out to the main input file is to place the changeable part in a separate file. The part of the main input file which specifies the material constitutive model may therefore be written as follows:

```
*Material, name=steel
*Include, input=material.inp
```

where the file `material.inp` takes the following form:

```
*Elastic
200000., 0.3
*Plastic
100., 0.
150., 1.
```

which contains the elastic constants (i.e., Young's modulus and Poisson's ratio) and the hardening curve. This section of the input file is changed during the minimization iterations.

The second part required for the inverse operator is the numerical solution of the minimization of the objective function given by Eq. (7.40). For this purpose, we may employ one of the available functions within MATLAB®. Here, specifically, we use the Trust Region algorithm. This function requires, as input, the objective function that outputs the vector of residuals that are to be minimized. In our case, the vector of residuals contains N differences in the indentation force computed by the objective function given by Eq. (7.40). Coding of this function represents the central part of the considered inverse problem. The function should perform the actions listed below.

1. Take the values of the material parameters as inputs and write them into the `material.inp` file.
2. Perform the ABAQUS analysis.
3. Load the results (i.e., the indentation curve) from the result file.
4. Load the experimental results and construct the vector of residuals.

In the problem under consideration the parameters that are passed to the function are the Young's modulus, first yield limit and the exponent of hardening. In this exercise, we assume that the value of the Poisson's coefficient is not changed (i.e., it is kept fixed). This simplification is reasonable as, for most steels, its value is almost identical.

Listing of the MATLAB® function in which residuals are computed for the present problem is given below.

```
function e=obj_fun(x)
% Function that computes vector of residuals
parE=x(1)*200000; % Denormalizing parameters
parY=x(2)*200; parN=x(3)*0.1;
%-----CHANGING THE INPUT FILE----------------------
elstr=parY/parE;
Nmb=0;
for strain=elstr:0.01:0.1
    Nmb=Nmb+1;
    ssc(Nmb,1)=parY*(parE*strain/parY)^parN;
    ssc(Nmb,2)=strain-ssc(Nmb,1)/parE;
end
newPL=['*Elastic',char(10)];
newPL=[newPL,num2str(parE),', 0.3',char(10)];
newPL=[newPL,'*Plastic',char(10)];
newPL=[newPL,num2str(ssc(1,1)),',',num2str(ssc(1,2))];
for i=2:size(ssc,1)
    line=[num2str(ssc(i,1)),',',num2str(ssc(i,2))];
    newPL=[newPL,char(10),line];
end
f_fil=fopen('material.inp','w'); % Writing changed input file;
fprintf(f_fil,newPL);
fclose(f_fil);
%-----PERFORMING ABAQUS RUN-------------------------
! abaqus j=inde_inp interactive
%-----LOADING RESULTS FROM ABAQUS .FIL FILE----------
c_curve=readfil('inde_inp.fil',0.150);
c_curve(:,2)=-c_curve(:,2);
peakC=find(c_curve(:,2)>=max(c_curve(:,2))); % Index of max force value
zeroC=find(c_curve(:,2)==0);
loadingC=c_curve(1:peakC,:);
unloadingC=c_curve(peakC:zeroC(1),:);
%-----LOADING EXPERIMENTAL RESULTS FROM FILE---------
expcur=load('expcur.txt');
peakX=find(expcur(:,2)>=max(expcur(:,2))); % Index of maxi force value
zeroX=find(expcur(:,2)==0);
loadingX=expcur(1:peakX,:);
unloadingX=expcur(peakX:zeroX(1),:);
%-----COMPARING THE TWO CURVES----------------------
NPL=50; % Number of interpolation points in the loading part of the curve
NPU=50; % Number of interpolation points in the unloading part of the curve
disBeg=max(min(loadingC(:,1)),min(loadingX(:,1)));
disEnd=min(max(loadingC(:,1)),max(loadingX(:,1)));
disZero=max(min(unloadingC(:,1)),min(unloadingX(:,1)));
```

```
45 pointsLO=disBeg:(disEnd-disBeg)/NPL:disEnd;
46 pointsUL=disEnd:-(disEnd-disZero)/NPU:disZero;
47 calL=interp1(loadingC(:,1),loadingC(:,2),pointsLO);
48 expL=interp1(loadingX(:,1),loadingX(:,2),pointsLO);
49 calU=interp1(unloadingC(:,1),unloadingC(:,2),pointsUL);
50 expU=interp1(unloadingX(:,1),unloadingX(:,2),pointsUL);
51 e=[(calL'-expL');(calU'-expU')];
```

Within the first two lines of the code, the parameters passed to the function are denormalized. The normalization of the parameters is rather important when the values of the sought parameters differ by a few orders of magnitude, as it is the case here. The three parameters are further used to build the stress vs. plastic strain curve (i.e., table) that is requited to specify the hardening in ABAQUS. The results are written in matrix `ssc`. The matrix is further used, together with the information about the Young's modulus to write the input file `material.inp` in code lines 21–23.

ABAQUS solver is invoked (code line 25), by specifying the main input file name. Upon finishing the analysis, results are loaded from the external ASCII file by using the MATLAB® function `readfil`. The structure of the ABAQUS resultant file is rather complex and detailed description of it is not within the focus of this book. For details, interested reader should refer to the ABAQUS User's manual [15]. The listing of the MATLAB® function that can be used to load the data in the form of reaction force vs. displacement is given below.

```
function [indcur]=readfil(file,ind_depth)
% Function that extract indentation curve from the abaqus result file
indcur=[];
steps=0; % Number of steps before the indentation step
RX=2; % Which reaction force is recorded in the curve (R1, R2 of R3)
f_fil=fopen(file); % Rading the .fil file.
s1=textscan(f_fil,'%c');  % This reading will get back one string data
fclose(f_fil);            % without any spaces in between, stored in s
s=s1{1}';
clear s1;
sizeS1=size(s);
sizeS=sizeS1(1,2); % Size of character vector
incr_string='I223I42000D'; % Code for increment string in ABAQUS .fil file
force_string='I16I3104I'; % Code for reaction force in ABAQUS .fil file
INCRM=strfind(s,incr_string);
firstINC=min(INCRM); % Taking the position of the first appearance
num=firstINC; % Inxed of running through the whole string s
nod=0;
curve=0; % index for curve matrix
% Writing load Vs displacement curve
while num<sizeS-10
    if s(num:num+10)==incr_string
        curve=curve+1;
        num=num+11;
        begg=num;
        foundD=0;
```

```
        while 2-foundD
            if s(num)=='D'
                foundD=foundD+1;
            end
            num=num+1;
        end
        endd=num-2;
        reldisp=str2num(s(begg:endd));
        if reldisp>1+steps
            indcur(curve,1)=1+1+steps-reldisp;
        else
            indcur(curve,1)=reldisp-steps;
        end
    end
    if s(num:num+8)==force_string
        digit=str2num(s(num+9)); % Read number of digits of node numerator
        nod=nod+1;
        num=num+8+1+digit+1+1;
        entry=0; % Number of entry in this block
        foundStar=0;
        while 1-foundStar
            foundD=0;
            begg=num;
            while 2-foundD-foundStar
                if s(num)=='D'
                    foundD=foundD+1;
                end
                if s(num)=='*'
                    foundStar=foundStar+1;
                end
                    num=num+1;
                end
                entry=entry+1;
                endd=num-2;
                if entry==RX
                    indcur(curve,2)=str2num(s(begg:endd));
                    searchingF=false;
                    break
                end
        end
            num=num+1;
            if num>sizeS-8
                break
            end
    end
    num=num+1;
end
indcur(:,1)=ind_depth*indcur(:,1);
```

In the last section of the `obj_fun` function, the two curves are compared over the region in which both of them exist. The curves are compared over 50 equally spaced points in the loading stage and over the same number of points along the unloading branch (specified within the code by variables NPL and NPU).

Upon finishing the coding of obj_fun, everything is set for the inverse analysis. The minimization part can be done by calling some of the existing MATLAB® functions. In the present problem, as previously mentioned, we make use of the Trust Region algorithm. The inverse problem is solved by executing the following script.

```
clear
guess=[0.5;0.5;0.5];
lb=[0;0;0];
ub=[2;2;2];
options=optimset('Display','iter','DiffMinChange',0.01,'TolFun',1e-4,...
    'DiffMaxChange',0.1);
result=lsqnonlin(@obj_fun,guess,lb,ub,options);
```

The above script performs a constrained minimization where the lower bound and the upper bound of the sought parameters are specified within vectors lb and ub. The MATLAB® function lsqnonlin employs the trust region algorithm (for details, see MATLAB® User's guide [16]) in order to minimize the objective function given as the argument of the function. Additional options related to the optimization algorithm are specified with argument options.

The developed codes represent computational tools that are required to solve an inverse problem. Before using them with real experimental results, it is useful to test the overall procedure with the so-called pseudo-experimental data (see Chapter 3). The verification of the procedure consists in the following steps. The numerical simulation is used to generate pseudo-experimental results that are subsequently treated as the real experimental data. If the inverse problem is well posed, it should lead to exactly the same parameter values that were employed to generate the pseudo-experimental data. Pseudo-experimental results offer a framework to test also the influence of various types of errors (i.e., by perturbing the pseudo-experimental values with selected levels of noise).

7.5 Summary

This chapter provided an overview of some of the key features related to the modeling of metals. In doing so, the constitutive representations previously introduced in Chapter 5 for elasto-plastic models were further expanded to provide a detailed outline of the steps involved in a three-dimensional numerical implementation. The concepts at the basis of stress return mapping and backward-Euler return algorithms were introduced to highlight their role in the solution process when evaluating the material response and to point out the elegant features at the basis of plasticity models.

A number of illustrative examples were presented to gain insight into the different yield functions and into the details of the implementation of a three-dimensional plasticity model. In general, the examples were kept as simple as possible to minimize the complexity of the coding.

The last part of the chapter focused on the parameter identification that needs to be performed when dealing with plasticity models. The numerical implementation of the inverse analysis presented here was particularly useful in gaining insight into the details of the required steps to be considered for a computer implementation. For a wider applicability, the direct analysis used for the inverse analysis approach made use of a finite element model developed with the ABAQUS commercial software. In this manner, it was possible to point out not only the computational aspects of the implementation but also to address all issues associated with the management of the identification process, and associated files and information, that need to be dealt with for a practical implementation.

References

[1] M.A. Crisfield, Non-linear Finite Element Analysis of Solids and Structures, vol. 1, John Wiley and Sons, New York, 2000.

[2] M.A. Crisfield, Numerical analysis of structures: developments in Thin-walled structures, Applied Science (1981) 235–284.

[3] S.W. Key, C.M. Stone, R.D. Krieg, A solution strategy for the quasi-static large deformation inelastic response of axissymmetric solids, in: Workshop on Nonlinear Finite Element Analysis in Structural Mechanics, Bochum, 1980.

[4] M. Ortiz, E.P. Popov, Accuracy and stability of integration algorithms for elastoplastic constitutive relations, International Journal of Numerical Methods in Engineering 21 (9) (1985) 1561–1576.

[5] M.L. Wilkins, Calculating elasto-plastic flow, in: B. Alder, et al. (Eds.), Methods of Computational Physics, vol. 3, Academic Press, 1964.

[6] R.D. Kreig, D.B. Kreig, Accuracy of numerical solution methods for the elastic-perfectly plastic model, Transactions of the ASME (1977) 510–515.

[7] J.C. Simo, R.L. Taylor, Consistent tangent operators for rate-independent elasto-plasticity, Computer Methods in Applied Mechanics and Engineering 48 (1985) 101–108.

[8] P. Jetteur, Implicit integration algorithm for elastoplasticity in plane stress analysis, Engineering Computations 3 (3) (1986) 251–253.

[9] J.C. Simo, R.L. Taylor, A return mapping algorithm for plane stress elastoplasticity, International Journal of Numerical Methods in Engineering 22 (1986) 649–670.

[10] T. Hughes, Finite Element Method. Linear Static and Dynamic Finite Element Analysis, Dover Publications INC, New York, 2000.

[11] G.R. Liu, S.S. Quek, The Finite Element Method: A Practical Course, Butterworth Heinemann, Oxford, 2003.

[12] G. Bolzon, G. Maier, M. Panico, Material model calibration by indentation, imprint mapping and inverse analysis, International Journal of Solids an Structures 41 (2004) 2957–2975.

[13] G. Bolzon, V. Buljak, B. Miller, G. Maier, Assessment of elastic-plastic material parameters comparatively by three procedures based on indentation test and inverse analysis, Inverse Problems in Science and Engineering 19 (6) (2011) 815–837.

[14] BS EN ISO 6508-1, Metallic materials. Rockwell hardness test. Test method, British Standards, 2016.

[15] ABAQUS / Standard, Theory and User's Manuals, Release 6.13, HKS Inc., Pawtucket, 2013.

[16] MATLAB. User's guide and optimization toolbox, Release 2016, The Mathd Works Inc., USA, 2016.

Chapter 8

Polymers

Polymers are natural or synthetic materials in which atoms are connected together to form long molecular chains. Since bonds between different molecular chains are significantly weaker than those that exist between atoms within each molecular chain, the chain represents the elementary building unit of a polymer. This distinguishing structure is at the basis of the physical properties exhibited by polymers and of their use in a wide range of applications that range from food packaging to automotive and aerospace industries.

From a mechanical viewpoint, the modeling of these materials is a rather challenging task given the wide range of behavioral features they can exhibit. In this context, the main focus of this chapter is to outline some of the aspects relevant to the selection of suitable constitutive models and to discuss strategies for their calibration. It is outside the scope of this chapter to provide a comprehensive description of the modeling of polymers' mechanical behavior for which reference should be made to dedicated textbooks (e.g., see [1], [2], or [3]).

It is difficult to classify the behavior of polymers, in general terms, as simply elastic, brittle, or plastic since their behavior is strongly affected by the material conditions. Specifically, the mechanical response depends mostly on the temperature and the rate of the application of loads. The same polymer can exhibit the features of a brittle solid, an elastic rubber or even a viscous liquid, depending on the temperature and the time scale of loading.

The mechanical behavior of polymers is predominantly governed by the presence of cross-links between their molecular chains. Polymers with almost no cross-links between the chains are denoted as *thermoplastics*. Polymers with relatively low cross-link density are referred to as *elastomers*, while polymers with many cross-links are called *thermosetting polymers* or *duromers*.

The nomination of the above groups in which different polymers are classified is related to their macroscopic thermo-mechanical behavior. Thermoplastic polymers soften when exposed to heat but they return to their original condition when cooled down. Because of this, these polymers can be hot-shaped to a wide range of different profiles, even repeatedly by heating and subsequent cooling. The elastomers, or rubbers, can be elastically deformed to fairly large strains (i.e., 100% or more) without suffering from any permanent deformation. Thermosetting polymers solidifies irreversibly (i.e., sets) when heated, and cannot be reshaped further. They are in general stronger than thermoplastics and hence better suited for high demanding, high temperature applications since they maintain their shape when heated.

Constitutive Modeling of Engineering Materials. https://doi.org/10.1016/B978-0-12-814696-5.00013-7

Both thermoplastic and thermosetting polymers can develop an irreversible plastic deformation, when loaded beyond their yield strength. The assumption of linear-elastic relationship between stress and strain prior to yielding may not be suitable for some polymers.

An important feature of the mechanical behavior of polymers is that both the elastic and plastic behaviors are time-dependent and, when modeling their mechanical behavior, visco-elastic and visco-plastic models should be considered [4]. This means that the adopted constitutive relation should, in general, contain time as a variable in addition to stress and strain, and the problem should be considered within a dynamic analysis framework.

The calibration of the adopted visco-elastic and/or -plastic model necessitates measurements collected from the dynamical response of material samples. For an adequate description of the mechanical behavior of a polymer, experimental data is required to be collected over a wide range of temperatures and loading rates. The following five classes of experimental measurements are commonly employed, separately or in combination, for the mechanical characterization of polymers (for details, see [1]): transient measurements, e.g., stress relaxation and creep tests; low-frequency vibration, e.g., free oscillation methods; high-frequency vibration, e.g., resonance method; forced vibration, e.g., non-resonance methods; and wave propagation velocity measurements, e.g., wave propagation methods.

We now give a closer look to the transient measurements (i.e., creep and relaxation test) since it is going to be used within the example of material characterization provided in this chapter. When a material that exhibits visco-elastic behavior is loaded by the constant amount of stress, the response of it in terms of deformation change in time follow qualitatively the one depicted on Fig. 8.1a: upon certain, instantaneous deformation, there will be an additional increase due to the viscous effects within the material. Similarly, if the same material is loaded by an instantaneous application of deformation, that is subsequently kept constant, the response in terms of stress change in time can be qualitatively described by the one depicted in Fig. 8.1b. The former experiment is commonly termed as the creep test, while the latter is called the relaxation test.

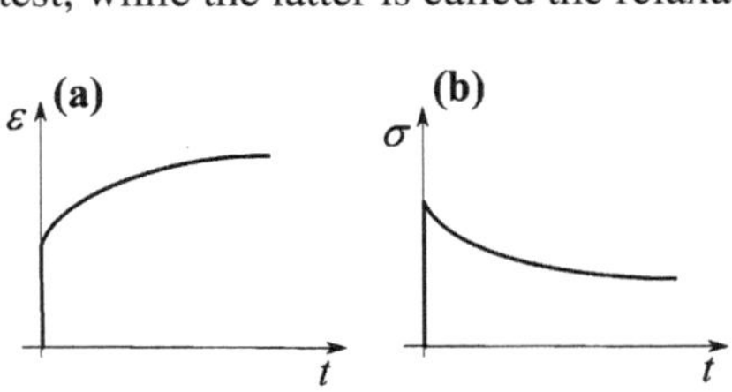

FIGURE 8.1 Time-dependent response of a visco-elastic material: (a) creep test and (b) relaxation test.

It is important to emphasize that the creep, as a deformation that changes in time, can be observed also as a consequence of visco-plastic behavior. In such a case, a slowly developed deformation due to creep will be an irreversible one. Its evolution in time in general can have more complex shape than the one depicted in Fig. 8.1a, occasionally without a settling value with constantly increasing deformation up to failure. For its proper modeling, rate dependent plasticity models should be employed.

In the following, we present simple illustrative examples to highlight the key steps involved in the inverse analysis procedure when applied to polymers. To keep the example simple, we focus only on the visco-elastic behavior and leave aside rate-dependent plastic behavior. In doing so, we also discuss the importance of selecting constitutive models that are sufficiently simple to enable effective parameter calibrations while minimizing the required number of tests and amount of experimental measurements to be collected. At the same time, the constitutive model needs to be sufficiently refined to well capture the mechanical response.

8.1 Visco-elastic response and model calibration

As introduced in Chapter 5, a material can be regarded as viscous when it undergoes time-dependent deformation when subjected to a stress maintained constant over time. The corresponding mechanical response is captured in Fig. 8.1a. It highlights how, after a certain, instantaneous deformation is applied, there is an additional increase in deformation due to viscous effects. If the same material is loaded by an instantaneous deformation maintained constant over time the response can be described in terms of stress changes over time as depicted in Fig. 8.1b.

It is common to model visco-elastic polymers with commercial finite element software. In such cases the material properties can be specified in a number of ways. One of the most used approaches is to rely on specialized subroutines available in these software that require, in input, the experimental measurements collected from standard tests, such as creep and/or relaxation tests. Such an approach is certainly easy to implement if the experimental data is available, even if it might not ensure that the characterization is appropriate for the modeling of loading scenarios different from those adopted during the experimental characterization tests. In such case, it may be preferable to specify the material properties by selecting a suitable model and by characterizing and providing the required model parameters.

Frequently used constitutive models to describe visco-elastic behavior relies on Prony series that formulates the model in terms of the relaxation moduli (tensile-compressive, shear, or bulk modulus). Such model representations require access to relaxation coefficients that could be obtained directly from relaxation tests. A difficulty related to the quantification of these coefficients is that the creep test is easier to perform than the relaxation test, and the assessment of the relaxation coefficients is not an easy task to perform when having access to only creep test data, as it is often the case.

We now consider an epoxy resin (i.e., thermosetting polymer) as a typical example of polymer material that exhibits visco-elastic behavior. When such a material is subjected to a constant strain over time, the material tends to relax due to viscous effects. Such behavior is commonly modeled by defining a suitable relaxation modulus function depending on the type of deformation induced to the material, e.g., tensile, compressive, shear, or bulk deformation. Usually, the relaxation modulus functions are defined so that viscous effects can model the deviatoric and volumetric deformation separately, usually denoted as $G(t)$ and $K(t)$, respectively.

For the purpose of this chapter, we further simplify the model by assuming that the bulk modulus is constant over time and that the viscous effects influence only the deviatoric behavior. Based on this simplification, commonly accepted for epoxy materials, the only relaxation function to be defined is $G(t)$. In finite element implementations the relaxation modulus functions can be described in terms of Prony exponential series (see [5]) and can be expressed as:

$$G(t) = G_\infty + \sum_{i=1}^{N} G_i e^{-\frac{t}{\tau_i}} \tag{8.1}$$

where parameter G_∞ represents the long-term shear modulus, N depicts the number of elements included in the Prony series and the parameters' values G_i and τ_i need to be quantified from material testing. It is difficult to associate specific physical meaning to the Prony series parameters as their quantitative values vary for the same material response, depending on the number of elements included in the series.

Visco-elastic materials are long-term elastic, in the sense that after having been subjected to a constant strain for a long period of time, they relax to a constant stress. This response is governed by the parameter G_∞, but the parameter is fairly difficult to measure. It is easier to measure an instantaneous response, governed by the instantaneous shear modulus G_0, responsible for the initial deformation of the material subjected to an instantaneous load. In view of this parameter the shear relaxation modulus can be written in a dimensionless form as follows:

$$g_R(t) = \frac{G(t)}{G_0} \tag{8.2}$$

based on which the Prony series expansion of the dimensionless relaxation modulus can be expressed in terms of the dimensionless Prony series parameters $\bar{g}_i$ and τ_i as:

$$g_R(t) = 1 - \sum_{i=1}^{N} \bar{g}_i e^{-\frac{t}{\tau_i}} \tag{8.3}$$

For illustrative purposes, we now consider an inverse analysis procedure aimed at the characterization of the visco-elastic response described by the Prony series expansion based on experimental data collected from a creep test.

In particular, we consider a simple test in which the material sample is subjected to an axial force maintained constant over time (see Fig. 8.2) and, during the test, we measure the sample elongation over time. Due to visco-elastic effects, the response of the elongation over time is expected to follow the trend depicted in Fig. 8.3.

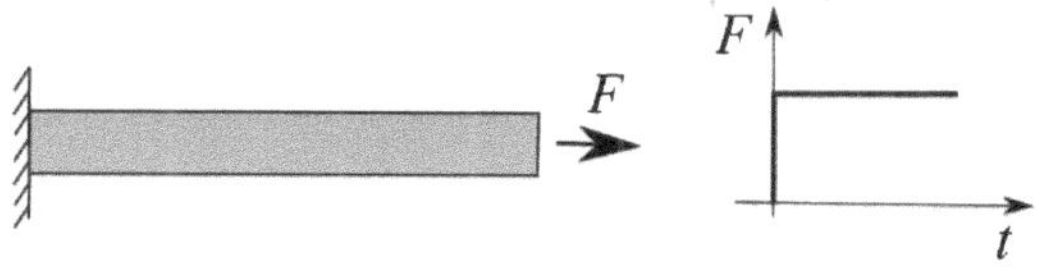

FIGURE 8.2 Creep test of an axially loaded bar.

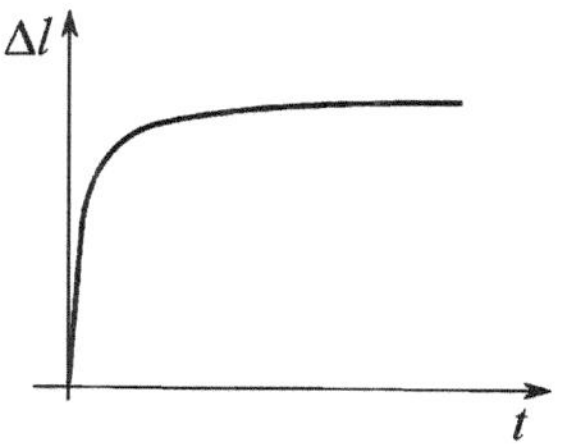

FIGURE 8.3 Typical elongation over time measured during a creep test.

For the purpose of this example, we simulate the creep test with the commercial code ABAQUS and perform a dynamic analysis for a period of 3600 seconds (i.e., 1 hour). The visco-elastic response is modeled by means of Prony series with $N = 4$ (see Eq. (8.3)). In this case, the number of parameters that need to be identified to capture the time-dependent response is 8: $\bar{g}_i$ and τ_i with $i = 1, ..., 4$.

The selection of the number of members to be included in the Prony series should be based on a trade-off based on the following considerations: (i) the larger the number, the more demanding is the quantification of the parameters through the inverse analysis, as the number of numerical simulations required to be carried out for each iteration of the inverse analysis is $P + 1$, with P being the number of sought parameters; and (ii) the use of a small number of members may lead to unrealistic oscillatory behavior in the model representation. Fig. 8.4 depicts the response of the time-dependent elongation by considering the response simulated with $N = 2$ and with $N = 4$. While the response modeled with 4 Prony members still exhibits some oscillations, the one modeled with only 2 members undergoes amplitudes of elongation that are even higher than those exhibited by the long-term elongation. It is worth noting that in reality certain oscillations in strain within the creep experiment can be observed (see [1] for details) due to rapid loading.

The experimental data to be used in this illustrative example is based on pseudo-experimental data, i.e., data generated from a numerical model. The

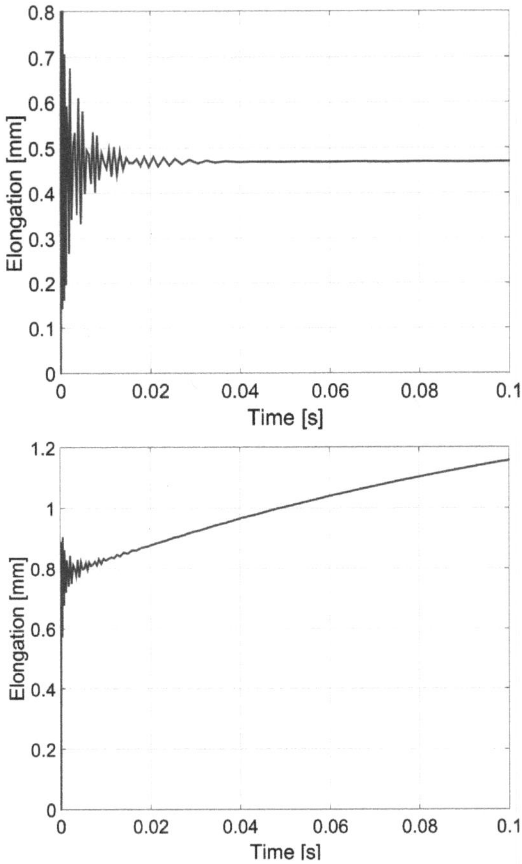

FIGURE 8.4 Elongation in time of an viscoelastic material modeled with 2 Prony series members (upper graph) and with 4 members (bottom graphs).

numerical model (required for the direct analysis of the inverse analysis and for the generation of the pseudo-experimental data) is here setup with a two-dimensional, plane stress model developed in ABAQUS. The model is setup to perform a dynamic analysis of the specimen subjected to a constant load for 3600 seconds. The results of the numerical model are written to an external file to be available to the inverse analysis procedure.

In the implementation of the inverse procedure, the identification process requires to change the input data of the ABAQUS model and this is carried out (as previously detailed in Chapter 7) by writing a changeable part of the ABAQUS input file in a separate file, i.e., `material.inp` file, that can take the form provided below.

```
*Elastic, moduli=INSTANTANEOUS
4000., 0.39
```

```
*Viscoelastic, time=PRONY
0.07, 0., 450.
0.15, 0., 0.065
0.31, 0., 0.0001
0.38, 0., 7E-07
```

where the Prony series parameters are specified after the ABAQUS command line `*Viscoelastic, time=PRONY`, with each row associated to a Prony series member. The number of lines should be equal to the adopted number of Prony series members. The first number in each line corresponds to the $\bar{g}_i$ value and the second number to the $\bar{k}_i$ value that are used to compute the bulk modulus relaxation, here taken as constant, while the last number corresponds to the parameter τ_i.

The input values specified above have been used to generate the pseudo-experimental results that have been plotted in Fig. 8.5.

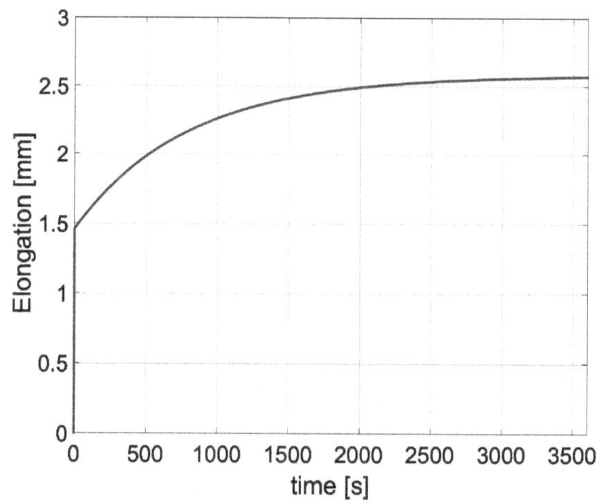

FIGURE 8.5 Simulated pseudo-experimental results: elongation *versus* time curve.

For simplicity, we assume the instantaneous material properties to be known a priori. In this manner, the identification process aims at the characterization of 8 parameters. Such an approach is acceptable by considering that the instantaneous properties could be identified through other simple experiments, e.g. three point bending test.

The second step of the inverse analysis procedure is to write a MATLAB® code that calculates the objective function. A number of tasks need to be performed within this code, and these are ordered in the following sequence: modify the input file as the inverse analysis progresses, perform the analysis by calling the ABAQUS solver, load the results from the external file (in this case the time-dependent elongation curve), and compare the computed results with the experimental ones (in this case to the pseudo-experimental set of values, depicted in Fig. 8.5). The structure of the MATLAB® code is identical to the one already introduced in Chapter 7 and, for this reason, the details of its implementation are not repeated here. The MATLAB® listing used in this example is provided below.

```
function e=obj_fun(x)
% Function that computes vector of residuals
g1=x(1)*0.07;
g2=x(2)*0.15;
g3=x(3)*0.31;
g4=x(4)*0.38;
t1=x(5)*450;
t2=x(6)*0.065;
t3=x(7)*0.0001;
t4=x(8)*7e-7;
%-----CHANGING THE INPUT FILE----------------------
newPL=['*Elastic, moduli=INSTANTANEOUS',char(10)];
newPL=[newPL,num2str(4000.),', 0.39',char(10)]; % Values are kept constant
newPL=[newPL,'*Viscoelastic, time=PRONY',char(10)];
newPL=[newPL,num2str(g1),',  0., ',num2str(t1),char(10)];
newPL=[newPL,num2str(g2),',  0., ',num2str(t2),char(10)];
newPL=[newPL,num2str(g3),',  0., ',num2str(t3),char(10)];
newPL=[newPL,num2str(g4),',  0., ',num2str(t4),char(10)];
f_fil=fopen('material.inp','w'); % Writting changed input file;
fprintf(f_fil,newPL);
fclose(f_fil);
%-----PERFORMING ABAQUS RUN------------------------
 ! abaqus j=creep_el_inp interactive
%-----LOADING RESULTS FROM ABAQUS .FIL FILE---------
c_curve=readfil('creep_el_inp.fil');
%-----LOADING EXPERIMENTAL RESULTS FROM FILE---------
expcur=load('expcur.txt');
%-----COMPARING THE TWO CURVES----------------------
start=max(c_curve(1,1),expcur(1,1));
points=start:0.5:3600;
cal=interp1(c_curve(:,1),c_curve(:,2),points);
exp=interp1(expcur(:,1),expcur(:,2),points);
e=cal'-exp';
```

The experimental curve and the computed one are compared over the time range from 0 to 3600 seconds, at intervals of 0.5 seconds (other intervals could have been considered, also varying in length over time). A dedicated MATLAB® function is used to read the results calculated with ABAQUS and saved in the ABAQUS result file. For completeness, the MATLAB® listing of this function is provided below. With respect to the function provided in Chapter 7 the one given here is modified to read the displacement over time.

```
function [indcur]=readfil(file)
% Function that extract indentation curve from the abaqus result file
indcur=[];
```

```
RX=1; % Which displacement component is recorded (U1, U2 of U3)
f_fil=fopen(file); % Rading the .fil file.
s1=textscan(f_fil,'%c');  % This reading will get back one string data
fclose(f_fil);            % without any spaces in between, stored in s
s=s1{1}';
clear s1;
sizeS1=size(s);
sizeS=sizeS1(1,2); % Size of character vector
incr_string='I223I42000D'; % Code for increament string in ABAQUS .fil file
disp_string='I15I3101I'; % Code for displacement in ABAQUS .fil file
INCRM=strfind(s,incr_string);
firstINC=min(INCRM); % Taking the position of the first appearance
num=firstINC; % Inxed of running through the whole string s
nod=0;
curve=0; % index for curve matrix
while num<sizeS-10
    if s(num:num+10)==incr_string
        curve=curve+1;
        num=num+11;
        begg=num;
        foundD=0;
        while 2-foundD
            if s(num)=='D'
                foundD=foundD+1;
            end
            num=num+1;
        end
        endd=num-2;
        reldisp=str2num(s(begg:endd));
        indcur(curve,1)=reldisp;
    end
    if s(num:num+8)==disp_string
        digit=str2num(s(num+9)); % Read number of digits of node numerator
        nod=nod+1;
        num=num+8+1+digit+1+1;
        entry=0; % Number of entry in this block
        foundStar=0;
        while 1-foundStar
            foundD=0;
            begg=num;
            while 2-foundD-foundStar
                if s(num)=='D'
                    foundD=foundD+1;
                end
                if s(num)=='*'
                    foundStar=foundStar+1;
                end
                    num=num+1;
            end
                entry=entry+1;
                endd=num-2;
                if entry==RX
```

```
                        indcur(curve,2)=str2num(s(begg:endd));
                        searchingF=false;
                        break
                    end
            end
                num=num+1;
                if num>sizeS-8
                    break
                end
        end
        num=num+1;
end
```

The inverse analysis procedure is the performed by running the MATLAB® script listed below.

```
clear
guess1=[0.80;0.85;0.86;1.1;1.15;0.97;1.1;1.05];
lb=[0.6;0.6;0.6;0.6;0.6;0.6;0.6;0.6];
ub=[1.8;1.8;1.8;1.8;1.8;1.8;1.8;1.8];
options=optimset('Display','iter','DiffMinChange',0.001,'TolFun',1e-4,...
    'DiffMaxChange',0.1);
result=lsqnonlin(@obj_fun,guess1,lb,ub,options);
```

For the problem considered, the sought parameters are normalized with respect to their target values, i.e., the values employed to generate the pseudo-experimental results. The minimization problem is solved as constrained, with suitable lower and upper bounds specified for all parameters. The limits adopted here are relatively tight, due to the fairly large influence that each of the parameters have on the measured response (expressed in the elongation versus time curve). This can be highlighted by comparing the curves generated by using the target parameter values (used to generate the pseudo-experimental results) and the initial "guess" values for the parameters that differ, in average, by about 15%. In this case, the two time-dependent elongation curves are quite different as shown in Fig. 8.6. This indicates the large sensitivity of the measured quantities (i.e., time-dependent elongations) to the sought parameters. By executing the MATLAB® script previously provided, the inverse analysis is capable of identifying the sought parameters after 10 iterations (the difference in each identified parameter with respect to its target value is less than 0.5%). Already after the 3rd iteration, the parameters are rather close to their target values as demonstrated by the vicinity of the two curves in Fig. 8.6.

8.2 Modeling and mechanical characterization of elastomers

Elastomers represent polymers with relatively low density of cross-links between molecular chains. Since their density is fairly low, they allow molecular chains to stretch relative to each other and, in doing so, give rise to the corresponding macroscopically observed large deformation. Differently from thermoplastics, the presence of cross-links in elastomers prevents relative sliding of

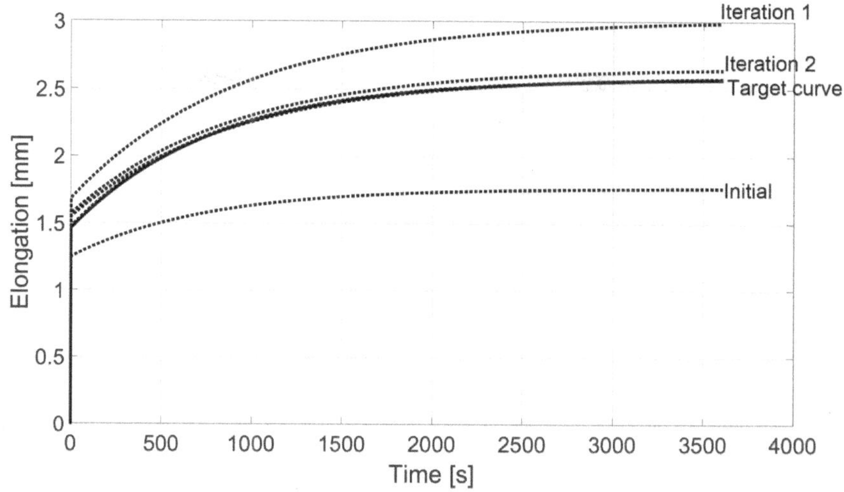

FIGURE 8.6 Time-dependent elongation curves identified for different iterations of the inverse analysis procedure.

molecular chains over each other and, therefore, the deformation is fully recoverable (i.e., elastic). Such elastic deformation is fairly large and can result in macroscopic strains of 100% and more. This unique property is called *hyperelasticity* (already introduced in Chapter 6). The most common example of this type of material is natural rubber. In the literature, materials with capability of developing large deformation are commonly denoted as *rubber-like materials.*

Mechanical modeling of elastomers and the characterization of this behavior can be challenging from the fact that experimentally observed stress–strain relations are highly nonlinear. Thus, a simple modulus of elasticity is not sufficient to describe this elastic behavior even when the material could be treated as isotropic. To describe the constitutive behavior of these materials, it is common to specify a strain energy density function and to use modeling techniques such as those described in Chapter 6.

There are several modeling functions that can be used to represent the strain energy density. Therefore the first task to be performed when modeling the response of rubber-like materials is to select a suitable strain energy density function based on the complexity of the problem considered and its expected deformation range. For the setting up of the material model, a mechanical characterization of the model parameters is then required to be carried out as briefly discussed in the following.

When selecting a strain energy density function, it is important to try to keep the model as simple as possible in order to minimize the complexity of the model calibration while ensuring a sufficient model refinement to achieve the sufficient accuracy required by the problem. For example, if the range of deformation is small (e.g., well below 100%), a simple strain energy function with few coefficients may be successfully used, even though it cannot capture well the typical "S" shape of a typical stress–strain curve as depicted in Fig. 8.7.

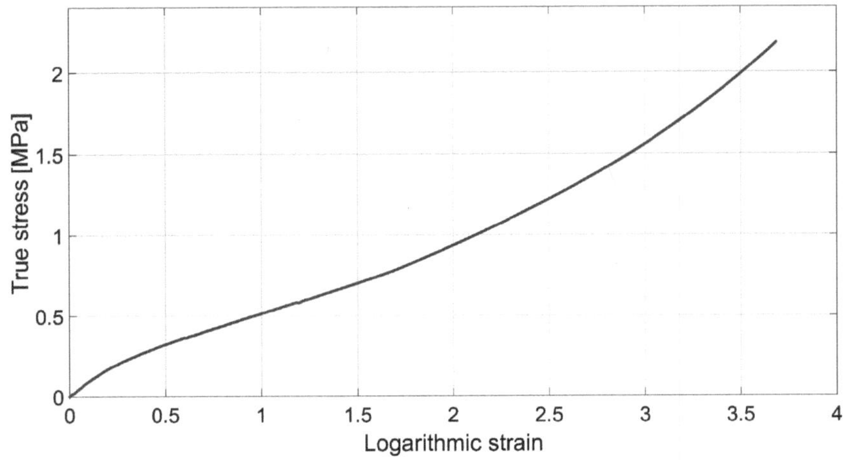

FIGURE 8.7 Typical true stress—logarithmic strain curve obtained from a tensile test for a silicone rubber sample.

Another important consideration to take into account is related to the availability of experimental data for the material parameter identification. In general, these models require a multitude of tests to be carried out to generate the sufficient experimental information required for the model calibration. As an indication, in the case we can perform only a simple test, such as an uniaxial test, it is preferable to select a simple model.

From a practical viewpoint, commercial FEM softwares available on the market provide access to a wide range of strain energy density functions. The very basic criterion for its selection is the range of the deformation relevant to the application of interest. In the case of a problem in which strains remain below 100% and if, in the deformation range, the material response does not exhibit an "S" shape (see Fig. 8.7), a Mooney–Rivlin model could be adopted for the modeling. In its incompressible form, the strain energy density function depends on only two parameters and reads:

$$U = C_{10}(I_{1C} - 3) + C_{01}(I_{2C} - 3) \tag{8.4}$$

where I_{1C} and I_{2C} are the first and the second invariant of the right Green–Cauchy deformation tensor (see Chapter 6).

We now illustrate this behavior by considering the example of a silicon rubber. Typically, the easiest way to obtain experimental data is to perform a tensile test (see Fig. 8.8) and to collect the elongation vs. applied force curve. To describe this behavior with a suitable constitutive model the measured quantities need first to be transformed into stress–strain relations. As the material undergoes large deformation, it is important to select an appropriate strain measure to

FIGURE 8.8 Tensile test of a silicone rubber.

adequately capture the material response (as discussed in Chapter 6). Arguably, the easiest strain measure to be used for this purpose is the logarithmic strain that can be evaluated as:

$$\varepsilon = \ln\left(\frac{l}{l_0}\right) \tag{8.5}$$

where l is the deformed length and l_0 is the initial one.

As far as the stress is concerned, the true (Cauchy) stress should be considered (i.e., the one that is computed on the basis of the deformed cross-sectional area) since the specimen undergoes large deformation and, by relying on its initial area, a significant error is introduced by which the stress exhibited by the material is underestimated. The evaluation of the true stress from the measured force requires access to the geometric changes of the cross-section that occur during the test. Such variation in geometry can be captured either through digital image correlation techniques or manually, by measuring the cross-sectional dimensions at different levels of load. A typical true stress—logarithmic strain curve that could be obtained from a tensile test of a silicone rubber sample is shown in Fig. 8.7.

The characterization of the material behavior can be carried out using an inverse analysis procedure. With this approach, the force-displacement curve obtained from the tensile test can be fed into the inverse analysis procedure. The tensile test can be simulated with the finite element method by employing, for example, the Mooney–Rivlin constitutive model with two parameters (i.e., parameters C_{10} and C_{01}) whose values are then sought through the identification process. It is expected that, with this model, the material behavior can be captured quite well up to strains of about 100%. Thus, only the part of the experimental curve up to the displacements that correspond to about this strain level should be used in the parameter identification process.

The numerical simulation of the tensile test here can be solved as a quasi-static problem. The MATLAB® files already introduced in previous examples can be used to solve this inverse problem, which leads to the following identified parameter values:

$$C_{10} = 0.139176; \quad C_{01} = 0.020004 \tag{8.6}$$

and the corresponding numerical and experimental curves are reported in Fig. 8.9.

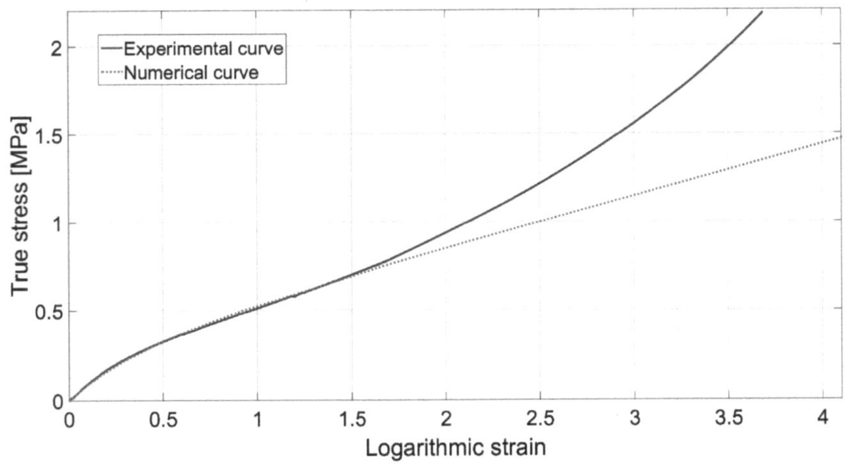

FIGURE 8.9 Numerical and experimental curves obtained for the tensile test of a silicone rubber sample.

The predicted numerical results obtained with the identified parameters show good agreement with the experimental curve for strains up to 150%. After this value, the curves start to separate as the Mooney–Rivlin model with only two parameters is not able to capture the "S-shape" response (as previously pointed out). In the case the modeling needs to capture the "S-shape" response of the material, it is required to consider material representation that are more refined that those based on low order models (such as the Mooney–Rivlin model).

One might expect that a better result can be obtained if, within an inverse analysis procedure, the whole curve is used, so that by averaging the error a better approximation is obtained also for the strains larger than 150%. In such a case, within the inverse analysis procedure, the parameters will tend to assume values resulting in unstable material behavior and the analysis will meet difficulties in convergence. Therefore if the analysis of interest includes stresses larger than 100%, a strain energy density function of higher order should be used instead.

Higher order models, as a rule, require larger amount of experimental data to be collected for calibration purposes. Biaxial and volumetric tests are commonly performed and combined with the results of uniaxial tests. In a biaxial test, two perpendicular forces are applied while the grade of the biaxiality (i.e., the ratio between the intensity of the two forces) can be adjusted. The specimen is usually of a cruciform shape, where the corner radius can be either relatively small or large, when compared to other specimen dimensions (see Fig. 8.10). In order to enrich the experimental data, multiple geometries may be considered in the preparation of the samples to be tested [6].

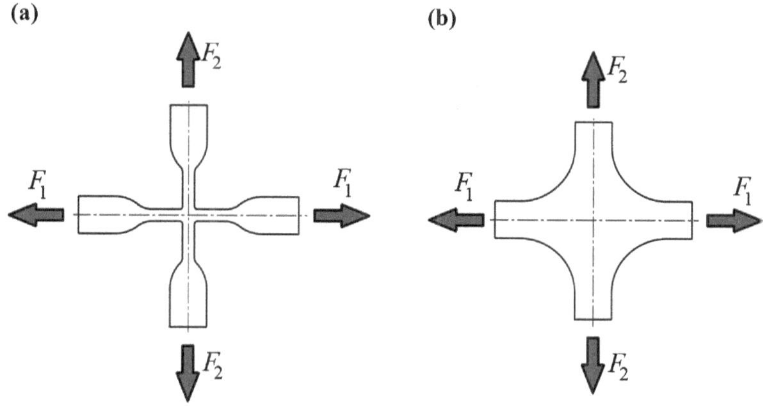

FIGURE 8.10 Specimen for biaxial test with (a) small radius or (b) large radius.

A volumetric compression test is an experiment that is performed under uniaxial strain conditions. The test equipment consists of a die in which the rubber sample is inserted (usually molded) and of a punch that is used to apply the loading pressure (see Fig. 8.11). During the test, the pressure vs. volume change ratio is measured (where the volume change is obtained based on the punch displacement) and this could then be used to characterize the material parameters.

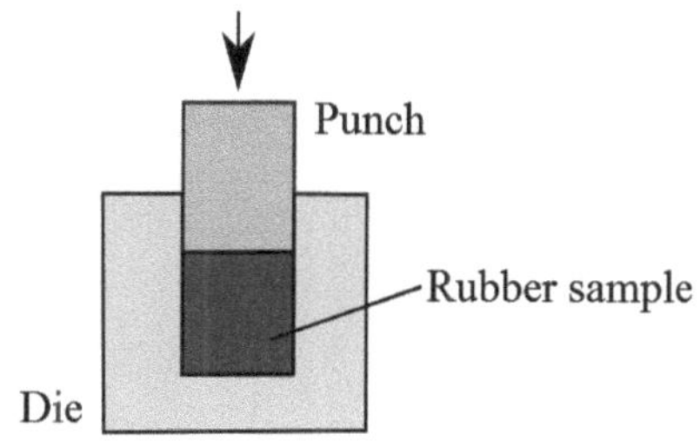

FIGURE 8.11 Schematic representation of a volumetric test.

8.3 Summary

In this chapter, we outlined some of the key aspects related to the modeling of the mechanical behavior of polymers. We placed particular attention on their time-dependent behavior because fundamental for the adequate representation of their response.

We introduced some of the main aspects related to the calibration of their material properties. In doing so, we pointed out the trade-off that needs to be considered when setting up a numerical model for this class of materials in which the constitutive model should be kept as simple as possible, to simplify the characterization of the material parameters, while accounting for the necessary refinement to adequately capture the material response.

Simple illustrative examples were considered to outline the key steps to be carried out in an inverse analysis procedure applied to polymers. For example, we performed a characterization of the material parameters describing a viscoelastic response based on experimental data collected from a creep test. In the second part of the chapter, a hyperelastic elastomer was considered. For the modeling of this type of material, the importance of representing the constitutive model in the form of a strain energy density function was discussed. While there are a wide range of different functions that can be used for this purpose, it was shown that, for applications in which the strain remains within 100%, lower-order models (such as the Mooney–Rivlin model that depends on only two parameters) can be successfully used. It was also noted that to capture the "S-shape," typical for stress–strain curves of rubber-like materials, higher order models need to be employed, even if these models require a large amount of experimental data for their calibration.

References

[1] I.M. Ward, J. Sweeney, Mechanical Properties of Solid Polymers, John Wiley and Sons, New York, 2013.

[2] Jorgen S. Bergstrom, Mechanics of Solid Polymers: Theory and Computational Modeling, Elsevier, San Diego, 2015.

[3] L. Tong, A.P. Mouritz, M.K. Bannister, 3D Fiber Reinforces Polymer Composites, Elsevier, Oxford, 2002.

[4] Joachim Roesler, Harald Harders, Martin Baeker, Mechanical Behavior of Engineering Materials, Springer, Berlin, 2007.

[5] J.F. Hauer, C.J. Demeuer, L.L. Scharf, Initial results in Prony analysis of power system response signals, IEEE Transactions on Power Systems 5 (1990) 80–89.

[6] M. Johlitz, S. Diebels, Characterisation of a polymer using biaxial tension test. Part I: Hyperelasticity, Archives of Applied Mechanics 81 (10) (2011) 1333–1349.

Chapter 9

Shape memory alloys

9.1 Introduction

This chapter provides an introduction to the behavior and modeling of shape memory alloys (SMAs) that have been gaining popularity over the last decades over a wide range of applications. The particularity of these materials relies on their ability to exhibit a direct coupling between the mechanical and thermal responses that can give rise, depending on the alloy composition, temperature, and stress conditions, to two distinctive behaviors commonly denoted as *shape memory effect* and *superelasticity* (or *pseudoelasticity*). In the case of the shape memory effect, the material is able to recover, after deformation, its initial shape when subjected to an increase in temperature, while superelasticity depicts the material's ability to recover its original shape after undergoing a hysteretic stress-deformation cycle that could occur, for example, through the application of a load cycle.

The unique response of shape memory alloys is induced by phase transformations that take place under different temperature and stress conditions between a high temperature phase, referred to as *austenite* (also known as the parent phase), and a low temperature phase, denoted as *martensite* (also known as the product phase). The austenite and the martensite have different crystal structure where the former is usually characterized by a body-centered cubic crystal structure and the latter by a face-centered cubic crystal structure (see Fig. 9.1). An overview of these phase transformations is introduced in Sections 9.2 and 9.3 when considering shape memory effect and superelasticity, respectively, while detailed information on their behavior and modeling can be found in specialized textbooks (e.g., [5,6]).

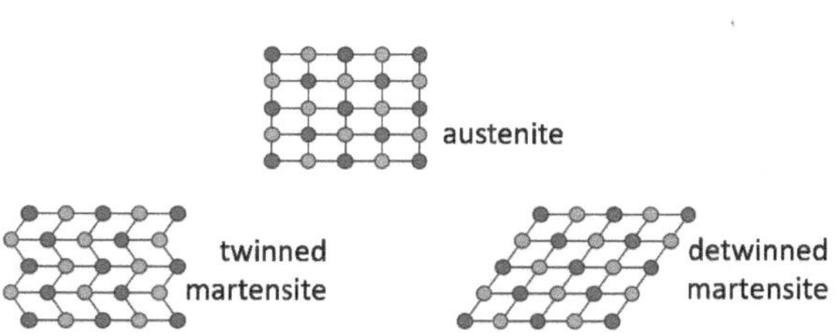

FIGURE 9.1 Schematic of the crystal structure of austenite, twinned martensite and detwinned martensite.

Constitutive Modeling of Engineering Materials. https://doi.org/10.1016/B978-0-12-814696-5.00014-9

A brief overview of the modeling of these behaviors is outlined in Section 9.4 and the use of the inverse analysis procedure for the material characterization is presented in Section 9.5.

9.2 Shape memory effect

The ability of shape memory alloys to exhibit shape memory effect (SME) has contributed to the extensive popularity of this material for a wide range of applications. One of the first popular alloys consisted of the Nickel-Titanium, commonly known as Nitinol (name originated by combining the abbreviations of its chemical composition with the initials of the laboratory, i.e., Naval Ordinance Lab, where it was first developed), will be used in the following for illustrative purposes, even if the main material features described below are applicable to shape memory alloys in general.

The shape memory effect is best introduced by considering an example of a SMA wire that is subjected to a particular loading and thermal cycle as depicted in Fig. 9.2. Let us assume that we start the cycle at point 1 where the material is at a sufficiently low temperature (for clarity, the temperature axis intersects the axes of the strain and stress at a value equal to the initial low temperature) and has not undergone any strain or stress histories. We then apply a load to the wire to induce tension along its length and, in this process, we produce an increase in strain and stress. In the initial part of the response (along the branch depicted by points 1 and 2), the material behaves in an elastic fashion. Under these conditions the material is assumed to be in its *twinned martensite* phase in which equilibrium of several *martensite variants* coexist (where the martensite variant represents an orientation direction that each martensite crystal can possess and, in twinned martensite, different variants can exists in the material).

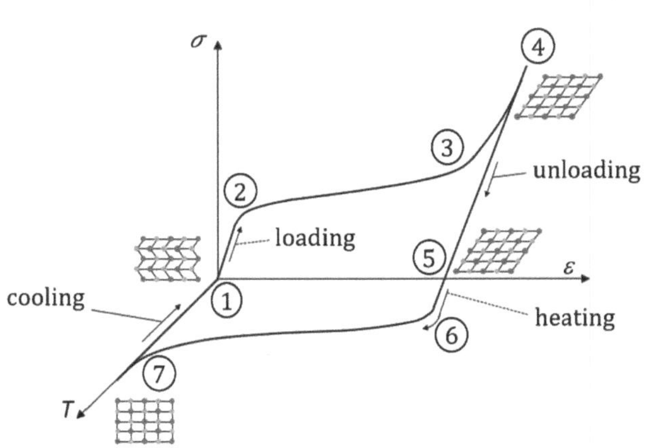

FIGURE 9.2 Qualitative overview of the shape memory effect.

Once the material reaches the stress level depicted at point 2, a reorientation of the martensite variants starts to occur and to give rise to a response that resembles a plastic behavior. The actual yield strength of the material occurs at a higher stress as outlined later. Once the reorientation has taken place (passed point 3), the material responds elastically again upon further loading (in the vicinity of point 4). The material is now in its detwinned martensite phase that has been achieved by reorientation of the different variants along a dominant variant. Upon unloading, the material responds in an elastic manner (i.e., along branch 4-5) and, once it is fully unloaded (i.e., no stress is resisted by the material), a residual deformation remains in the SMA wire (as depicted at point 5). It is possible to recover this residual deformation by heating the wire. Once a certain temperature is reached (depicted by point 6), the material begins a phase transformation from detwinned martensite to austenite. At point 7, the material is in its austenite phase and has fully recovered its original shape. By cooling down the material to the initial low temperature, the material does not change in shape and transforms back to twinned martensite, therefore closing the loop (i.e., back to point 1).

The different phase transformations that characterize the shape memory effect can also be outlined by referring to the phase diagram of the material (see Fig. 9.3). In the initial undeformed and unloaded condition, the material is considered to be twinned martensite (bottom-left part of the diagram) as long as the temperature of the material is below the temperature A_s as shown on the horizontal axis of Fig. 9.3. In this case, A_s defines the temperature at which the phase transformation from martensite to austenite starts, while A_f depicts the temperature at which the transformation is completed and the material is in the austenite phase (region on the right-hand side of the diagram).

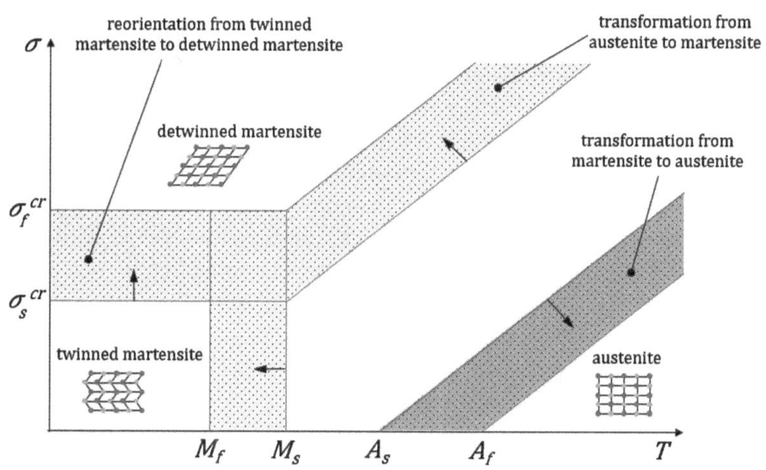

FIGURE 9.3 Qualitative overview of the phase diagram of the shape memory alloy.

Let us now assume that we start to load a SMA wire in tension. A reorientation of the martensite variants starts to take place once the stress resisted by the material reaches the *detwinning start stress* σ_s^{cr}. The reorientation is completed once the material stress equals the *detwinning finish stress* σ_f^{cr}. At this point the material consists of detwinned martensite (denoted by the top-left region of the diagram in Fig. 9.3). Stresses σ_s^{cr} and σ_f^{cr} are also denoted as critical stresses at the start and finish of the transformation from twinned to detwinned martensite.

By completely unloading the material, no further phase transformation or martensite reorientation takes place as the material structure remains detwinned martensite. In fact, the reorientation takes place only during the loading process and the material does not convert back to the twinned martensite upon unloading. At this point of the loading cycle the material is expected to exhibit a residual deformation (see point 5 in Fig. 9.2).

By heating up the material, it is possible to induce a new phase transformation from martensite to austenite when the temperature range in the material is between A_s and A_f. Once the temperature is equal or greater than A_f, the material has recovered its original shape. In the case the temperature of the SMA is reduced, the material starts a further phase transformation from austenite to twinned martensite at a temperature M_s. This phase transformation is then completed at a temperature M_f.

The temperature at which the phase transformations between the martensite and austenite phases tends to increase as the stress in the material grows. This is depicted by the positive gradient of the lines representing the start and finish temperatures of martensite and austenite for increasing stress levels (i.e., slope of the phase transformation strip).

When considering a SMA unloaded at a temperature below A_s or resisting some degree of stress σ below the temperature $A_s(\sigma)$ (where the latter depicts the varying temperature of the austenite start as a function of the applied stress to capture the inclined slope of the curve defining the beginning of the transformation from martensite to austenite as shown in the phase diagram of Fig. 9.3), the stress–strain curve observed in an uniaxial tensile loading can be of the form illustrated in Fig. 9.4. The initial deformation plateau is generated by the reorientation from twinned to detwinned martensite structure that can take place in the material. The shape memory effect can recover this deformation produced during the martensite reorientation (exhibited in the stress range between σ_s^{cr} and σ_f^{cr}) that remains in the material upon complete unloading by means a suitable thermal cycle. This is not the case for the deformation produced at the higher level of stress (the second plateau in the stress–strain curve) that consists of plastic and permanent deformations.

9.3 Superelasticity

The superelastic (or pseudoelastic) response of SMA has been widely exploited in a wide range of applications. This material characteristic enables the material to recover large deformations that are produced in a stress–strain hysteresis. Pseudo-elasticity can occur when the material has a temperature that enables it to be in its austenite phase (e.g., above A_f in its unstressed conditions).

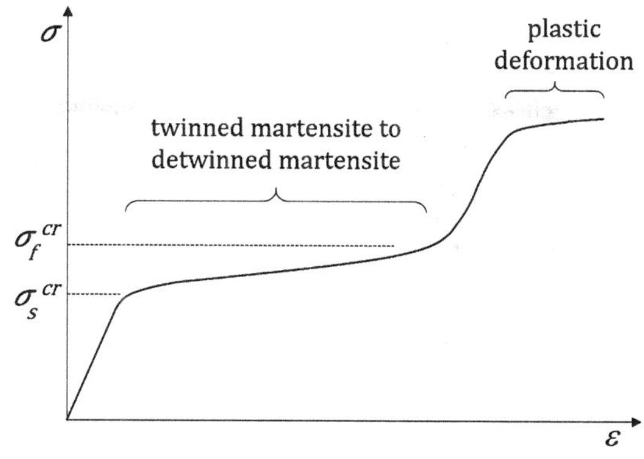

FIGURE 9.4 Qualitative stress–strain response highlighting occurrence of martensite reorientation and plastic deformation for SMA at temperature below $A_S(\sigma)$.

We will now outline this behavior by considering the hysteretic stress-deformation cycle shown in Fig. 9.5 that depicts the response of a SMA wire loaded in tension. In particular, the material at point 1 is assumed to be undeformed and unloaded. Upon application of a stress, the material responds in an elastic manner until it reaches the stress level at point 2. At this particular level of stress, the austenite starts to transform to detwinned martensite. The transformation is completed around point 3. If the material is further loaded beyond this point, the material deforms elastically in its detwinned martensite state and, eventually, it starts to undergo plastic deformation. An elastic response is observed when the unloading of the material after its transformation to detwinned martensite is completed, as depicted along branch 3-4. The phase transformation back to austenite takes place between points 4 and 5. Further unloading passed point 5 occurs with the material being in the austenite phase. Upon full removal of the load, the material is back to point 1. The recovery of the strain occurs under the assumption that no plastic deformation is exhibited by the material during the loading stage beyond point 3.

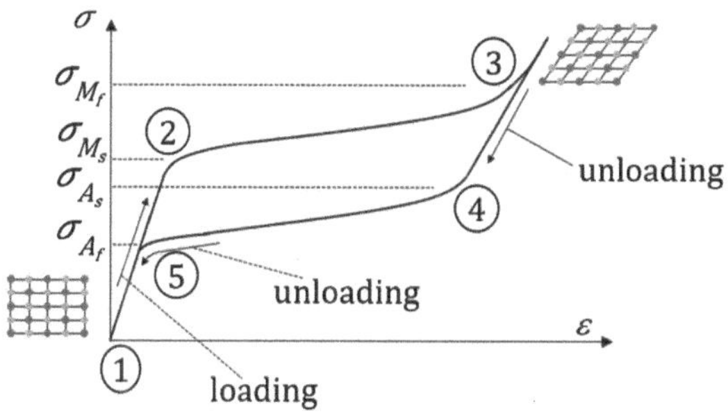

FIGURE 9.5 Qualitative overview of the superelastic response.

The phase transformations (i.e., from austenite to detwinned martensite and from detwinned martensite back to austenite) can be visualized in the phase diagram of Fig. 9.3 by assuming that we move from a point on the horizontal axis (i.e., subjected to no stress) at a temperature greater than A_f upwards until we pass through the region through which the austenite to martensite transformation occurs. In the unloading process, the material passes through the reverse phase transformation.

9.4 Material modeling

Over the last decades, extensive research has been carried out to develop constitutive models capable of representing the response of shape memory alloys considering both shape memory effect and superelasticity. In this section, we intend to provide only a brief introduction to SMA modeling to highlight how the deformations induced in the material by means of phase transformations can be included as an inelastic deformation in material constitutive representation. A detailed treatment of constitutive models for SMA materials can be found in dedicated textbooks, e.g., [5,6], or research publications.

The material behavior is dependent upon a number of variables, including the stress, the temperature and the crystal structure that is a function of the loading and thermal histories. To capture this response an internal variable is commonly introduced in SMA material models to capture the current condition of the material expressed in terms of a martensite material fraction and denoted as ξ. A material that is in its martensite phase only is depicted by $\xi = 1$, while the case of fully austenite phase is described by $\xi = 0$. To gain insight into the material response, it is useful to further distinguish within the martensite material fraction between the twinned martensite and the detwinned martensite. This distinction can be expressed as follows:

$$\xi = \xi_S + \xi_T \tag{9.1}$$

in which ξ_S and ξ_T define the material fractions of the detwinned and twinned martensite, respectively. The subscripts "S" and "T" represent the fact that these material conditions are also referred to as stress-induced and temperature-induced martensite, respectively. As this subscript representation is widely used in the literature, this has been maintained in this chapter as well.

For the purpose of this chapter, we consider a simple 1D SMA model that has been widely used in the literature and that has been proposed by Brinson [2]. The constitutive model can be described as follows (while recent model refinements related to the phase transformation process can be found in [1,4])):

$$\sigma - \sigma_0 = E(\xi)\varepsilon - E(\xi_0)\varepsilon_0 + \Omega(\xi)\xi_S - \Omega(\xi_0)\xi_{S0} + \Theta(T - T_0) \tag{9.2}$$

where σ depicts the current stress resisted by the material, $E(\xi)$ defines the elastic modulus of the material based on its current crystal structure (specified by ξ), ε is the strain exhibited by the material, Θ describes the thermal response of the material, while the variables with the subscript "0," i.e., σ_0, ξ_0, ξ_{S0}, ε_0, and T_0, depict the initial conditions of the material. The term $\Omega(\xi)$ describes the phase transformation that takes place in the material and it can be determined with the following expression:

$$\Omega(\xi) = -\varepsilon_L E(\xi) \tag{9.3}$$

where ε_L defines the maximum residual strain that the material exhibits in its stress-induced martensite state. From a modeling viewpoint the effect of this term resembles the plastic strain with the only difference that, in the SMA response, the deformation associated to the $\Omega(\xi)$ term can be recovered under certain conditions as described in following examples.

The elastic modulus of the material $E(\xi)$ is calculated by linearly interpolating between the elastic moduli of the austenite phase E_A and of the martensite phase E_M as a function of the martensite material fraction ξ:

$$E(\xi) = E_A + \xi(E_M - E_A) \tag{9.4}$$

while other interpolation functions could be used in the evaluation of $E(\xi)$.

In this model representation the values for ξ_S and ξ_M depend on the stress σ, temperature T and martensite material fraction ξ and can be calculated based on the phase transformation exhibited by the material. These occur when the material falls within a shaded area of the phase diagram of Figs. 9.3 and 9.6. No phase transformation are assumed to occur outside the shaded areas.

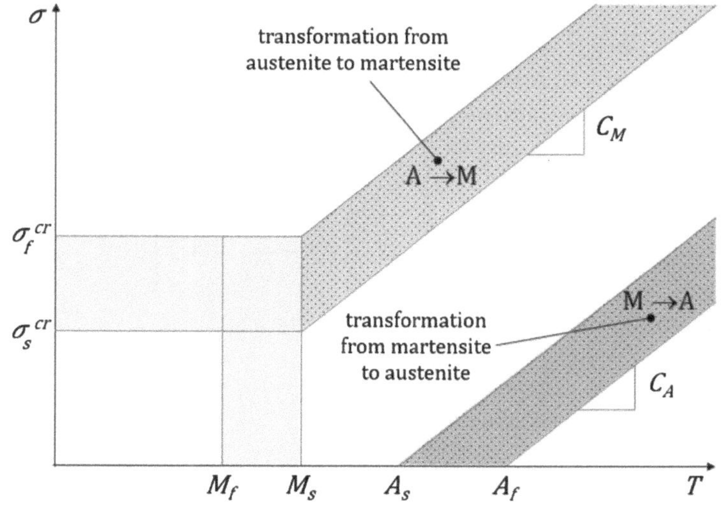

FIGURE 9.6 Qualitative overview of transformation strips for $T > M_S$.

When the material is moving to its detwinned martensite state (i.e., passing through the gray-shaded area with dots depicted by A→M in Fig. 9.6), the following expressions can be used to determine the material response:

- for $T > M_s$ and $\sigma_s^{cr}+C_M(T-M_s) < \sigma < \sigma_f^{cr}+C_M(T\text{-}M_s)$:

$$\xi_S = \frac{1-\xi_{S0}}{2} cos\left[\frac{\pi}{\sigma_s^{cr}-\sigma_f^{cr}}[\sigma-\sigma_f^{cr}-C_M(T-M_s)]\right]+\frac{1+\xi_{S0}}{2} \quad (9.5)$$

$$\xi_T = \xi_{T0} - \frac{\xi_{T0}}{1-\xi_{S0}}(\xi_S-\xi_{S0}) \quad (9.6)$$

- for $T < M_s$ and $\sigma_s^{cr} < \sigma < \sigma_f^{cr}$:

$$\xi_S = \frac{1-\xi_{S0}}{2} cos\left[\frac{\pi}{\sigma_s^{cr}-\sigma_f^{cr}}(\sigma-\sigma_f^{cr})\right]+\frac{1+\xi_{S0}}{2} \quad (9.7)$$

$$\xi_T = \xi_{T0} - \frac{\xi_{T0}}{1-\xi_{S0}}(\xi_S-\xi_{S0}) + \Delta_{T\xi} \quad (9.8)$$

where $\Delta_{T\xi}$ is calculated based on the following expression when $M_f < T < M_s$ and $T < T_0$:

$$\Delta_{T\xi} = \frac{1-\xi_{T0}}{2}\left[cos[a_M(T-M_f)]+1\right] \quad (9.9)$$

and

$$a_M = \frac{\pi}{M_s - M_f} \quad (9.10)$$

or, otherwise, $\Delta_{T\xi}$ is taken equal to zero in all other cases.

The phase transformation from martensite to austenite is described in Fig. 9.6 by the gray-shaded area with dots depicted by M→A. In this case, the material response that occurs for $T > A_s$ and $C_A(T-A_f) < \sigma < C_A(T-A_s)$ can be evaluated as follows:

$$\xi = \frac{\xi_0}{2}\left[cos\left(a_A(T-A_s-\frac{\sigma}{C_A})\right)+1\right] \quad (9.11)$$

$$\xi_S = \xi_{S0} - \frac{\xi_{S0}}{\xi_0}(\xi_0-\xi) \quad (9.12)$$

$$\xi_T = \xi_{T0} - \frac{\xi_{T0}}{\xi_0}(\xi_0-\xi) \quad (9.13)$$

and

$$a_A = \frac{\pi}{A_f - A_s} \quad (9.14)$$

For ease of notation, in the following examples we will consider only material responses that, in the phase diagram, are described by temperatures greater than M_s (see Fig. 9.6). We will also make use of a simplification introduced in [3] that enables to rewrite Eq. (9.2) as follows:

$$\sigma = E(\xi)(\varepsilon - \varepsilon_L \xi_S) \tag{9.15}$$

in which the thermal strain has been neglected because significantly smaller than the transformation strain.

We will now consider an example where a SMA wire is subjected to an axial loading and unloading cycle assumed to be performed at the same temperature. The stress and temperature history adopted for this case is illustrated in the phase diagram of Fig. 9.7. The key values of stress and temperature are presented in Table 9.1, while the material properties used in the simulations are summarized in Table 9.2 [2]. In this example, we assume to start from an unstressed austenite material (point A_1 in Fig. 9.7) and to apply an axial tension up to a stress level that is beyond the A→M transformation strip (to point A_2 in Fig. 9.7) to ensure that the material is fully transformed to detwinned martensite. The material is then assumed to be unloaded to point A_3 and, in this process, to undergo the transformation M→A. The corresponding stress–strain representation of this loading cycle is illustrated in Fig. 9.8 to highlight the superelastic response.

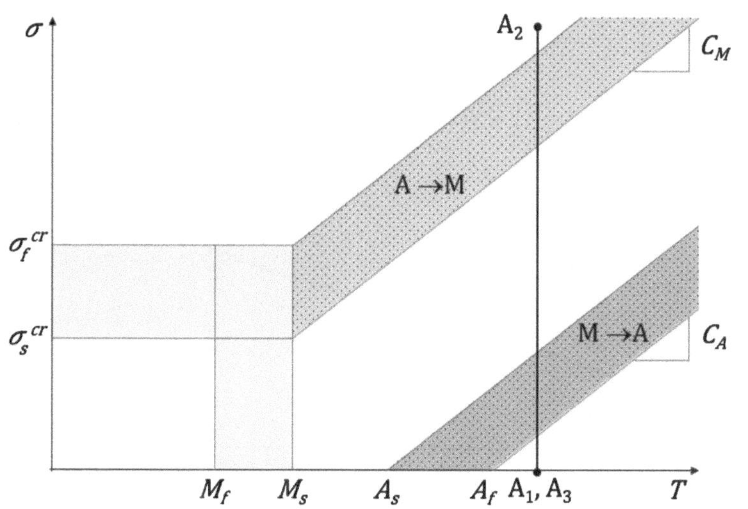

FIGURE 9.7 Outline of stress and temperature history for numerical example 1.

The second example that we consider covers a slightly more complex loading and temperature history as shown in Fig. 9.9. After the loading process between points B_1 and B_2 along which the material transforms from austenite to detwinned martensite (identical to the previous step A_1-A_2), the material is slightly cooled down to point B_3. This temperature reduction is introduced to

TABLE 9.1 Description of the variations of σ ad T for example 1.

	A1	A2	A3
T	50/°C	50/°C	50/°C
σ	0 MPa	440 MPa	0 MPa

TABLE 9.2 Material properties specified in the numerical examples.

Elastic moduli	Thermal effects and maximum residual strain
$E_A = 67{,}000$ MPa	$\Theta = 0.55$ MPa/°C
$E_M = 26{,}300$ MPa	$\varepsilon_L = 0.067$
Transformation temperatures	**Transformation constants**
$M_f = 9$°C	$C_M = 8$ MPa/°C
$M_s = 18.4$°C	$C_A = 13.8$ MPa/°C
$A_s = 34.5$°C	$\sigma_s^{cr} = 100$ MPa
$A_f = 49$°C	$\sigma_f^{cr} = 170$ MPa

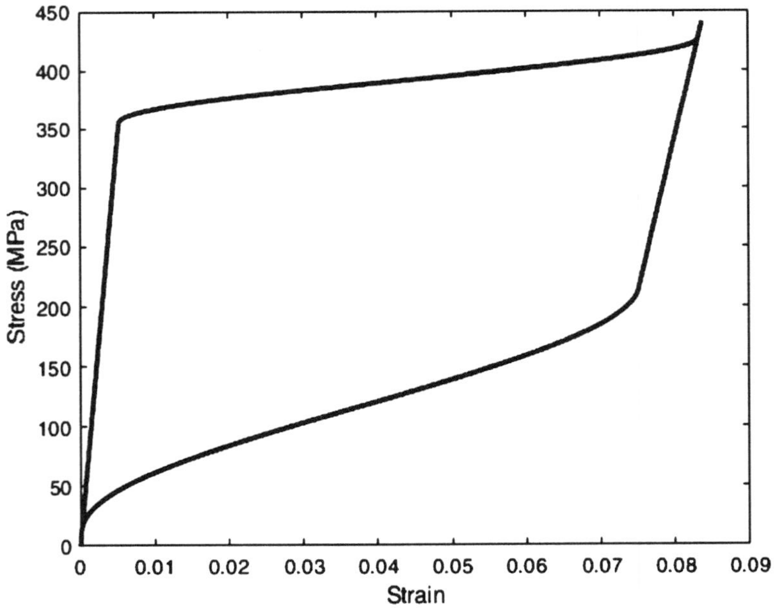

FIGURE 9.8 Example 1: stress–strain curve.

ensure that, at the end of the unloading process (when moving from point B_3 to B_4), the material cannot completely transform back to the austenite phase. This is depicted by the fact that final point B_4 falls within the M→A transformation strip. The relevant values of the stress and temperature defining the load cycle

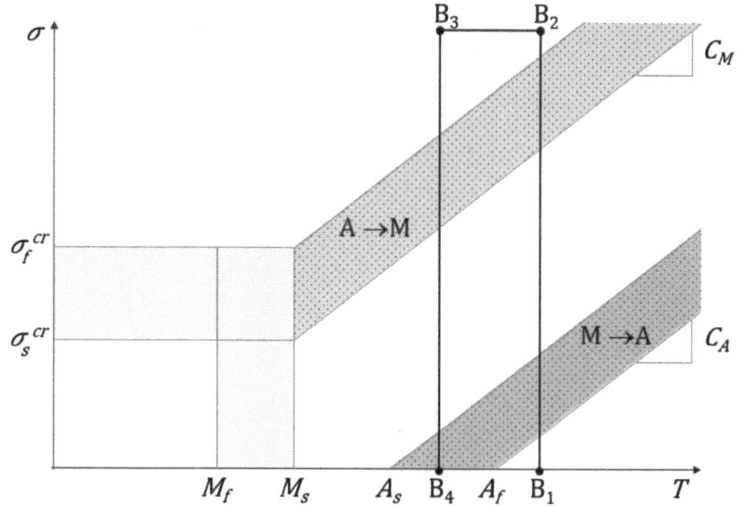

FIGURE 9.9 Outline of stress and temperature history for numerical example 2.

are depicted in Table 9.3 and the material properties are given in Table 9.2. The corresponding stress–strain curve taking place during this load and temperature cycle is presented in Fig. 9.10. In this case the stress versus deformation up to the peak stress is identical to the one of Fig. 9.8. The unloading is different because the M→A transformation starts at a slightly lower stress and does not reach its completion upon unloading, therefore causing a residual strain to remain in the unloaded condition as shown in Fig. 9.10.

TABLE 9.3 Description of the variations of σ ad T for example 2.

	B1	B2	B3	B4
T	50/°C	50/°C	40/°C	40/°C
σ	0 MPa	440 MPa	440 MPa	0 MPa

The load and temperature cycle of the third example considered is shown in Fig. 9.11 and follows points C_1, C_2, C_3, C_4, and C_5. The relevant values for σ and T are depicted in Table 9.4 while the material properties are specified in Table 9.2. The cycle adopted in this example extends the one of example 2 as, while point C_4 falls on the left of the transformation strip (i.e., the unloaded SMA is in its detwinned martensite phase), the SMA has a final heating step (between points C_4 and C_5) that enables it to fully recover its original shape, therefore reflecting the shape memory effect. This response can also be observed in the stress–strain representation shown in Fig. 9.12.

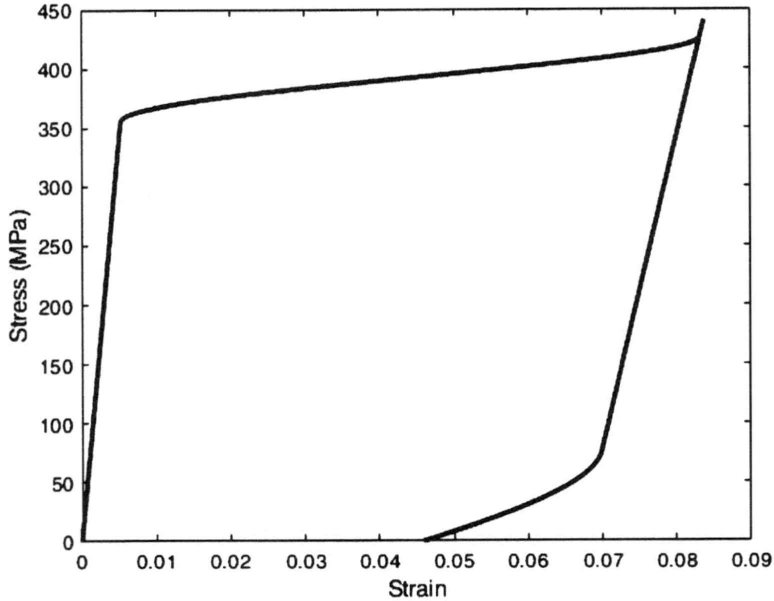

FIGURE 9.10 Example 2: stress–strain curve.

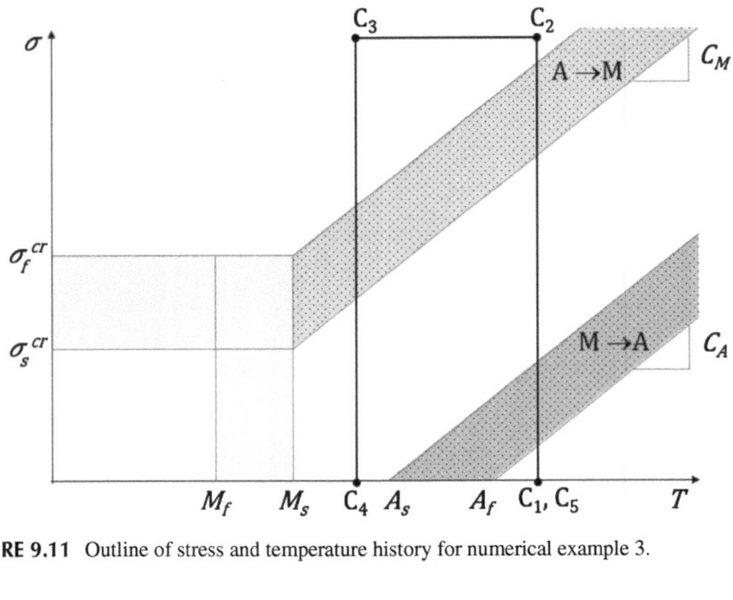

FIGURE 9.11 Outline of stress and temperature history for numerical example 3.

9.5 Characterization of material properties

The identification of the material properties of SMA materials is typically based on well established experimental tests and laboratory procedures. Constitutive

TABLE 9.4 Description of the variations of σ ad T for example 3.

	C1	C2	C3	C4	C5
T	50/°C	50/°C	30/°C	30/°C	50/°C
σ	0 MPa	440 MPa	440 MPa	0 MPa	0 MPa

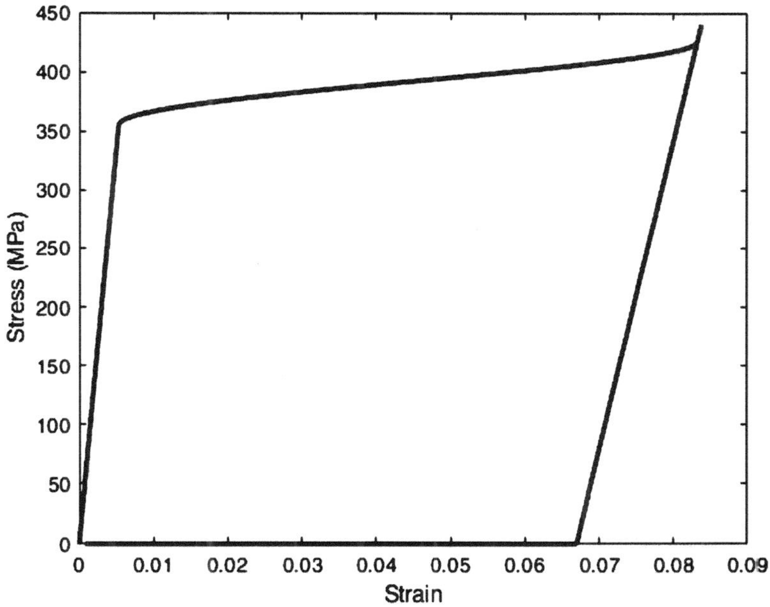

FIGURE 9.12 Example 3: stress–strain curve.

models available in the literature tend to rely on material parameters that have a physical meaning or that can be easily obtained from experimental characterizations. In this section, we provide simple examples in which inverse analysis procedures can be implemented for the identification, for example, of the slopes of the phase transformation strips between martensite and austenite depicted in the SMA phase diagrams. Such inverse analysis techniques can complement current approaches based on the post-processing of experimentally measured curves. For example, an advantage of implementing inverse analysis procedure could lead to a reduction in the number of tests that need to be carried out, especially when dealing with the more complex constitutive models that range from 1D to 3D representations of the material response. Also, inverse analysis strategies lend themselves to complement the characterization of material tested with complex experimental setups from which the material parameters cannot be directly extrapolated.

For the purpose of this section, we consider a SMA wire subjected to an axial tensile loading. In terms of experimental measurements, we make use of the stress–strain diagram generated by the load and temperature cycle induced to the sample. The examples presented in the following are generated by considering pseudo-experimental data obtained by specifying the material properties of Table 9.2 in a direct analysis of the problem.

The objective function considered for the inverse analysis approach is kept as simple as possible. In this case, it compares variations in the stress–strain curves obtained from the pseudo-experiment and in the computed values based on the following expression:

$$f(x) = \frac{1}{2}\sum_{i=1}^{n_R} r_i^2(x) \tag{9.16}$$

where the residual r_i depicts the difference between the pseudo-experimental strain and the computed strain at a particular level of stress and temperature that is calculated as follows:

$$r_i = \varepsilon_i^{com} - \varepsilon_i^{exp} \tag{9.17}$$

where ε_i^{exp} depicts the measured strain and ε_i^{com} denotes the computed deformation. The number of points selected can influence the resolution of the identified parameters produced by the inverse analysis procedure.

The first identification example that we consider aims at the characterization of the slope of the phase transformation strip defined by the term C_M (Fig. 9.6). In order to perform the inverse analysis procedure, we need to select a testing protocol for the direct analysis of the characterization process. For simplicity, we start by considering the use of the experiment presented in the previous section in Fig. 9.7 (denoted as example 1). In Fig. 9.13, we present the variation of the objective function for different values of C_M where it can be seen that the minimum of the function is well defined and located at the position of the sought parameter value (i.e., position of 1 on the horizontal axis that represents a normalized value obtained by dividing the sought value by the pseudo-experimental result). In such a scenario, it is clear that, at least in the range considered here, the problem appears to be well posed. For the purpose of this chapter, we will not consider the influence of noise on the pseudo-experimental data to resemble the variability in the measurements that can occur in real-world situations and that can influence the effectiveness of the inverse analysis problem. The stepping through the convergence iterations of the identification process obtained for one initialization (that included an error of 5%) is illustrated in Fig. 9.14. It can be seen that the convergence is obtained quickly. Similar results could be obtained by adopting other testing protocols, such as those used in examples 2 and 3. The MATLAB® scripts for the inverse analysis are based on those presented in Chapter 3 and are not reproduced here.

We will now apply the same inverse analysis procedure to the identification of the slope of the phase transformation strip from martensite to austenite depicted by the term C_A. By relying on the load and temperature cycle of example

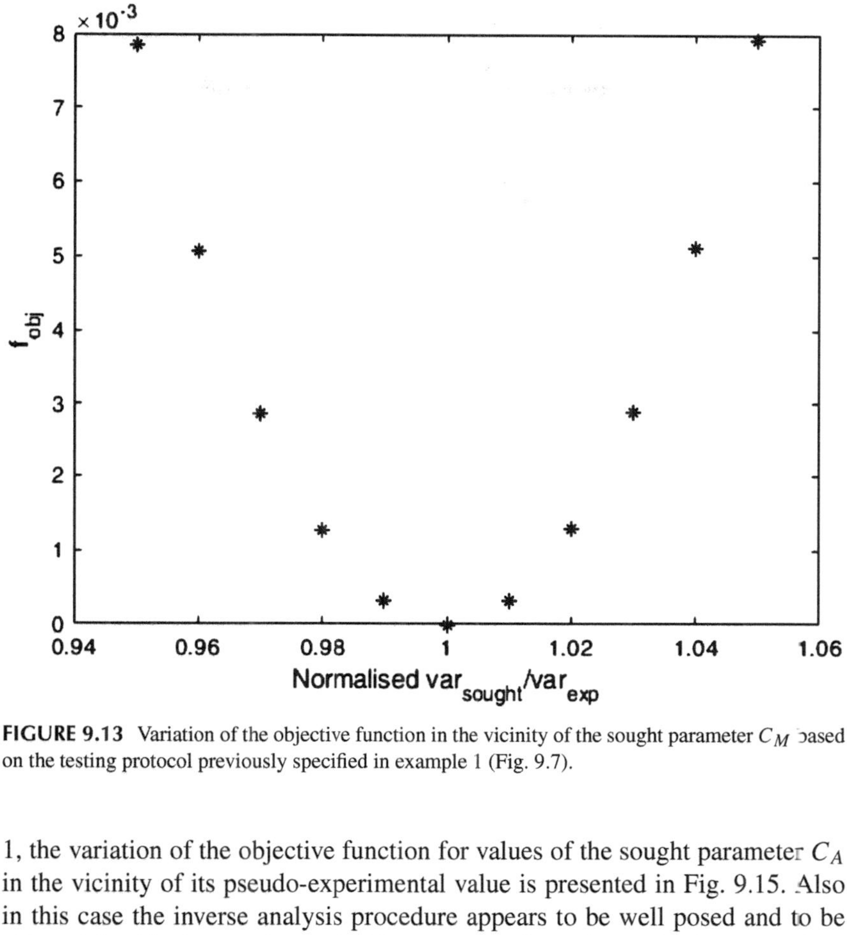

FIGURE 9.13 Variation of the objective function in the vicinity of the sought parameter C_M based on the testing protocol previously specified in example 1 (Fig. 9.7).

1, the variation of the objective function for values of the sought parameter C_A in the vicinity of its pseudo-experimental value is presented in Fig. 9.15. Also in this case the inverse analysis procedure appears to be well posed and to be able to identify the sought result. The stepping through an initialization of the problem is shown in Fig. 9.16.

It is interesting to point out that, when seeking to identify C_A, the use of the load and temperature cycle of example 3 (Fig. 9.11) would not enable the characterization of C_A as depicted by the trends of the objective function in the vicinity of the value of the pseudo-experimental data shown in Fig. 9.17. The reason for the lack of variation of the objective function for different values of C_A is simply related to the fact that the cycle adopted in example 3 does not intersect the phase transformation strip from martensite to austenite at stress levels different from zero and, therefore, it is not able to detect the impact of the slope of this phase transformation strip on the calculated strain values.

For illustrative purposes, we consider the identification of both parameters C_A and C_A based on the testing procedures of examples 1 and 3. Figs. 9.18 and 9.19 depict the variation of the objective function in the vicinity of the value of the pseudo-experimental values. These figures highlight how the objective

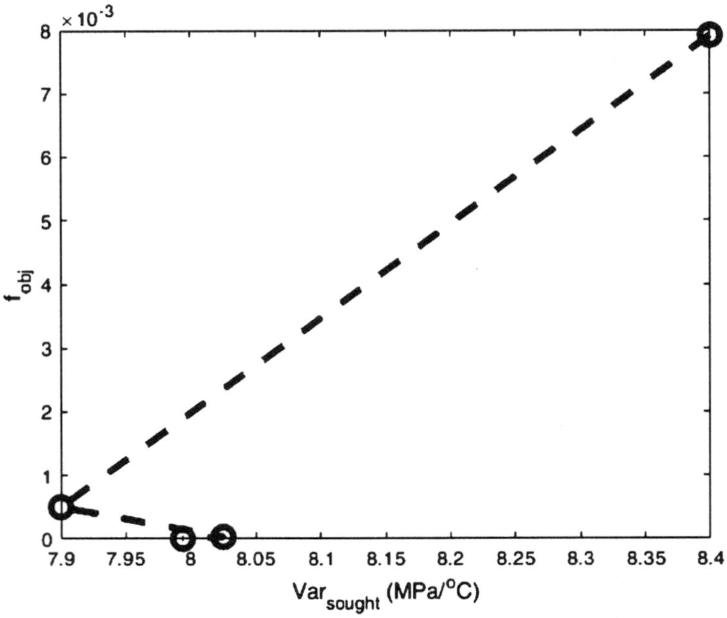

FIGURE 9.14 Convergence iterations during the identification of the sought parameter C_M based on the testing protocol previously specified in example 1 (Fig. 9.7).

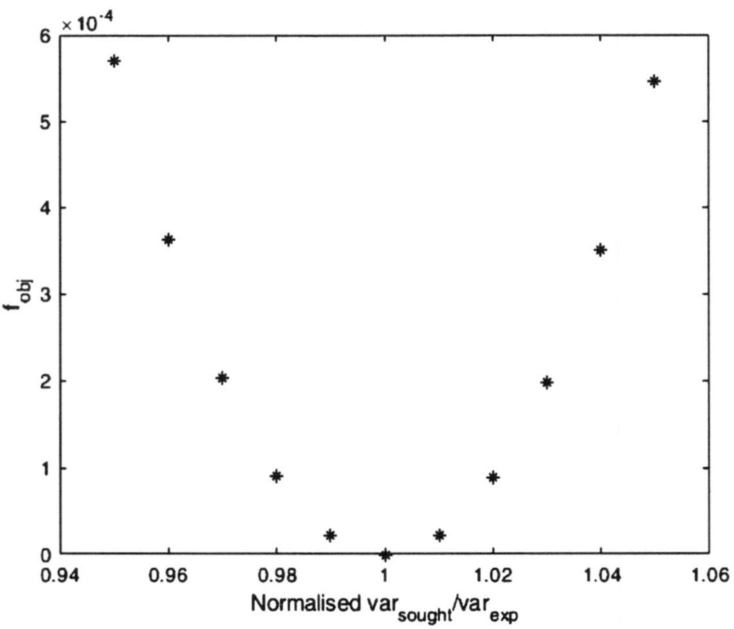

FIGURE 9.15 Variation of the objective function in the vicinity of the sought parameter C_A based on the testing protocol previously specified in example 1 (Fig. 9.7).

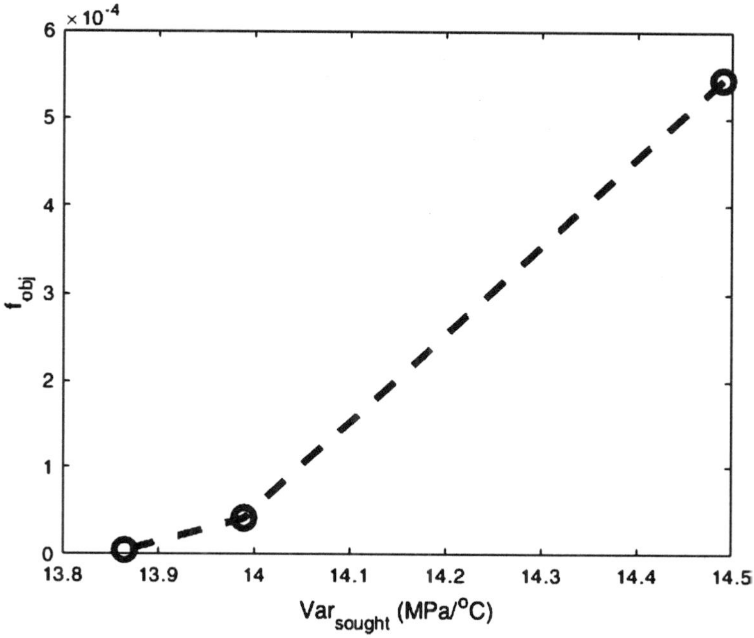

FIGURE 9.16 Convergence iterations during the identification of the sought parameter C_A based on the testing protocol previously specified in example 1 (Fig. 9.7).

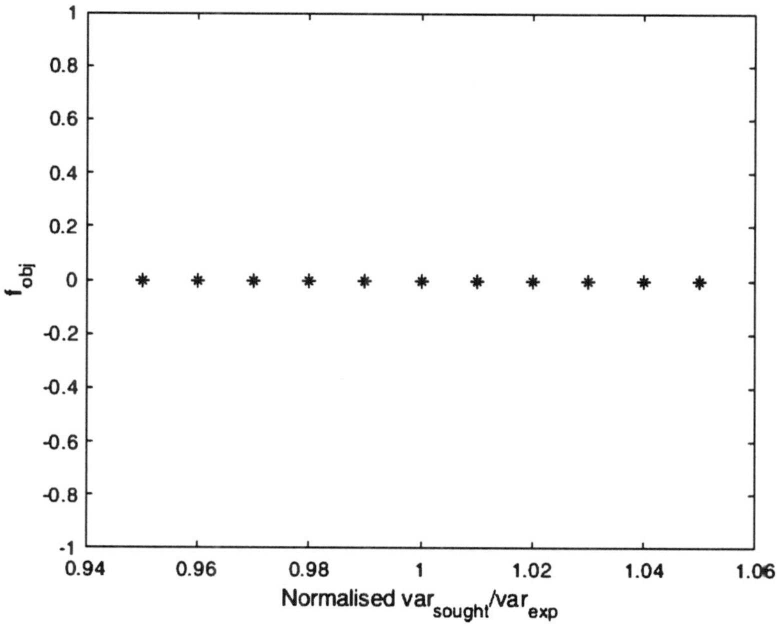

FIGURE 9.17 Variation of the objective function in the vicinity of the sought parameter C_A based on the testing protocol previously specified in example 3 (Fig. 9.11).

function has a stronger concave up shape for variations of C_A when compared to the variations exhibited for C_M. Such trends, especially the softer concave up variation for C_M, would need to be further investigated by introducing different levels of noise in the pseudo-experimental data to account for the variability expected to occur in real measurements. The inverse analysis procedure carried out for an initialization with an error of 5% is presented in Fig. 9.20.

For completeness, the variation of the objective determined with the test cycle of example 3 is presented in Fig. 9.21. These results confirm the previously observed conclusion that example 3 cannot be used to predict C_A because its load and temperature cycle does not excite this material parameter in the direct analyses.

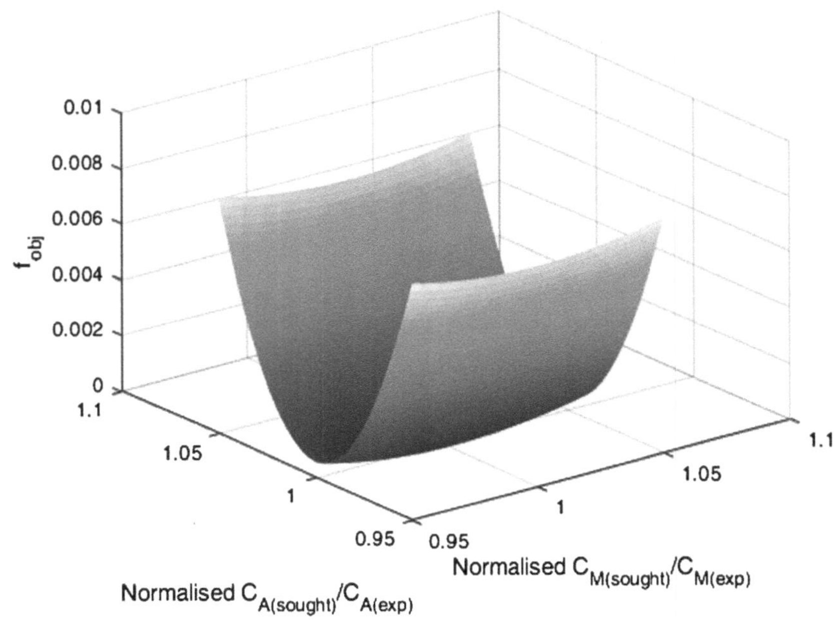

FIGURE 9.18 Variation of the objective function (surface plot) in the vicinity of the sought parameters C_M and C_A based on the testing protocol previously specified in example 1 (Fig. 9.7).

The examples presented in this section intend to highlight the opportunities that the use of inverse analysis procedures can provide for the characterization of the material parameters describing the response of SMA materials. Particular attention needs to be specified in the selection of the load and temperature cycle to be used for the experimental test as it needs to provide sufficient information to capture the sought material response and parameters.

9.6 Summary

This chapter provided a brief introduction to the behavior, modeling, and material characterization of shape memory alloys (SMAs). In the initial part of the

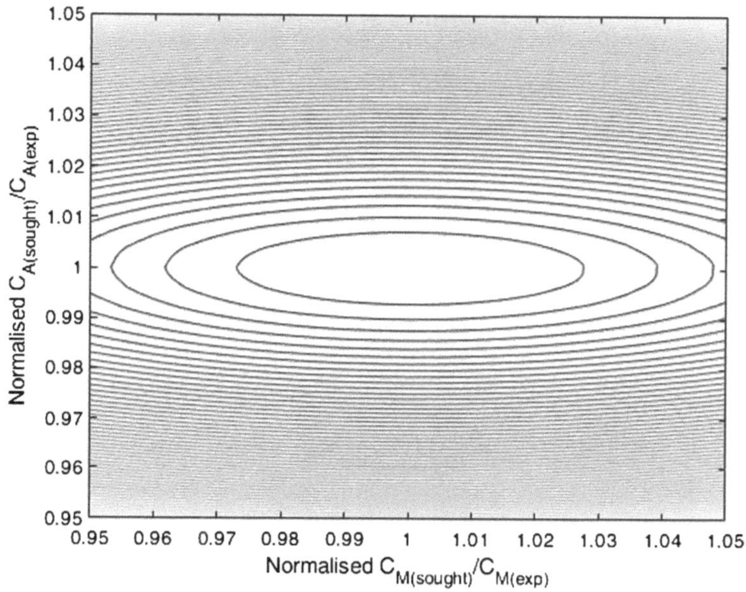

FIGURE 9.19 Variation of the objective function (contour plot) in the vicinity of the sought parameters C_M and C_A based on the testing protocol previously specified in example 1 (Fig. 9.7).

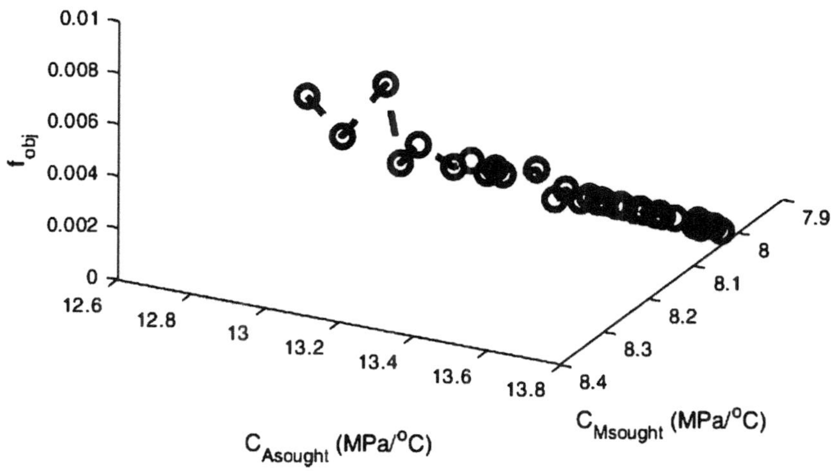

FIGURE 9.20 Convergence iterations during the identification of the sought parameters C_M and C_A based on the testing protocol previously specified in example 1 (Fig. 9.7).

chapter, the key features that distinguish the response of SMA materials, such as shape memory effect and superelasticity, were presented. These behaviors originate from the ability of the material to undergo phase transformations between austenite and martensite phases depending on its stress and temperature conditions as well as material composition. The particularity of the shape mem-

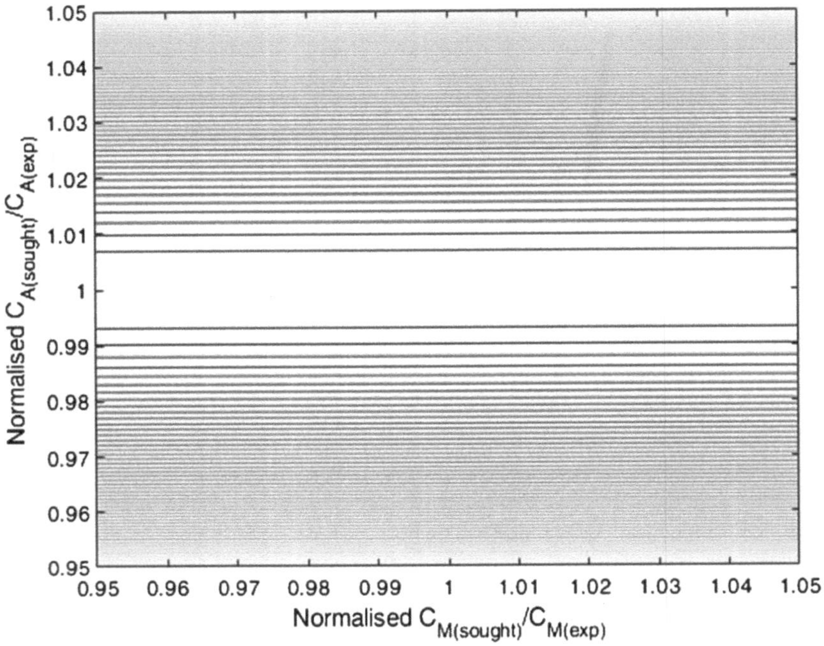

FIGURE 9.21 Variation of the objective function in the vicinity of the sought parameters C_M and C_A based on the testing protocol previously specified in example 3 (Fig. 9.11).

ory effect relies on the ability of the material to recover its initial shape upon an application of an increase in temperature. In this case the material regains its original shape when transforming from the martensite phase (oriented in its detwinned condition) to austenite. In the case of the superelastic response the material is able to undergo hysteretic stress-deformation cycles without permanent deformations remaining upon completion of the load cycle in its unstressed condition. In this case, the material is in the austenite phase in its unstressed state and, during loading, it transforms to detwinned martensite while, upon unloading, the material returns to austenite. The actual material condition for different levels of stress and temperature can be well described by means of phase diagrams as outlined for simple selected examples.

A wide range of constitutive models are available in the literature to simulate the response of SMA that range for their complexity as well as from 1D to 3D representations. For the purpose of this chapter, we introduced a simple model to illustrate the key features that can be included in a constitutive model to describe the shape memory effect and superelasticity. For further details on the wide range of constitutive models available, reference should be made to dedicated textbooks, e.g., [5,6].

In the final part of the chapter, an introduction to the use of inverse analysis approaches to identify the material parameters defining the SMA response was presented. Such approach could complement available characterization strategies that were developed over the years for SMA materials. Simple identification problems were presented to highlight some of the key features that require attention when implementing characterization procedures by means of inverse analysis techniques.

References

[1] A. Bekker, L.C. Brinson, Phase diagram based description of the hysteresis behavior of shape memory alloys, Acta Materialia 46 (1998) 3649–3665.
[2] L.C. Brinson, One-dimensional constitutive behavior of shape memory alloys: thermomechanical derivation with non-constant material functions and redefined martensite internal variable, Journal of Intelligent Material Systems and Structures 4 (1993) 229–242.
[3] L.C. Brinson, M.S. Huang, Simplifications and comparisons of shape memory alloy constitutive models, Journal of Intelligent Material Systems and Structures 7 (1996) 108–114.
[4] X.J. Gao, R. Qiao, L.C. Brinson, Phase diagram kinetics for shape memory alloys: a robust finite element implementation, Smart Material and Structures (2007) 2102–2115.
[5] D.C. Lagoudas, Shape Memory Alloys, Springer, 2008.
[6] L. Lecce, A. Concilio, Shape Memory Allow Engineering, Butterworth-Heinemann, 2015.

Index

Y

Z

Kolibri 360 GmbH
Ferdinand-Jühlke-Straße 7
99095 Erfurt, Deutschland
produktsicherheit@kolibri360.de